*NATURAL SCIENCES IN AMERICA*

# NATURAL SCIENCES IN AMERICA

*Advisory Editor*
KEIR B. STERLING

*Editorial Board*
EDWIN H. COLBERT
EDWARD GRUSON
ERNST MAYR
RICHARD G. VAN GELDER

# The Species Problem

*Edited by*
ERNST MAYR

**ARNO PRESS**
A New York Times Company
New York, N. Y. • 1974

Reprint Edition 1974 by Arno Press Inc.

Copyright © 1957 by The American Association for the Advancement of Science.

Reprinted by permission of The American Association for the Advancement of Science.

NATURAL SCIENCES IN AMERICA
ISBN for complete set: 0-405-05700-8
See last pages of this volume for titles.

Manufactured in the United States of America

---

Library of Congress Cataloging in Publication Data

American Association for the Advancement of Science.
    The species problem.

    (Natural sciences in America)
    Papers presented at a symposium held at the Atlanta meeting of the American Association for the Advancement of Science, Dec. 28-29, 1955.
    Reprint of the 1957 ed. published in Washington, D. C. as Publication no. 50 of the American Association for the Advancement of Science.
    Includes bibliographies.
    1. Species--Congresses. I. Mayr, Ernst, 1904- ed. II. Title. III. Series: IV. Series: American Association for the Advancement of Science. Publication no. 50.
QH83.A73 1974     575      73-17831
ISBN 0-405-05749-0

REF
QH
83
.A73
1974
Cop. 1

# The Species Problem

A Symposium Presented at the Atlanta Meeting of the American Association for the Advancement of Science, December 28-29, 1955

*Edited by*

ERNST MAYR

Publication No. 50 of the
AMERICAN ASSOCIATION FOR THE ADVANCEMENT OF SCIENCE
Washington, D.C., 1957

© 1957
THE AMERICAN ASSOCIATION FOR THE
ADVANCEMENT OF SCIENCE

Library of Congress Catalog
Card Number 57-11321

# PREFACE

Few biological problems have remained as consistently challenging through the past two centuries as the species problem. Time after time attempts were made to cut the Gordian knot and declare the species problem solved either by asserting dogmatically that species did not exist or by defining, equally dogmatically, the precise characteristics of species. Alas, these pseudo-solutions were obviously unsatisfactory. One might ask: "Why not simply ignore the species problem?" This also has been tried, but the consequences were confusion and chaos. The species is a biological phenomenon that cannot be ignored. Whatever else the species might be, there is no question that it is one of the primary levels of integration in many branches of biology, as in systematics (including that of microorganisms), genetics, and ecology, but also in physiology and in the study of behavior. Every living organism is a member of a species, and the attributes of these organisms can often best be interpreted in terms of this relationship. This is particularly true in comparative studies.

The continued interest in the species problem thus requires no apology. Indeed the interest in the species is perhaps greater now than it has been at any other time during the last hundred years. One reason for this is the increase in autecological investigations, which include a joint study of the physiological and ecological properties of species that has resulted in an active contact between physiology and ecology, with the species the area of overlapping interest. A second reason is the current tendency in population genetics to study the interaction, rather than the action, of genes, and thus to lead to a study of gene pools of which the species is the largest. A third is the emergence of a rejuvenated new systematics that devotes much of its attention to the intimate study of the population structure of species, a new school that is strongly represented in the southeastern United States. It is not surprising therefore that the species problem received more votes than any other subject when Dr. J. G. Carlson, Chairman of Section F (Zoology) of the American Association for the Advancement of Science, together with the officers of the Association of Southeastern Biologists, asked for suggestions on topics for a symposium at the 1955 meeting of the AAAS in Atlanta. These two organizations therefore decided to sponsor a symposium on the species problem and invited me to serve as chairman. The Association's Section G (Botany), the American Society of Parasitologists, and the Society of Systematic Zoology agreed to cosponsor the symposium. The strong backing of the symposium by these organizations greatly contributed to its success.

When selecting participants in the symposium one of my considerations was to avoid duplication of similar earlier symposia. Verne Grant, John Imbrie, and Hampton L. Carson, who had not recently spoken on the species problem, were invited to discuss the species problem in plants, in paleontology, and in genetics. Their fresh approach was supplemented by treatment of groups of organisms or topics that had not before been related to the species problem: John Langdon Brooks agreed to speak on freshwater organisms, T. M. Sonneborn on protozoans, C. Ladd Prosser on the physiological aspects of the species problem, and John A. Moore on the species problem as seen by an embryologist.

The papers are published essentially as given at Atlanta, with one exception. Dr. Sonneborn, when preparing his contribution, realized the need of a comprehensive survey of the species problem in the protozoans and particularly in the ciliates with their interesting and versatile modes of reproduction. He found that the nature of their population structure, whether they are inbreeders or outbreeders, is responsible for the specific aspects of much of their reproductive specialization. To gather the material with which to substantiate this new interpretation necessitated a review of the widely scattered literature and could not be brought to completion until well after the Atlanta meeting. Dr. Sonneborn's masterly synthesis, a fundamental contribution to the protozoan literature, was not available for the chairman's summation.

It would be too much to expect this symposium to solve the species problem, yet it made a solid contribution toward its solution. It broadened the base on which to discuss the problem by utilizing new organisms. It led to a delimitation of the areas of general agreement among biologists. It presented a clear statement of the various species concepts and frankly stated and enumerated difficulties in their application to different types of natural populations. Finally, it illuminated certain aspects of the ageless species problem that had been neglected previously, and it attempted a statement of still controversial issues. From these papers it should be evident that the species problem is still one of the important issues in biology.

ERNST MAYR

# CONTRIBUTORS

John Langdon Brooks, Osborn Zoological Laboratory, Yale University, New Haven, Connecticut

Hampton L. Carson, Department of Zoology, Washington University, St. Louis, Missouri

Verne Grant, Rancho Santa Ana Botanic Garden and Claremont Graduate School, Claremont, California

John Imbrie, Department of Geology, Columbia University, New York, New York

Ernst Mayr, Museum of Comparative Zoology, Harvard University, Cambridge, Massachusetts

John A. Moore, Department of Zoology, Barnard College and Columbia University, New York, New York

C. Ladd Prosser, Physiology Department, University of Illinois, Urbana, Illinois

T. M. Sonneborn, Department of Zoology, Indiana University, Bloomington, Indiana

# CONTENTS

Species Concepts and Definitions
    ERNST MAYR                                                      1
The Species as a Field for Gene Recombination
    HAMPTON L. CARSON                                              23
The Plant Species in Theory and Practice
    VERNE GRANT                                                    39
The Species Problem in Freshwater Animals
    JOHN LANGDON BROOKS                                            81
The Species Problem with Fossil Animals
    JOHN IMBRIE                                                   125
Breeding Systems, Reproductive Methods, and Species
        Problems in Protozoa
    T. M. SONNEBORN                                               155
An Embryologist's View of the Species Concept
    JOHN A. MOORE                                                 325
The Species Problem from the Viewpoint of a Physiologist
    C. LADD PROSSER                                               339
Difficulties and Importance of the Biological Species
    ERNST MAYR                                                    371
Index                                                             389

# SPECIES CONCEPTS AND DEFINITIONS

ERNST MAYR: MUSEUM OF COMPARATIVE ZOOLOGY,
HARVARD COLLEGE, CAMBRIDGE, MASSACHUSETTS

The importance of one fact of nature is being recognized to an ever increasing extent: that the living world is comprised of more or less distinct entities which we call species. Why are species so important? Not just because they exist in huge numbers, and because each species, when properly studied, turns out to be different from every other, morphologically and in many other respects. Species are important because they represent an important level of integration in living nature. This recognition is fundamental to pure biology, no less than to all subdivisions of applied biology. An inventory of the species of animals and plants of the world is the base line of further research in biology. Whether he realizes it or not, every biologist—even he who works on the molecular level—works with species or parts of species and his findings may be influenced decisively by the choice of a particular species. The communication of his results will depend on the correct identification of the species involved, and thus, on its taxonomy.

Yet, when I was first approached by the Chairman of the Division of Zoology of the American Association for the Advancement of Science to organize a symposium on the species problem I was, to put it mildly, hesitant. Much discussion of this subject in recent years suggested that there was perhaps no need for such a symposium. Ensuing correspondence, however, convinced me otherwise, and certain publications showed clearly that further thinking on this subject is welcome, if not necessary. The species problem continues to be one of the most disputed subjects in biology, in spite of the intense preoccupation with it during the past two hundred years. The recent publications by Spurway, Burma, and Arkell attest this. This symposium can be considered a success if it throws light on some of the disputed questions or even if it does nothing more than lead to a more precise phrasing of the basic points of disagreement.

One way of laying a foundation for such an investigation is to recall some of its history. Who was the first to realize that there is a species problem and what was his proposed solution? What were the subsequent developments? Time does not permit a thorough coverage of the field, but even a glance at the high lights is revealing. If we open a history of biology, the two names mentioned most prominently under the heading of "Species" will be Linnaeus and Darwin. Linnaeus will be cited as the champion of two characteristics of the species, their constancy and their sharp delimitation (their "objectivity"). One of the minor tragedies in the history of biology has been the assumption during the hundred and fifty years after Linnaeus that constancy and clear definition of species are strictly correlated and that one must make a choice of either believing in evolution (the "inconstancy" of species) and then having to deny the existence of species except as purely subjective, arbitrary figments of the imagination, or, as most early naturalists have done, believing in the sharp delimitation of species but thinking that this necessitated denying evolution. We shall leave the conflict at this point and merely anticipate the finding made more than a hundred years after Linnaeus that there is no conflict between the fact of evolution and the fact of the clear delimitation of species in a local fauna or flora.

The insistence of Linnaeus on the reality, objectivity, and constancy of species is of great importance in the history of biology for three reasons. First, it meant the end of the belief in spontaneous generation as far as higher organisms are concerned, a belief which at that time was still widespread. Lord Bacon and nearly all leading writers of the pre-Linnaean period, except Ray, believed in the transmutation of species and the Linnaean conception "of the reality and fixity of species perhaps marks a necessary stage in the progress of scientific inquiry." (See Poulton, 1903, pp. lxxxiv-lxxxvii for further references on the subject). "Until about 1750 almost no one believed that species were stable. Linnaeus had to show that species were not erratic and ephemeral units before organic evolution as we know it could have any meaning." (Conway Zirkle in litt.) The idea that the seed of one plant could occasionally produce an individual of another species was so widespread that it died only slowly. We all know that it raised its ugly head once more during the past ten years. In spite of Redi's and Spallanzani's experiments spontaneous generation was still used in 1851 by the philosopher Schopenhauer as an explanation for the origin of higher categories. Linnaeus thus did for the higher organisms what Pasteur did one hundred years later for the lower.

A second reason why his emphasis was important is that it took the species out of the speculations of the philosophers who approached the species problem in the spirit of metaphysics and stated, for instance, that "only individuals exist. The species of a naturalist is nothing but an illusion" (Robinet, 1768). We shall return later to the point why species are more than merely an aggregate of individuals.

A third reason why the insistence on the sharp delimitation of species in the writings of Linnaeus is of historical importance is that it strengthened the viewpoint of the local naturalist and established the basis for an observational and experimental study of species in local faunas and floras, of which Darwin took full advantage.

Linnaeus was too experienced a botanist to be blind to the evidence of evolutionary change. Greene (1912) gathered numerous citations from his writings which clearly document Linnaeus' belief in the common descent of certain species, and Ramsbottom (1938) and Sirks (1952) have traced how Linnaeus expressed himself more and more freely on the subject, as his prestige grew. Paradoxically, Linnaeus did more, perhaps, to lay a solid foundation for subsequent evolutionary studies by emphasizing the constancy and objectivity of species than if he, like Darwin, had emphasized the opposite.

Darwin looked at the species from a viewpoint almost directly opposite to that of Linnaeus. As a traveler naturalist and particularly because of his studies of domesticated plants and animals he was impressed by the fluidity of the species border and the subjectivity of their delimitation. The views of both Linnaeus and Darwin underwent a change during the life of each. With Linnaeus the statements on the constancy of species became less and less dogmatic through the years. In Darwin, as the idea of evolution became firmly fixed in his mind, so grew his conviction that this should make it impossible to delimit species. He finally regarded species as something purely arbitrary and subjective. "I look at the term species as one arbitrarily given for the sake of convenience to a set of individuals closely resembling each other, and that it does not essentially differ from the term variety which is given to less distinct and more fluctuating forms. . . . The amount of difference is one very important criterion in settling whether two forms should be ranked as species or variety." And finally he came to the conclusion that "In determining whether a form should be ranked as a species or a variety, the opinion of naturalists having sound judgment and wide experience seems the only guide to follow" (Darwin, 1859). Having thus eliminated the species as a concrete unit of nature, Darwin had also neatly eliminated the problem of the multiplication of species. This explains why he made no effort in his classical work to solve the problem of speciation.

The seventy-five years following the publication of the *Origin of Species* (1859) saw biologists rather clearly divided into two camps, which we might call, in a somewhat oversimplified manner, the followers of Darwin and those of Linnaeus. The followers of Darwin, which included the plant breeders, geneticists, and other experimental biologists minimized the "reality" or objectivity of species and considered individuals to be the essential units of evolution. Characteristic for this frame of mind is a symposium held in the early Mendelian days, which endorsed unanimously the supremacy of the individual and the nonexistence of species. Statements made at this symposium (Bessey, 1908) include the following: "Nature produces individuals and nothing more. . . . Species have no actual existence in nature. They are mental concepts and nothing more. . . . Species have been invented in order that we may refer to great numbers of individuals collectively."

Taxonomists, one of the speakers claimed, did not merely name the species found in nature but actually "made" them. "In making a species the guiding principle must be that it shall be recognizable from its diagnosis." A leftover from this period is the statement of a recent author: "Distinct species must be separable on the basis of ordinary preserved material."

It is a curious paradox in the history of biology that the rediscovery of the Mendelian laws resulted in an even more unrealistic species concept among the experimentalists than had existed previously. They either let species saltate merrily from one to another, as did Bateson and DeVries, defining species merely as morphologically different individuals, or they denied the existence of species altogether except as intergrading populations. Whether these early Mendelians considered species as continuous or discontinuous units, they all agreed in their arbitrariness and artificiality. There is an astonishing absence of any effort in this school to study species in nature, to study natural populations.

A study of natural populations had become the prevailing preoccupation in an entirely independent conceptual stream, that of the naturalists, which ultimately traces back to Linnaeus. The viewpoint of the naturalist was particularly well expressed by Jordan (1905), who stated "The units of which the fauna of a region is composed are separated from each other by gaps which, at a given place, are not bridged by anything. This is a fact which can be checked by any observer. Indeed, the activity of a local naturalist begins with the searching out of these units which with Linnaeus we call species." (For a more detailed discussion see Mayr, 1955.) Although this was the prevailing viewpoint among taxonomists, it was completely ignored by the general biologists by whom, as a result of Darwin's theory, "Species were mostly regarded merely as arbitrary divisions of the continuous and ever changing series of individuals found in nature . . . of course, active taxonomists did not overlook the existence of sharply and distinctly delimited species in nature—but as the existence of those distinct units disagreed with the prevailing theories, it was mentioned as little as possible" (Du Rietz, 1930). The two streams of thought are still recognizable today even though most geneticists, under the leadership of Dobzhansky, Huxley, Ford, and others, have swung into Jordan's camp. The principal opponents of the concept of objectively delimitable species are today found among philosophers and paleontologists. Publications maintaining this viewpoint are those of Gregg (1950), Burma (1949, 1954), Yapp (1951), and Arkell (1956). These are only the most recent titles in a vast literature, some of which is cited in the bibliography.

The point which is perhaps most impressive when one studies these voluminous publications is the amount of disagreement that has existed and still exists. The number of possible antitheses that have been established in this field may be characterized by such alternate views, to mention only a few, as follows:

Subjective *versus* objective;
Scientific *versus* purely practical;
Degree of difference *versus* degree of distinctness;
Consisting of individuals *versus* consisting of populations;
Only one kind of species *versus* many kinds of species;
To be defined morphologically *versus* to be defined biologically.

To give a well-documented history of the stated controversies would fill a book. As interesting as this chapter in the history of human thought is, the detailed presentation of the gropings and errors of former generations would add little to the task before us. Let us concentrate therefore on the gradual emergence of the ideas which we, today, consider as central and essential. Three aspects are stressed in most modern discussions of species, that (1) they are based on distinctness rather than on difference and are therefore to be defined biologically rather than morphologically, (2) they consist of populations, rather than of unconnected individuals, a point particularly important for the solution of the problem of speciation, (3) they are more succinctly defined by isolation from non-conspecific populations than by the relation of conspecific individuals to each other. The crucial species criterion is thus not the fertility of individuals, but rather the reproductive isolation of populations. Let us try to trace the emergence of these and related concepts.

It is not surprising that species were considered merely "categories of thought" by many writers in periods so strongly dominated by idealistic philosophy as were the eighteenth and nineteenth centuries. Thoughts as that expressed in the above quoted statement of Robinet were echoed by Agassiz, Mivart, and particularly among those paleontologists who considered their task merely the classification of "objects" ($=$ fossil specimens). In opposition to this, an increasingly strong school developed which considered species as "definable," "objective," "real." Linnaeus was, of course, the original standard bearer of this school to which also belonged Cuvier, De Candolle, and many taxonomists in the first half of the nineteenth century. They supported their case sometimes by purely morphological arguments such as Godron (1853) who stated: "c'est un fait incontestable que toutes les espèces animales et végétables se séparent les unes des autres par de caractères absolues et tranchées." Others used a more biological argument, as I will discuss below.

What is unexpected for this pre-Darwinian period, however, is the frequency with which "common descent" is included in species definitions. When such an emphatically anti-evolutionary author as v. Baer (1828) defines the species as "the sum of the individuals that are united by common descent," it becomes evident that he does not refer to evolution. What is really meant is more apparent from Ray's species definition (1686) or a statement by the Swedish botanist Oeder (1764) that it characterizes species "dass sie aus ihres gleichen entsprungen seien und wieder ihres gleichen erzeugen." Expressions like "community of origin" or "individus descendants des parents communs" (Cuvier) are frequent in the literature. These are actually attempts at reconciling a typological species concept (with its stress of constancy) with the observed morphological variation. Constancy was a property of species taken very seriously not only by Linnaeus and his followers but curiously enough also by Lamarck and by Darwin himself: "The power of remaining constant for a good long period I look at as the essence of a species" (letter to Hooker, Oct. 22, 1864). Such constancy in time was the strongest argument in favor of a morphological species concept, but it could be proved only by the comparison of individuals of different generations. Different morphological "types" that are no more different than mother and daughter or father and son can safely be considered as conspecific. They are "of the same blood." It is obvious that this early stress of descent was essentially the consequence of a morphological species concept. Yet this consideration of descent eventually led to a genetic species definition.

Virtually all early species definitions regarded species only as aggregates of individuals, unconnected except by descent, as is evident not only from the writings of Robinet, Buffon, and Lamarck, but also of much more recent authors (e.g., Britton, 1908; Bessey, 1908). The realization that these individuals are held together by a supraindividualistic bond, that they form populations, came only slowly. Illiger (1800) spoke of species as a community of individuals which produce fertile offspring. Brauer (1885) spoke of the "natural tie of blood relationship" through which the "individuals of a species are held together," and which "is not a creation of the human mind . . . if species were not objective, it would be incomprehensible that even the most similar species mix only exceptionally and the more distant species never." Plate (1914) was apparently the first to state explicitly the nature of this bond: "The members of a species are tied together by the fact that *they recognize each other as belonging together* and reproduce only with each other. The systematic category of the species is therefore entirely independent of the existence of Man." Finally, in the language of current population genetics this community becomes the "co-adapted gene pool," again stressing the integration of the members of the population rather than the aggregation of individuals (a viewpoint which is of course valid only for sexually reproducing organisms).

The growth of thinking in terms of populations went hand in hand with a growing realization that species were less a matter of difference than of distinctness. "Species" in its earlier typological version meant merely "kind of." This, as far as inanimate objects are concerned, is measured in terms of difference. But one cannot apply this same standard to "kinds of" organisms, because there are various biological "kinds." Males and females may be two very different "kinds" of animals. Jack may be a very different "kind" of a person from Bill, yet neither "kind" is a species. Realization of the special aspects of biological variation has led to a restriction in the application of the term species to a very particular "kind," namely the kind that would interbreed with each other. The first three authors found * by me who state this clearly are Voigt (1817), "Man nennt Spezies . . . was sich fruchtbar mit einander gattet, fortpflanzt"; Oken (1830), "Was sich scharet und paaret, soll zu einer Art gerechnet werden"; and Gloger (1833), "What under natural conditions regularly pairs, always belongs to one species." (He stated that by stressing "regularly" he wanted to eliminate the complications due to occasional hybridization.) Gloger later (1856) gave a different, but similar definition: "A species is what belongs together either by descent or for the sake of reproduction." It is interesting how completely all these definitions omit any reference to morphological criteria. They are obviously inapplicable to asexually reproducing organisms.

This is an exceedingly short outline of some of the trends in the development of a modern species concept. More extensive treatments can be found in the publications of Geoffroy St. Hilaire (1859), Besnard (1864), de Quatrefages (1892), Bachmann (1905), Plate (1914), Uhlmann (1923), Du Rietz (1930), Kuhn (1948), and other authors cited in the bibliography. Several conclusions are self-evident. One is that biological or so-called modern species criteria were already used by authors who published more than one hundred years ago, long before Darwin. Another is that a steady clarification is evident, yet that there is still much uncertainty and widespread divergence of opinion on many aspects of the species problem. It is rather surprising that not more agreement has been reached during the past two

* Still earlier statements can no doubt be found in the extensive literature, particularly on hybridization.

hundred years in which these questions have been tossed back and forth. This certainly cannot be due to lack of trying, for an immense amount of time and thought has been devoted to the subject during this period. One has a feeling that there is a hidden reason for so much disagreement. One has the impression that the students of species are like the three blind men who described the elephant respectively as a rope, a column, or a giant snake when touching its tail, its legs, and its trunk.

Perhaps the disagreement is due to the fact that there is more than one kind of species and that we need a different definition for each of these species. Many attempts have been made during the last hundred years to distinguish these several kinds of species, among the most recent being those of Valentine (1949) and Cain (1953). Camp and Gillis (1943) recognized no less than twelve different kinds of species. Yet, a given species in nature might fit into several of their categories, and in view of this overlap no one has adopted either this elaborate classification or any of the simpler schemes proposed before or afterwards.

**Species Concepts**

An entirely different approach to the species problem stresses the kaleidoscopic nature of any species and attempts to determine how many different aspects a species has. Depending on the choice of criteria, it leads to a variety of "species concepts" or "species definitions." At one time I listed five species concepts, which I called the practical, morphological, genetic, sterility, and biological (Mayr, 1942). Meglitsch (1954) distinguishes three concepts, the phenotypic, genetic, and phylogenetic, a somewhat more natural arrangement. Two facts emerge from these and other classifications. One is that there is more than one species concept and that it is futile to search for *the* species concept. The second is that there are at least two levels of concepts. Such terms as "practical," "sterility," "genetic" signify concrete aspects of species which lead to what one might call "applied" species concepts. They specify criteria which can be applied readily to determine the status of discontinuities found in nature. Yet they are secondary, derived concepts, based on underlying philosophical concepts, which might also be called primary or theoretical concepts. I believe that the analysis of the species problem would be considerably advanced, if we could penetrate through such empirical terms as phenotypic, morphological, genetic, phylogenetic, or biological, to the underlying philosophical concepts. A deep, and perhaps widening gulf has existed in recent decades between philosophy and empirical biology. It seems that the species problem is a topic where productive collaboration between the two fields is possible.

An analysis of published species concepts and species definitions indicates that all of them are based on three theoretical concepts, neither more nor less. An understanding of these three philosophical concepts is a prerequisite for all attempts at a practical species definition. And all species criteria or species definitions used by the taxonomist in his practical work trace back ultimately to these basic concepts.

*The Typological Species Concept.* This is the simplest and most widely held species concept. Here it merely means "kind of." There are languages, as for instance German, where the term for "kind" (*Art*) is also used for "species." A species in this concept is "a different thing." This concept is very useful in many branches of science and it is still used by the mineralogist who speaks of "species of minerals" (Niggli, 1949) or the physicist who speaks of "nuclear species." This simple concept of everyday life was incorporated in a more sophisticated manner in the philosophy of Plato. Here, however, the word *eidos* (*species*, in its Latin translation) acquired a double meaning that survives in the two modern words "species" and "idea" both of which are derived from it. According to Plato's thinking objects are merely manifestations, "shadows," of the eidos. By transfer, the individuals of a species, being merely shadows of the same type, do not stand in any special relation to each other, as far as a typologist is concerned. Naturalists of the "idealistic" school endeavor to penetrate through all the modifications and variations of a species in order to find the "typical" or "essential" attributes. Typological thinking finds it easy to reconcile the observed variability of the individuals of a species with the dogma of the constancy of species because the variability does not affect the essence of the eidos, which is absolute and constant. Since the eidos is an abstraction derived from individual sense impressions, and a product of the human mind, according to this school, its members feel justified in regarding a species "a figment of the imagination," an idea. Variation, under this concept, is merely an imperfect manifestation of the idea implicit in each species. If the degree of variation is too great to be ascribed to the imperfections of our sense organs, more than one eidos must be involved. Thus species status is determined by degrees of morphological difference. The two aspects of the typological species concept, subjectivity and definition by degree of difference, therefore depend on each other and are logical correlates.

The application of the typological species concept to practical taxonomy results in the morphologically defined species, "degree of morphological difference" is the criterion of species status. Species are defined on the basis of their observable morphological differences. This concept has been carried to the extreme where mathematical formulas were proposed (Ginsburg, 1938) that would permit an unequivocal answer to the question whether or not a population is a different species.

Most systematists found this typological-morphological concept inadequate and have rejected it. Its defenders, however, claim that all taxonomists, when classifying the diversity of nature into species, follow the typological method and distinguish "archetypes." At first sight there seems an element of truth in this assertion. When assigning specimens either to one species or to another, the taxonomist bases his decision on a mental image of these species that is the result of past experience with the stated species. The utilization of morphological criteria is valuable and productive in the taxonomic practice. To assume, however, that this validates the typological species concept overlooks a number of important considerations. To begin with, the mental construct of the "type" is subject to continuous revision under the impact of new information. If it is found that two archetypes represent nothing more than two "kinds" within a biological species, they are merged into a single one. It was pointed out above that males and females are often exceedingly different "kinds" of animals. Even more different are in many animals the larval stages, or in plants sporophyte and gametophyte, or in polymorph populations the various genotypes. A strictly morphological-typological concept is inadequate to cope with such intraspecific variation. It is equally incapable of coping with another difficulty, namely an absence of visible morphological differences between natural populations which are nevertheless distinct and reproductively isolated, and therefore to be considered species. The frequent occurrence of such "cryptic species" or "sibling species" in nature has been substantiated by various genetic, physiological, or ecological methods. They form another decisive argument against defining species on a primarily morphological basis. Any attempt in these two situations to define species "by degree of difference" is doomed to failure. Degree of difference can be specified only by a purely arbitrary decision.

More profound than these two essentially practical considerations is the fact that the typological species concept treats species merely as random aggregates of individuals which have the "essential properties" of the "type" of the species and "agree with the diagnosis." This static concept ignores the fact that species are not merely classes of objects but are composed of natural populations which are integrated by an internal organization and that this organization (based on genetic, ethological, and ecological properties) gives the populations a structure which goes far beyond that of mere aggregates of individuals. Even a house is more than a mere aggregate of bricks or a forest an aggregate of trees. In a species an even greater supraindividualistic cohesion and organization is produced by a number of factors. Species are a reproductive community. The individuals of a species of higher animals recognize each other as potential mates and seek each other for the purpose of reproduction. A multitude of devices insures intraspecific reproduction in all organisms. The species is an ecological unit which, regardless of the individuals of which it is composed, interacts as a unit with other species in the same environment. The species, finally, is a genetic unit consisting of a large, intercommunicating gene pool whereas each individual is only a temporary vessel holding a small portion of this gene pool for a short period of time. These three properties make the species transcend a purely typological interpretation or the concept of a "class of objects."

The very fact that a species is a gene pool, with numerous devices facilitating genic intercommunication within and genic separation from without, is responsible for the morphological distinctness of species as a byproduct of their biological uniqueness. The empirical observation that a certain amount of morphological difference between two populations is normally correlated with a given amount of genetic difference is undoubtedly correct. Yet, it must be kept in mind at all times that the biological distinctness is primary and the morphological difference secondary. As long as this is clearly understood, it is legitimate and indeed very helpful to utilize morphological criteria. This caution has been exercised, consciously or unconsciously, by nearly all proponents of the morphological species concept. As pointed out by Simpson (1951) and Meglitsch (1954), they invariably abandon the morphological concept when it comes in conflict with biological data. This was true for Linnaeus himself and for his followers to the present day.

The typological species concept has a certain amount of operational usefulness when applied to inanimate objects. Ignoring the population structure of species, however, and incapable of coping with the facts of biological variation, it has proved singularly inadequate as a conceptual basis in taxonomy. Much of the criticism directed against the taxonomic method was provoked by the application of the typological concept by taxonomists themselves or by other biologists who mistakenly considered it the basis of taxonomy.

*The Second Species Concept.* This is sometimes called the nondimensional species concept and has no generally accepted designation. The essence of this concept is the relationship of two coexisting natural populations in a nondimensional system, that is, at a single locality at the same time (sympatric and synchronous). This is the species concept of the local naturalist. It was introduced into the biological literature by the English naturalist John Ray and confirmed by the Swedish naturalist Linnaeus. It is based not on difference but on distinction, and this distinction in turn is characterized by a definite mutual relationship, namely that of reproductive isolation. The word "species" is here best defined in combination with the word "different." The relationship of two "different species" can be objectively defined as reproductive isolation. We have, thus, an objective yardstick for this species concept, something that is absent in all others. Philosophers have objected to the use of the terms "objective" or "real" for species, and it may be more neutral to use the terms arbitrary or nonarbitrary (Simpson, 1951). Presence or absence of interbreeding of two populations in a nondimensional system is a completely nonarbitrary criterion. Since the nondimensional species concept is based on a relationship, the word species is here equivalent to words like, let us say, the word brother, which also has a meaning only with respect to a second phenomenon. An individual is a brother only with respect to someone else. Being a brother is not an inherent property as hardness is a property of a stone. Describing a presence or absence relationship makes this species concept nonarbitrary.

This species concept seems so self-evident to every naturalist that it is only rarely put in words. That the species is more than an aggregate of individuals, held together by a biological bond, has long been realized, as was pointed out in the historical survey above. The interbreeding within the species is more conspicuous, and it was thus more often emphasized than is the reproductive isolation against other species. Eimer, as early as 1889 (p. 16) defined species as "groups of individuals which are so modified that successful interbreeding [with other groups] is no longer possible." The first author, however, who stated the nondimensional species concept in its full extent and implication was Jordan (1905).

In spite of its theoretical superiority, the nondimensional species has a number of serious drawbacks (which will be discussed later), particularly its limitation to sexually reproducing species and to such without the dimensions of space and time.

Yet, as a basic, nonarbitrary yardstick, this is the species concept on which we have to fall back whenever we encounter a borderline situation.

*The Third Species Concept.* This is a concept of an entirely different kind, it is the concept of the polytypic or multidimensional species. In contradistinction to the other two concepts, of which one is based on a degree of difference, the second one on the completeness of a discontinuity, this concept is a collective one. It considers species as groups of populations, namely such groups as interbreed with each other, actually or potentially. Thus this species concept is a concept of the same sort as the higher categories, genus, family, or order. Like all collective categories it faces the difficulty, if not impossibility, of clear demarcation against other similar groupings. What this species gains in actuality by the extension of the nondimensional situations in space and time, it loses in objectivity. As unfortunate as this is, it is inevitable since the natural populations, encountered by the biologist, are distributed in space and time and cannot be divorced from these dimensions. Thus, this species concept likewise has its good and its bad points.

## Species Definitions

All our reasoning in discussions of "the species" can be traced back to the stated three primary concepts. As concepts, of course, they cannot be observed directly, and we refer to certain observed phenomena in nature as "species," because they conform in their attributes to one of these concepts or to a mixture of several concepts. From these primary concepts, just discussed, we come thus to secondary concepts, based on particular aspects of species. We have already mentioned the so-called morphological species concept, which, in most cases, is merely an applied typological concept, using morphological criteria. The case of the so-called genetic species concept shows that all three of the basic concepts can be expressed, on this level, in genetic terms. Some geneticists, for instance, subscribed to the typological concept and defined species by the degree of genetic difference as did Lotsy or DeVries; others stressed the genetic basis of the isolating mechanisms between species thereby endorsing the nondimensional species concept; still others finally emphasized the gene flow among interbreeding populations in a multidimensional system, thus adopting the multidimensional collective species concept. All three groups of geneticists thought they were dealing with a uniquely "genetic species concept," yet they were merely observing secondary manifestations of the primary concepts.

It is evident from the analysis of the morphological and genetic species concepts, that such derived concepts are attempts to deal directly with the discontinuities in nature. In the past, almost every taxonomist worked with his own personal yardstick based on a highly individual mixture of elements from the three basic concepts. As a consequence one taxonomist might call species every polymorph variant, a second one every morphologically different population, and a third one every geographically isolated population. Such lack of standards, which is still largely characteristic for the taxonomic literature, has been utterly confusing to taxonomists and other biologists alike. It has therefore been the endeavor of many specialists within recent decades to find a standard yardstick, on which there could be general agreement. A historical study of species definitions indicates clearly a trend toward acceptance of a synthetic species definition, often referred to as "biological species" definition. It is essentially based on the nondimensional ("reproductive gap") and the multidimensional ("gene flow") species concepts. Nearly all species definitions proposed within the last fifty years incorporate some elements of these two concepts. This is evident from the species definitions of Jordan (Mayr, 1955), Stresemann (1919), and Rensch (1929). Du Rietz (1930) called the species "a syngameon . . . separated from all others by . . . sexual isolation." Dobzhansky (1935) was apparently the first geneticist to define species in the terms customary among naturalists and taxonomists, namely interbreeding and reproductive isolation; other recent definitions are variants of the same theme. Mayr (1940) defined species as "groups of actually or potentially interbreeding natural populations which are reproductively isolated from other such groups." Simpson (1943) gave the definition "a genetic species is a group of organisms so constituted and so situated in nature that a hereditary character of any one of these organisms may be transmitted to a descendent of any other," and Dobzhansky (1950) defined the species as "the largest and most inclusive . . . reproductive community of sexual and cross-fertilizing individuals which share in a common gene pool."

It might be useful to mention some qualifications which are often included in species definitions but needlessly so. Anything that is equally true for categories above and below species rank should be omitted, since there is no sense burdening a species definition with features which do not help discrimination between species and infraspecific populations.

1. Species characters are adaptive. This component of Wallace's (1889) species definition was correctly rejected by Jordan (1896). Adaptiveness is not diagnostic for species characters and not even necessarily true. Not every detail of the phenotype needs to be adaptive as long as the phenotype as a whole is adaptive and as long as the genotype itself is the result of selection.

2. Species are evolved and evolving. Again this is true for the entire organic world from the individual to the highest categories and adds nothing to the species definition.

3. Species differ genetically. This is only the morphological species concept expressed in genetic terms. It does not permit discriminating species from infraspecific populations or from individuals.

4. Species differ ecologically. This qualification is unnecessary and misleading for the same reasons as the genetic one. Ecological differences exist for all ecotypes within species and in general for all geographical isolates. Conspecific populations are sometimes more different ecologically than are good species.

A yardstick, such as the biological species concept, is not automatic. To apply it properly requires skill and experience. This is particularly true in the recognition of situations where it cannot be applied, for one reason or another, and where the worker has to fall back on the criterion of "degree of difference." It will be one of the tasks of this symposium to investigate to what extent the diversity of animal and plant life permits application of the standard yardstick of the biological species concept. When can it not be applied and for what reasons? What types of difficulties are there? And finally, are there perhaps other basic concepts in addition to the stated ones? By approaching the species problem with new questions and new material, perhaps this symposium can make a contribution to its solution.

## REFERENCES

Arkell, W. J. 1956. The species concept in paleontology. *Systematics Assoc. Publ. No. 2*, pp. 97-99.

Bachmann, H. 1905. Der Speziesbegriff. *Verhandl. schweiz. naturforsch. Ges.*, **87**, 161-208.

Baer, K. E. von. 1828. *Entwickelungs-Geschichte der Thiere*. Königsberg.

Besnard, A. F. 1864. Altes und Neues zur Lehre über die organische Art (Spezies). *Abhandl. zool. mineral. Ver. Regensburg*, **9**, 1-72.

Bessey, C. E. 1908. The taxonomic aspect of the species question. *Am. Naturalist*, **42**, 218-24.

Brauer, F. 1885. Systematisch-zoologische Studien. *Sitzber. Akad. Wiss. Wien*, **91** (Abt. 1), 237-413.

Britton, N. L. 1908. The taxonomic aspect of the species question. *Am. Naturalist*, **42**, 225-42.

Burma, B. H. 1949a. The species concept: A semantic review. *Evolution*, **3**, 369-70.

Burma, B. H. 1949b. The species concept: Postscriptum. *Evolution*, **3**, 372-73.

Burma, B. H. 1954. Reality, existence, and classification: A discussion of the species problem. *Madroño*, **7**, 193-209.

Cain, A. J. 1953. Geography, ecology and coexistence in relation to the biological definition of the species. *Evolution*, **7**, 76-83.

Camp, W. H., and C. L. Gillis. 1943. The structure and origin of species. *Brittonia*, **4**, 323-85.

Darwin, C. 1859. *On the Origin of the Species by Means of Natural Selection*. London.

Dobzhansky, T. 1935. A critique of the species concept in biology. *Phil. Sci.*, **2**, 344-55.

Dobzhansky, T. 1950. Mendelian populations and their evolution. *Am. Naturalist*, **84**, 401-18.

Doederlein, L. 1902. Über die Beziehungen nahe verwandter "Thierformen" zu einander. *Z. Morphol. Anthropol.*, **26**, 23-51.

Dougherty, E. C. 1955. Comparative evolution and the origin of sexuality. *Systematic Zool.*, **4**, 145-69.

Du Rietz, G. E. 1930. The fundamental units of botanical taxonomy. *Svensk. Bot. Tidsskr.*, **24**, 333-428.

Eimer, G. H. T. 1889. *Artbildung und Verwandtschaft bei Schmetterlingen.* Jena, Vol. II, p. 16.

Geoffroy Saint Hilaire, I. 1859. Histoire naturelle génerale des règnes organiques, **2**, 437, Paris.

Ginsburg, I. 1938. Arithmetical definition of the species, subspecies and race concept, with a proposal for a modified nomenclature. *Zoologica*, **23**, 253-86.

Gloger, C. L. 1833. *Das Abändern der Vögel durch Einfluss des Klimas.* Breslau.

Gloger, C. L. 1856. Ueber den Begriff von "Art" ("Species") und was in dieselbe hinein gehört. *J. Ornithol.*, **4**, 260-70.

Godron, D. A. 1853. *De l'espèce et des races dan les êtres organisés et specialment de l'unité de l'espèce humaine.* Paris, 2 vols.

Greene, E. L. 1912. Linnaeus as an evolutionist. In *Carolus Linnaeus*, pp. 73-91. C. Sower & Co., Philadelphia, Pa.

Gregg, J. R. 1950. Taxonomy, language and reality. *Am. Naturalist*, **84**, 419-35.

Huxley, J. 1942. *Evolution, the Modern Synthesis.* Allen and Unwin, London.

Illiger, J. C. W. 1800. *Versuch einer systematischen vollständigen Terminologie für das Thierreich und Pflanzenreich.* Helmstedt.

Jordan, K. 1896. On mechanical selection and other problems. *Novit. Zool.*, **3**, 426-525.

Jordan, K. 1905. Der Gegensatz zwischen geographischer und nichtgeographischer Variation. *Z. wiss. Zool.*, **83**, 151-210.

Kuhn, E. 1948. Der Artbegriff in der Paläontologie. *Eclogae Geolog. Helv.*, **41**, 389-421.

Lorkovicz, Z. 1953. Spezifische, semispezifische und rassische Differenzierung bei *Erebia tyndarus* Esp. *Rad. Acad. Yougoslave*, **294**, 315-58.

Mayr, E. 1940. Speciation phenomena in birds. *Am. Naturalist*, **74**, 249-78.

Mayr, E. 1942. *Systematics and the Origin of Species*. Columbia University Press, New York, N.Y.

Mayr, E. 1949. The species concept: Semantics versus semantics. *Evolution*, **3**, 371-72.

Mayr, E. 1951. Concepts of classification and nomenclature in higher organisms and microorganisms. *Ann. N.Y. Acad. Sci.*, **56**, 391-97.

Mayr, E. 1955. Karl Jordan's contribution to current concepts in systematics and evolution. *Trans. Roy. Entomol. Soc. London*, pp. 45-66.

Mayr, E. 1956. Geographical character gradients and climatic adaptation. *Evolution*, **10**, 105-8.

Mayr, E., and C. Rosen. 1956. Geographic variation and hybridization in populations of Bahama snails (*Cerion*). *Am. Museum Novit.*, **1806**, 1-48.

Mayr, E., E. G. Linsley, and R. L. Usinger. 1953. *Methods and Principles of Systematic Zoology*. McGraw-Hill Book Co., New York, N.Y.

Meglitsch, P. A. 1954. On the nature of the species. *Systematic Zool.*, **3**, 49-65.

Niggli, P. 1949. *Probleme der Naturwissenschaften* (Der Begriff der Art in der Mineralogie). Basel.

Plate, L. 1914. Prinzipien der Systematik mit besonderer Berücksichtigung des Systems der Tiere. In *Die Kultur der Gegenwart*, III (iv, 4), pp. 92-164.

Poulton, E. B. 1903. What is a species? *Proc. Entomol. Soc. London* for 1903, pp. lxxvii-cxvi.

Quatrefages, A. de. 1892. Darwin et les précurseurs français. Paris.

Ramsbottom, J. 1938. Linnaeus and the species concept. *Proc. Linnean Soc. London*. (150 session), pp. 192-219.

Ray, J. 1686. *Historia Plantarum*, p. 40.

Rensch, B. 1929. *Das Prinzip geographischer Rassenkreise und das Problem der Artbildung*. Bornträger Verl., Berlin.

Schopenhauer, A. 1851. *Parerga und Paralipomena: kleine philosophische Schriften*. Vol. 2, pp. 121-22. Berlin.

Simpson, G. G. 1943. Criteria for genera, species, and subspecies in zoology and paleozoology. *Ann. N.Y. Acad. Sci.*, **44**, 145-78.

Simpson, G. G. 1951. The species concept. *Evolution*, **5**, 285-98.

Sirks, M. J. 1952. Variability in the concept of species. *Acta Biotheoretica*, **10**, 11-22.

Spring, A. F. 1838. *Ueber die naturhistorischen Begriffe von Gattung, Art und Abart und über die Ursachen der Abartungen in den organischen Reichen.* Leipzig.

Spurway, H. 1955. The sub-human capacities for species recognition and their correlation with reproductive isolation. *Acta XI Congr. Intern. Orn.*, Basel, 1954, pp. 340-49.

Stresemann, E. 1919. Über die europäischen Baumläufer. *Verhandl. Orn. Ges. Bayern*, **14**, 39-74.

Sylvester-Bradley, P. C. 1956. The species concept in paleontology. Introduction. *Systematics Assoc. Publ. No. 2.*

Thomas, G. 1956. The species concept in paleontology. *Systematics Assoc. Publ. No. 2*, pp. 17-31.

Uhlmann, E. 1923. Entwicklungsgedanke und Artbegriff in ihrer geschichtlichen Entstehung und sachlichen Beziehung. *Jena. Z. Naturw.*, **59**, 1-114.

Valentine, D. H. 1949. The units of experimental taxonomy. *Acta Biotheoretica*, **9**, 75-88.

Voigt, F. S. 1817. *Grundzüge einer Naturgeschichte als Geschichte der Entstehung und weiteren Ausbildung der Naturkörper.* Frankfurt a.M.

Wallace, A. R. 1889. *Darwinism: An Exposition of the Theory of Natural Selection, with Some of Its Applications.* London.

Yapp, W. B. 1951. Definitions in biology. *Nature*, **167**, 160.

# THE SPECIES AS A FIELD FOR GENE RECOMBINATION

HAMPTON L. CARSON: DEPARTMENT OF ZOOLOGY, WASHINGTON UNIVERSITY, ST. LOUIS, MISSOURI

The impact of genetics has reached into every corner of biology. Nowhere has its influence been stronger than in the related fields of taxonomy, speciation, and evolution. In fact, around the fresh insight and new techniques provided by genetics has grown a new science, which might be called experimental evolution. Basically, its substance is derived from population genetics.

Huxley's *The New Systematics* (1940) chronicled the effects of genetic discoveries on taxonomy, ecology, paleontology and geographical distribution. At the same time the groundwork was laid for the revitalization of work on speciation. The same period, that is, just before and during the early years of World War II, witnessed the appearance of an extraordinary series of broad works on evolution. Thus the books of Dobzhansky (1937), Mayr (1942), Huxley (1942), and Simpson (1944) came in rapid succession. What is remarkable about these works is not their differences of opinion but rather their essential agreement on basic principles. A modern synthetic theory of evolutionary cause had apparently been crystallized.

These key syntheses drew their freshness of approach largely from the attention given to genetic phenomena; their function was to integrate widely in evolutionary thought the major principles established by the more strictly genetic and mathematical fundamentals elaborated by the earlier works of Fisher (1930), Haldane (1932), and Wright (1932). It takes time for a synthesis of this enormity, even in its barest outlines, to be evaluated and digested and thus serve as a guide to future biological work. One could hardly expect immediate wide dissemination of such a major generalization, let alone the years of contemplation and testing which precede general acceptance of any such idea. In the present instance, an initial delay was imposed by World War II. Only now is it just beginning to be possible for the new generation of evolutionists to start critical evaluation of what appears to have been the most spectacular era in basic biological thought about evolution since the years following the publication of the *Origin of Species*.

The most immediate effect of the syntheses mentioned above was to stimulate a rash of investigations designed to answer new questions about the species and especially about its populations. These new questions arise primarily from purely theoretical consideration of the consequences of Mendelian inheritance in interbreeding communities of organisms. A geneticist, in most instances, can work only with organisms that will interbreed freely so as to produce abundant offspring. His work is thus of necessity usually at or below the species level. It will therefore be a foregone conclusion that the evolutionarily inclined geneticist will have developed his own perspective—his own particular way of looking at the species. The geneticist who is oriented toward populations is really attempting the experimental study of heredity on a level above that of the individual. The present effort will be a somewhat personal picture—one geneticist's view of the accomplishments of this approach in the last fifteen years and the effects that it has had on the species concept.

## Genotype, Phenotype, and Wild Type

The term genotype is sometimes used in a strictly operational way in experimental genetics to refer to the genic condition under observation at a single locus or sometimes at a relatively small number of loci (e.g., Aa or Aa Bb Cc Dd). In the sense used in this paper, however, the word refers to the entire genetic constitution or the sum total of the genes of an individual organism. This point requires emphasis because, as will be pointed out, from the evolutionary point of view, it is unrealistic to select a few genes for special attention. We are being forced more and more to emphasize the total genotype.

The geneticist is continually trying to see in behind the façade of the phenotype to the genotypic basis that is the hereditary endowment of an individual. The genotype is truly a property of an individual organism, a kind of personal information which crosses the narrow hereditary bridge from one generation to the next. Sexual reproduction generates diverse genotypes, and each genotype confers individuality. It is because of this individuality of the genotype that the suffix "type" in the word conveys an erroneous idea. The geneticist is wholly unable to set up types with properties of the types of the taxonomist or, in fact, to work with types of any sort. Indeed, the major emphasis of this article arises from the fact that one of the most important corollaries of our modern understanding of sexual reproduction is that types are absolutely foreign, in fact diametrically opposed, to the whole concept.

The term "wild type" has run into similar difficulties. It, again, was originally used in an operational sense to refer to an allelic condition at a single locus on a chromosome. An unjustified assumption of uniformity from individual to individual is made when the term wild type is extended to cover many loci not under direct experimental observation. Thus, in some of the earlier studies of genetics there was a tendency for natural populations of organisms to be interpreted as if the individuals comprising them were essentially homozygous for "wild-type" genes, with only a few scattered recessive mutant alleles here and there at various loci. This "classical" hypothesis is now giving way to what Dobzhansky (1955) refers to as the "balance" hypothesis. This latter designation refers to the fact that recent studies of Mendelian populations under natural conditions have revealed that a fantastic amount of genetic variability is concealed in recessive form in the individuals of these populations.

The mechanisms which maintain a high level of genic variability in natural populations are only just now becoming clear. It is not within the scope of this short paper to review them, and the reader is referred to the review of Dobzhansky (1955). The very existence of this extensive variability, however, is the real focal point in the unique view of the species that the geneticist is beginning to take. The statement may safely be made that in all probability, no two individuals in a sexual population are genotypically identical. The geneticist thus sees, in every population, arrays of individually different genotypes. He sees the group as a whole, as a statistically varying array. The newer methods of population genetics are attempting to deal directly with the laws governing the behavior and fate of genotypes in populations.

**The Species as a Gene Pool**

Those who have been bold enough to advance genetical definitions of species are on safest ground when they deal with species which can be shown to be sexually reproducing, cross-fertilizing arrays of individuals. When a species has such a breeding structure, it is tempting to view the hereditary material possessed by the species as a whole as a gigantic pool of genes. Reproducing individuals may be viewed as throwing random assortments of their genes, packaged in gametes, into the pool. The individuals of the next generation are thus genetically chance recombinants drawn from this gene pool. Dobzhansky (1950) has utilized Wright's term "Mendelian population" for the reproductive community of individuals who share in a common gene pool.

Such a formulation as the above should serve only to direct thinking along the general path followed by the population geneticist. If it is taken more seriously than this, it is likely to be misleading. Simple concepts of gene frequency and the consequent related predictions of the Hardy-Weinberg law may be demonstrated in a very simple model of a gene pool. A hundred poker chips, 30 red and 70 white can be pooled in a fishbowl. The two colors represent the two alternative alleles at one locus. Drawing out chips in pairs thus represents zygote formation for a single locus; the homozygotes will be red-red and white-white and the heterozygotes red-white. The relationship of gametic frequency to zygotic frequency, predicted by the Hardy-Weinberg law, may thus be directly observed.

This model may be elaborated somewhat; for instance, one may intuitively grasp some of the enormous statistical consequences of Mendelian heredity in sexual reproduction by drawing zygotes from two or more fishbowls simultaneously. One may attempt to diagram by this method the long-range effects of selection on certain zygotic types. The elimination of a deleterious gene, for instance, may be observed by extending selection over a number of generations. In short, it is possible through such a model to gain a crude measure of realization of how evolutionary changes in gene frequency can occur through the media of mutation, recombination and differential reproduction.

Such models, however, have serious drawbacks. In many ways, for instance, it is unsatisfactory to verbalize a mathematical concept like the above. The naturalist will also object to the pictorial analogy, even for emphasis of a point of view, of the species population to chips in a fishbowl. It is so crude as to make it necessary to point out immediately the important ways in which species populations *do not* behave like poker chips in a fishbowl. The employment of the word "pool" in the expression above, is nevertheless tremendously important to stress the point that groups of individual organisms really possess a corporate genotype, transcending that of the individuals. No other word in common usage appears to come closer than "pool" in expressing the nature of the phenomena involved. At this juncture it seems unwise to invent a new one; rather it is better to tell what a species pool is really like.

## The Local Population

The term "gene pool" has also been used in a number of senses. Thus, as is done very widely, one may refer to the total gene resources, both past and present, of a species as its gene pool. Whenever the term "gene flow" is employed, as is often glibly done when neither actual genes nor actual flow are under experimental observation, this broad view of the gene pool in both space and time is implied. This concept is most useful for the student of the species who has an essentially historical approach. He attempts to interpret the mosaic of present-day populations of a single species in terms of the action of evolutionary directive factors on the totality of gene resources which have been available in its ancestral populations.

In yet another sense, the parameter of time may be omitted and the contemporary genetic material of a species may be viewed as constituting a pool of genes. Now in this case, the emphasis is on the total genic material currently found throughout the geographical range of the species. This gene pool is thus the actual legacy willed by selection to the present-day series of generations which a single human investigator may observe. This type of pool is the province of the animal and plant breeder, the eugenicist, and the evolutionist who may be tempted to predict what may happen next.

Neither of the above "pools," however, is the really crucial one from the point of view of the population geneticist. His attention is focused on a gene pool which can be defined much more rigorously and which may be studied with precise observational and experimental methods. This is the gene pool found in a single contemporary local breeding population, made up of randomly, or almost randomly, interbreeding individuals. Wright has encouraged the use of Gilmour and Gregor's (1939) term deme to refer to this microgeographical unit, momentarily suspended in time, which lies at the base of the hierarchal array of species populations.

Because breeding populations of species are entities distributed in space, the individual genotypes formed in local populations in any one generation are never drawn at random from the entire gene pool of the species. Despite dispersal and outcrossing mechanisms, organisms tend to breed with those that are close to them, in a physical sense, in their environment. This gives reality to the deme and makes it the unit which most closely resembles the fishbowl model.

Of central importance is the fact that the local population is the only part of the larger gene pool which is active from the evolutionary point of view in any one generation, that is, at any one time. Although transitory, the local population is always the unit in which any new character first makes its appearance. It is the only place where gene recombination can be effective as a directive factor in evolution. Indeed, the local population, at any one time level, always represents the point of contact between the hereditary material of a species and its environment. It is at this point that new genotypes are generated and this is where they meet the immediate acceptance or rejection of the omnipotent environment.

The local population is not equivalent to any of the intraspecific designations of the taxonomist, such a subspecies, local race, or variety. The local population is not only suspended in time, but it is usually much smaller than any of these. Furthermore, even a local race may have a population structure which divides it genetically into a series of partially isolated interbreeding populations. A local population can, for example, be a hybrid swarm; in such a case we are observing the release of an enormous amount of variability through the recombination of genomes which have been separated historically. The local population may indeed have any one of quite a variety of past histories, which will affect the amount of genetic recombination which occurs. The fate of the genetic variability and of the continuing composition of the gene pool nonetheless is determined by the inexorable laws of population genetics and the directive factors of evolution working at a local level.

## Recombination in Populations

Genetics is really the study of sexual reproduction. An enormous amount of research on meiosis and chromosome cycles, perhaps the bulk of it, was engaged in when only the vague outlines of the evolutionary and genetic meanings of these chromosome behaviors were clear. The details of sexual phenomena in protozoa, for example, have been filled in with great elaborateness as an adjunct to the ancient problems of taxonomic relationships and phylogenetic interpretations. The chromosome cycles that one observes in the life history of an organism, however, are not merely historical relics to be studied with the techniques and attitudes of comparative anatomy and phylogeny —they are rather dynamic genetic systems, concerned in most cases with the recombination of genes. They should be studied for what they are—basic mechanisms which determine the genetic makeup of the individuals which comprise species populations.

One of the major consequences of the particulate nature of the hereditary material is the potentiality which automatically exists for recombination. The sexual mode of reproduction, as had been brought out so brilliantly in the work of Darlington (1939) and White (1954), is biologically meaningful only when viewed basically as a process of recombination of hereditary particles. The analytical methodology of the whole science of genetics is rooted in recombination; in fact, perhaps the most satisfactory definition of a gene is a hereditary entity which acts as a unit in recombination.

Given a certain amount of hereditary diversity provided by gene mutation, the sexual process, through its essential feature of recombination, can generate an enormous number of diverse genotypes. These genotypes, or recombinants, are exposed to the action of natural selection in local populations. In the above context, recombination occupies a focal position in modern theory of evolutionary cause. For this reason, the devices whereby sexual reproduction accomplishes recombination deserve close scrutiny.

Recombination in sexual reproduction is a function of three processes. These are: crossing over between homologous chromosomes, random segregation of chromatids from bivalent chromosomes, and chance recombination of gametes. Thus the greatest amount of recombination will result in an organism which has intensive and extensive crossing over, high chromosome number and a high degree of interbreeding between relatively unrelated individuals (outcrossing). Conversely, if exchanges between homologous chromosomes are few, if the chromosome number is small and if a considerable degree of inbreeding occurs, the amount of recombination that can be produced may be relatively much smaller.

As natural selection operates with gene combinations, it follows that where the hereditary material of the species is faced with a new environmental challenge, the success with which the species meets this challenge depends to a very great extent on its ability to form new gene combinations. In other words, the possession of a high potential for recombination carries with it a correspondingly high potential for rapid evolution, whereas general inability to recombine the hereditary material implies evolutionary stagnation and greatly restricts ability to alter the composition of the gene pool.

If one wants to interpret the recent past history of a species or to predict how rapid and far-reaching an evolutionary change might occur in it, it is of the greatest importance to know the specific system by which it reproduces. If the system is a sexual one, then some knowledge of the potential recombination which might be accomplished in a local population of this species is imperative. Thus, the geneticist views the species population in part with a sort of x-ray vision—looking within to such matters as details of meiosis—as well as assessing the importance of the more extrinsic factors which affect recombination, such as the mating system.

An important point of emphasis is that estimates of the ability to undergo recombination express only a potential ability—a potential with which the species might respond should the conditions arise which demand evolutionary aggression. If the conditions do not arise, the status quo is likely to be maintained. The potential may be there but never actually realized. The possession of a high recombination potential by a species is no more an automatic guarantee of further progress than is a high mutation rate.

To return to the point of view of the local population. As a single generation is bridged by a population, the increment of change possibly depends entirely on the wealth of gene combinations which are formed in the local pools. The widespread geographical nature of most species means that only a small portion of the total pool of genes in the species is drawn into combination at any one time. Thus it is that the geneticist finds himself forced, like the members of all other disciplines which deal with species, to consider both geographical and historical matters.

## The Species in Space and Time

As we have seen previously, the local population occupies a really pivotal position for the genetic nature of the species. Its composition depends not only on the past history of the local population itself but also on the history of the populations near it, and from which it may have inherited genes. Thus the parameters of time and space in the study of local populations are so interrelated that it is often difficult to separate their effects. The words "gene flow" are sometimes used to describe the process whereby the local population obtains influx of hereditary material from the outside. This is, in most cases, an extremely slow, gradual process, usually extending over many generations.

In a larger sense, then, the gene pool of the species is distributed in time as well as in space. The local population where the actual recombination trials are taking place may owe its contemporary composition to very complex interactions with the populations in its immediate ancestry. In certain cases, cytogenetic studies enable us to prove that this past history involved polyploidy or hybridization or both. In other cases, it is possible to interpret the present genetic peculiarities of a species population as being traceable to recent strong isolation of a local population. In any event, the net result is an alteration of the content of the gene pool and thus both qualitatively and quantitatively of the raw materials with which recombination operates. Thus, for instance, a newly formed hybrid swarm, such as one that has resulted from the breakdown of ecological barriers by human disturbance of the habitat, may be looked upon primarily as a process resulting in the sudden release of recombinations consequent upon the original wide outcross. Hybridization between species with high chromosome numbers which maintain high frequencies of distributed crossovers is, in fact, the condition which would appear to be able to release the greatest amount of hereditary variability in the shortest possible time.

## How Much Recombination?

The amount of recombination permitted in a given species by the genetic system of that species is best looked upon as an adaptive property which is under the control of natural selection. Too much recombination is inadaptive because it tends to break up adaptive complexes of genes which have been welded together by natural selection. This reduces the immediate fitness of the organism. In order to survive, a species must maintain high immediate fitness and it is not surprising that devices limiting or preventing recombination are common in species populations. Not infrequently, as will be brought out below, the process appears to have been carried to its logical conclusion and recombination is inhibited completely. At this point, sexual reproduction ceases to have significance as a biologically important process.

Organisms which have completely discarded sexual reproduction, and reproduce exclusively by asexual means are definitely in the minority in nature. Thus it appears that too little recombination jeopardizes the ability of the species to meet drastic changes in environmental conditions over very long periods of time (see Thoday, 1953). Most organisms which have survived to the present day display a balance between these two forces. There is neither too much recombination on the one hand, which tends to break up adaptive complexes, or too little, which leads to evolutionary rigidity, specialization, inability to change, and eventually extinction in the face of an environmental challenge.

## Open, Restricted, and Closed Recombination Systems

If we refer to the sum total of all the devices leading toward recombination as the "recombination system," we may attempt to assess the capacity of a given species to generate recombinations in a typical contemporary local population. The effectiveness of the recombination system obviously varies enormously from one species to the next. Such variation is almost certain to be continuous in such a way that hard and fast categories could not conceivably be applied. Thus the attempt which follows is designed merely to give a very rough outline for this formulation of the problem. The point should again be stressed that it is in almost every case necessary to speak in terms of potential rather than actual recombination. Obviously, too, recombination can be effective only where there are differences which can be recombined. Precise methods do not as yet exist for estimating the total amount of recombinable heterozygosity in a species or a population. Heterozygosity has its ultimate origin in the mutation process, but its existence in any given population will depend on many different factors such as possible recent wide outcrossing, including species hybridization, a recent or past high rate of mutation, or recent drastic fluctuations in population size.

A recombination system wherein a high degree of recombination is permitted may be referred to as open. The human species, for example, displays all the ingredients which may be associated with such a system. The chromosome number ($2n = 48$) is high, so that a substantial amount of recombination occurs by the segregation of whole chromosomes alone, quite apart from crossing over. The number of chromosomally different gametes, which can potentially be formed by an individual which carries one pair of allelic differences per pair of homologous chromosomes is given by the simple formula $2^n$, where $n$ is the number of chromosome pairs concerned. In the present example, the number is $2^{24}$, more than 16 million, produced by a single individual. Perhaps even more important, however, there is evidence that crossing over between homologues is extensive and unhampered throughout the genome. This system is superimposed on the foregoing one. If the assumption is made that a moderate number of allelic differences exist per chromosome pair, a fantastically high recombination potential is provided on an individual basis.

Even under rather close local inbreeding systems, the purely chromosomal devices mentioned above have high potential effectiveness. But superimposed on these internal mechanics are the intricate and far-reaching influences of the choice of mates, the mating system, which is the external and ecologically affected phase of genetic recombination. Through it, chance recombination of gametes is effected. In human populations, the mating systems are intricate in the extreme. Suffice it only to say here that we are witnessing in a few generations the rapid recombination of large segments of the human gene pool which were once separate.

Flies of the genus *Drosophila* may serve as an example of a recombination system which is much more restricted than the above example (Patterson and Stone, 1952). The highest chromosome number in the group is $2n = 12$, and most species have suffered a diminution in whole chromosome recombination potential by further reduction of the effective chromosome number by centromere fusions. Secondly, this genus of flies, together with a large segment of the higher Diptera have no crossing over in the male sex. This condition, despite the distributed crossing over in females, reduces the effective recombination due to crossing over by one-half compared to organisms which have distributed crossovers in both sexes. In certain species, the accumulation of inverted sections in the chromosomes in natural populations results essentially in the isolation of blocks of the germ plasm from effective genetic recombination. This system is particularly evident and important in many endemic species of *Drosophila*. The outcrossing potential of *Drosophila* is also relatively low. Many species crosses can be made in the laboratory, but, with rather insignificant exceptions, do not occur under natural conditions.

*Drosophila* as a genus shows an interesting variety of recombination systems. In some species, inversions are absent, or nearly so, from natural populations, whereas in others inversions are abundant. Even within an individual species, there are some populations which are highly heterozygous for inversions and others which are essentially homozygous. For instance, Carson (1954, 1955) has devised a means of comparing quantitatively the amount of genetic recombination which occurs in different populations of *Drosophila robusta* of the eastern United States. Marginal populations of this species display a much greater potential for genetic recombination than do populations from the central part of the range. Marginal populations thus appear to have a recombination system that is especially well suited to evolutionary aggression and pioneering.

In any organism in which reproduction is solely by asexual means, the recombination system may be considered closed. The cases most easily interpreted in this regard are those in which the asexual method of reproduction has clearly come to be irrevocably and completely substituted for the sexual method. Such a situation exists, for example, in certain of the obligate apomicts of the genus *Crepis* (Stebbins, 1950). Adaptation, as it becomes increasingly perfect, may be increasingly interfered with by recombination in such a way that selection will tend to bypass and then gradually eliminate entirely the very process that makes adaptation possible. Organisms which have eliminated sexual reproduction may be looked upon as representing a closed system, a dead end in evolution.

To the student of genetic recombination and its importance in evolution it would appear likely that all organisms which reproduce exclusively by asexual means have acquired this property secondarily and originally reproduced by sexual reproduction. It is perhaps not generally realized in biology that recent findings in population genetics have thrown new light on the basic biological meaning of sex. It may well be that many of the older concepts on the origin of sexual reproduction will have to be reexamined in detail.

**Conclusion and Summary**

The thesis is advanced that the most meaningful approach to the sexually reproducing species that the geneticist may make is by studying and describing it as a system of recombining genes. Such a system is often spoken of as a "gene pool." The breeding structure of the species is such that there is an almost continuous compartmentation of this larger gene pool into smaller and smaller units of recombination, with varying degrees of isolation from one another. At the bottom of the ladder and in a focal position is the local population, which at any one time level is the functional gene pool from which recombinations are being drawn. These combinations, encapsulated as individual organisms, are confronted by natural selection at the local level. In a broader sense, however, the gene pool includes all pools of the species that are separated geographically and temporally from one another.

The pattern of sexual reproduction existing in a particular local population constitutes its recombination system at that point in time and space. This may be a relatively open or closed system, depending on the chromosomal mechanisms and mating system that prevail. Thus a species may have extensive genetic recombinations available for selection in any one generation or series of successive generations, or the potentiality for recombination may be partially closed or even locked up and unavailable. These characteristics will be of very great importance in understanding (1) the past history of a species, (2) its present variability, and (3) its potentiality for future evolutionary change.

## REFERENCES

Carson, H. L. 1954. Variation in genetic recombination in natural populations. *J. Cellular Comp. Physiol.*, **45** (suppl. 2), 221-36.

Carson, H. L. 1955. The genetic characteristics of marginal populations of *Drosophila*. *Cold Spring Harbor Symposia Quant. Biol.*, **20**, 276-87.

Darlington, C. D. 1939. *The Evolution of Genetic Systems*. The University Press, Cambridge, England.

Dobzhansky, T. 1937. *Genetics and the Origin of Species*. Columbia University Press, New York, N.Y.

Dobzhansky, T. 1950. Mendelian populations and their evolution. *Amer. Naturalist*, **84**, 401-18.

Dobzhansky, T. 1955. A review of some fundamental concepts and problems of population genetics. *Cold Spring Harbor Symposia Quant. Biol.*, **20**, 1-15.

Fisher, R. A. 1930. *The Genetical Theory of Natural Selection*. The Clarendon Press, Oxford, England.

Gilmour, J. S. L., and J. W. Gregor. 1939. Demes: A suggested new terminology. *Nature*, **144**, 333.

Haldane, J. B. S. 1932. *The Causes of Evolution*. Harper & Bros., New York, N.Y.

Huxley, J. 1940. *The New Systematics*. The Clarendon Press, Oxford, England.

Huxley, J. 1942. *Evolution, the Modern Synthesis*. Allen and Unwin, London.

Mayr, E. 1942. *Systematics and the Origin of Species*. Columbia University Press, New York, N.Y.

Patterson, J. T., and W. S. Stone. 1952. *Evolution in the Genus Drosophila*. The Macmillan Company, New York, N.Y.

Simpson, G. G. 1944. *Tempo and Mode in Evolution*. Columbia University Press, New York, N.Y.

Stebbins, G. L., Jr. 1950. *Variation and Evolution in Plants*. Columbia University Press, New York, N.Y.

Thoday, J. M., 1953. The components of fitness. *Symposia Soc. Exptl. Biol.*, **7**, 96-113.

White, M. J. D. 1954. *Animal Cytology and Evolution*. The University Press, Cambridge, England.

Wright, S. 1932. The roles of mutation, inbreeding, crossbreeding and selection in evolution. *Proc. 6th Intern. Congr. Genetics*, **1**, 356-66.

# THE PLANT SPECIES
# IN THEORY AND PRACTICE

VERNE GRANT: RANCHO SANTA ANA BOTANIC GARDEN AND CLAREMONT GRADUATE SCHOOL, CLAREMONT, CALIFORNIA

## Biological Units

A philosophical tendency in modern biology, which has been in process of formulation for a century or more, is the view that the world of life is composed of a series of units organized into successively higher levels of complexity. The elementary biological units, gene, chromosome, cell, individual, deme (or local population), society, species, biotic community, represent different levels of organization of biological materials. All these units, in spite of numerous differences, share three common properties: an internal integration, the power of reproduction, and the ability to become modified in time. It would be correct to describe them all as integrated wholes, as reproductive units, and as entities capable of development.

The process of intellectual criticism has sharpened the concept of each one of the biological units. The progress of thinking follows a pattern. The unit in question is discovered and its autonomy and unity are frequently overemphasized by the earlier investigators in the field. With the growth of knowledge a reaction sets in. Some biologist in a later period is certain to maintain that the entity has no well-marked boundaries and hence has no definable unity, or even that it has no objective existence at all. This heresy forces a reexamination of the concept hitherto held and of the evidence upon which it was based. A controversy over "the problem of the gene" or "the concept of the association" ensues, which may eventually culminate in a deeper understanding of the nature of the particular unit.

One of the most common adjustments which has to be made

in the development of a unit concept in biology is the recognition that the unit does not enjoy an independent and unconditioned existence but that it forms a constituent part of a higher organization. Neither the extreme position that the unit is autonomous (and therefore objectively defined), nor the opposite position that it is merely an integral part of a more complex unit (which alone is objectively demarcated), is likely to prove correct in the final synthesis. The truth is more likely to lie in the middle ground that the elementary unit possesses an objective existence which, however, is subject to certain limitations due to its integration into a higher system.

Thus in nineteenth century biology the adherents of the cell theory were forced to retreat from their original position under the pressure of the knowledge that the organism as a whole constitutes a physiological unit and not a mere aggregation of cells. The compromise solution embodied in the now generally accepted organismal theory did not, however, negate, but merely qualified, the unitary nature of the cell. The individual in turn is now recognized as an elementary component of a more complex unit, the breeding population, in cross-fertilizing organisms. The facts of modern population genetics which make this conclusion inescapable can be accepted without denying the existence of the individual as a subordinate but real unit in its own right. The acceptance of these facts, moreover, is fatal to the opposite extreme view, still held in some quarters, that the individual is the ultimate biological unit.

A second adjustment which is frequently necessary in the formulation of a satisfactory concept of a biological unit is the redrawing of the boundaries. A unit as originally defined may require subdivision in the light of subsequent evidence. The concept of the gene seems to be going through this process at the present time. There is accumulating evidence that the Mendelian gene, which is the unit of crossing over, is not equivalent to the unit of biochemical activity in the chromosomes. The theoretical difficulty of delimiting individuals in clonally reproducing organisms has led to the recognition of two units in place of one: the individual, as defined by physiological autonomy, and the clone, defined by descent from a single zygote. Similarly, various authors have been forced to recognize different kinds of species.

The task of finding objective criteria for the definition of the biological units, that is, criteria for marking off their boundaries which can be applied by different observers with equivalent results, has in general proved easier in the case of the microscopic units than with the macroscopic ones. Thus the chromosome and the cell are clearly definable units; there is every reason to expect that the recombination gene and the functional gene can be satisfactorily defined; but an inordinate amount of controversy has been devoted to the species and the biotic community. As between the latter two entities, moreover, it is apparent that biologists are much closer at the present time to an understanding of the species question than they are to finding a basis for the definition of the community. This largest and most complex of the biological units presents the greatest difficulties of definition and delimitation of them all.

## Evolution of the Species Concept

The original species concept was that of primitive man. His idea of species was based on the observation of discrete units in the local fauna and flora. He had a specific name for each kind of animal and plant to which he paid any attention; these names corresponded rather well to objectively existing populations in nature. Essentially the same species concept was adopted by the local naturalist. Thus John Ray in 1686 defined the species as the unit which breeds true within its own limits (cf. Darlington, 1940).

The primitive species concept was not equipped for coping with the phenomenon of geographical variation, which only became apparent when the stage of man's activities expanded from the restricted home territory to an area equal in extent to the distribution range of many species. A species which was clear-cut in the local community might become the center of a "species problem" when a series of collections from different localities was assembled and compared. To accommodate the expanded knowledge of individual species gained since the nineteenth century an expanded concept was required. Criteria had to be sought that would lead to the demarcation of species not only within a local community but also on the wider stage of the world fauna and flora. The result of this search, conducted by numerous students during the past seventy years, is the biological species concept, which defines species on the basis of reproductive isolation or the presence of barriers to gene exchange.

The period between primitive man and the modern evolutionists, that is, the period during which geographical variation and the intergradation of morphological characters were known, but the idea of reproductive isolating mechanisms had not yet been formulated to explain the limits of this variation, was marked by an interregnum in the species concept. The Linnaean or typological species concept, which emphasized morphological difference rather than discontinuity as the criterion of species and recognized species universally throughout the plant and animal kingdoms, prevailed for two centuries during this interregnum, and it has inevitably left its stamp on modern systematics.

In its original conception the Linnaean species was a fixed and immutable entity. In the nineteenth century, with the increase in knowledge concerning natural variation and the rise of the theory of evolution, the fixed category became regarded more often as an arbitrarily determined stage above the level of the variety. The morphologically defined species, which seemed to be an objective reality in the period of Linnaeus and Cuvier, became frankly recognized as a subjective concept in the nineteenth and early twentieth centuries.

## Critique of the Species Concept

The problem of defining the species in plants has been complicated by the fact that different botanists have applied the name of species to entities ranging in magnitude from homozygous biotypes (Lotsy) to species groups (Engler). A common practice among botanists is to recognize as a species any population exhibiting distinctive morphological characters combined with a definite geographical range. The result, of course, is that many named species are equivalent to the geographical races of a polytypic species. There are also a few botanists who adopt the so-called biological species concept.

The opinion held by many botanists today that the species is undefinable is in part a heritage from the interregnum of the typological species concept. This agnosticism is written into the *International Code of Botanical Nomenclature* which through four editions has studiously refrained from defining the species. If one is operating within the framework of a typological species concept, this attitude is certainly justified.

The species which is defined on the basis of a certain number and degree of morphological differences is indeed a subjectively defined entity. No criterion of the amount of morphological difference between two forms that marks them as species, rather than as subspecies, sections, or genera, etc., and that is capable of application generally has ever been proposed. It follows that different observers confronted with the same range of variation in a series of specimens will frequently be unable to agree on the taxonomic disposition of the case, in so far as their decisions are predicated on a typological species concept. That such differences of opinion are commonplace is evident to anyone who has compared different taxonomic treatments of the same group prepared by different but equally well qualified authors.

It is generally agreed by the adherents of both the typological and the biological species concepts that the species of the typological concept can be defined only as that entity which a competent systematist regards as a species. This conclusion constitutes the strongest argument against retaining the typological species concept in taxonomy, now that an objectively defined concept is available to take its place. Unfortunately, the advocates of the subjectivity of the species as a unit usually ignore the existence of the biological definition. Thus Mason writes that, "I have seen no putative definition of a taxonomic category so worded as to be incapable of application either to the next higher or the next lower category of the taxonomic structure. That which is a species to one taxonomist may be a subspecies to another . . ." (Mason, 1950). The presence of reproductive isolating mechanisms is a feature of species, but not of subspecies; the smallest population possessing such reproductive barriers is the species, and not the polytypic section or genus.

The difficulties of the typological species concept have led some students to adopt, not the biological, but still another, nominalistic concept of the species. The proponents of this concept stipulate that the species is (are) an "empty category," "mental units rather than biological units," "highly abstract fictions," an "abstract category in the taxonomic structure" which "has no foundation in reality and obviously cannot be objectively defined." (Mason, 1950; Davidson, 1954; Burma, 1954).

The nominalistic concept is logically sound although scientifically barren. If all that is implied by the word "species" is the category in the taxonomic hierarchy corresponding to the specific name in the binomial system of nomenclature, there can be no disagreement as to its artificiality. In this sense of the word, the unicorn, if properly described according to the Rules of Nomenclature, would be as real a species as the horse.

According to Mason again (1950), "The wisdom of past experience has dictated that the taxonomist purposely refrain from defining these categories [species, genus, family, order, etc.] in any way that will impose restrictions on the freedom with which he may express the interrelationships that he construes to exist." With regard to the genus, family, and order the absence of an objective definition is not so much a matter of wisdom as of necessity. With regard to the species the attempt to preserve a wide elasticity of usage, besides being unnecessary, is also unwise, since it leads to a chaotic condition in taxonomic work which could be avoided. Every branch of natural science is sooner or later compelled to define its basic units in objective terms. It would be remarkable if plant taxonomy were to prove unique in this respect. In the meantime, as so aptly pointed out by Camp (1951), the adherence to a subjective criterion of its basic unit is responsible for the fact that plant taxonomy, in many of its branches, is not yet a science but an art.

The opinion that the species is not a definite entity has been arrived at in three different ways. We have seen that the typological species definition leads to the conclusion that the species is a subjective entity. The nominalistic species concept frankly eschews the search for an objective criterion of species. It remains to note that the biological species concept itself denies the existence of definite species units in certain plant groups. In fact, a large part of the species problem, as it currently exists in botany, arises from the attempt to apply the species concept to groups in which biological species do not exist. The view of many botanists that the species is undefinable is, in such groups, justified by the facts.

If one wishes to use the term species everywhere throughout the plant and animal kingdoms, one must adopt the typological species concept, since morphological similarity and convenience in classification are the only criteria that can be applied universally. The general usage of the species category is a traditional practice in taxonomy. It has even become a dogma, sanctioned by a pronouncement in the Code of Nomenclature that all plants belong to species. Neither the species concept of primitive man, nor that of the naturalist, nor that of Linnaeus, however, was based on a consideration of organisms representing all the types of life cycle now known to modern biology. The historical grounds for applying the species concept to all forms of life are consequently not compelling.

The issue between alternative concepts of the species can and should be decided on the basis of the respective merits of the contending viewpoints rather than on purely historical considerations. The chief advantage of the typological species concept is its universality; this meritorious feature is offset by the corresponding disadvantage of subjectivity. The biological concept, on the other hand, enables us to define the species objectively in the numerous and dominant class of sexual organisms, but it is not applicable outside such organisms.

To the present author, the privilege of applying a subjective species definition to all organisms does not represent a very important gain; whereas the price that must be paid for this privilege, namely the giving up of an objectively defined species concept applicable to the majority of higher organisms, does represent a very important loss. If in any given group of organisms we do not find an evolutionary development into reproductively isolated populations, we can, after all, fall back upon a morphological criterion and call the various forms by some other name than species, such as *taxon* or *binom*. Biologists who reserve the category of species for morphologically similar groupings of individuals, on the other hand, have not left any alternative designation for the reproductively isolated system of breeding populations.

The reproductively isolated population, whether it is called a species or some other name, undoubtedly represents a major biological unit, which must be investigated and discussed. This task will be facilitated if the unit in question can simply be referred to as "the species" or, if any qualification is needed, as "the biological species." For purely operational purposes, if for no other reason, the adoption of the biological definition of the species is justified in discussions of the origin, nature, and behavior of reproductively isolated population systems.

The substitution of a biological for a morphological definition of the species does not provide a panacea for the solution of every complex situation in nature. The naive assumption that it should do so, more often tacitly implied than explicitly stated, is responsible for the disillusionment in the biological species concept felt by some biologists whenever they run into difficulties of interpretation. Such difficulties should be expected, considering the wide diversity of life cycles, breeding systems, and population structures existing among sexual organisms. All that can be claimed for the biological species concept is that it provides a framework that works better than any other yet proposed for the analysis of species problems.

The remainder of this paper will be devoted to a consideration of certain problems involved in the recognition and demarcation of biological species in the higher plants. This inquiry will give attention to the difficulties encountered in setting the limits of species, the causes of those difficulties, and the biological significance of the causes.

## Criteria of Reproductive Isolation

Relatively few groups of organisms have been studied from enough points of view with sufficient thoroughness to reveal their whole pattern of species-separating mechanisms. In most groups the presence and distribution of isolating mechanisms will have to be inferred from indirect evidence of various sorts. Among the most important of these are the finding of discontinuities in the variation pattern, marginal overlapping of distribution areas, complete sympatry, or hybrid sterility in the laboratory or experimental garden.

If these criteria of reproductive isolation are consistent among themselves in any given group, there will be no species problem. It may happen, however, that one or more criteria will fail to coincide with the rest. Under these circumstances an automatic application of any single criterion in drawing species boundaries might easily lead to a false picture of the natural species. The inconsistency indicates that a critical appraisal of the criteria themselves is always needed.

A reproductively isolated cross-fertilizing population should reveal its genetic separation from other such populations by a gap in the pattern of variation. Some plant species in every genus do show the expected morphological discontinuity, but others for various reasons do not. If applied uncritically, the criterion of discontinuity would often result in treating as species populations belonging to other categories. In a polytypic species with a disjunct distribution, the subspecies may be separated by morphological gaps, i.e., *Gilia leptantha*. The morphologically discrete units in *Agoseris* seem to be the species groups; in *Ceanothus* they frequently correspond to the sections; in the California oaks one of the discrete entities is a subgenus; and in *Allophyllum* it is the entire genus, which has, in fact, until recently been treated as a single species.

The overlapping range of variation of three biologically well isolated species of *Gilia* is shown in Fig. 1. Each symbol represents a population. The coordinates give the measurements of the two most important diagnostic characters for these three species,

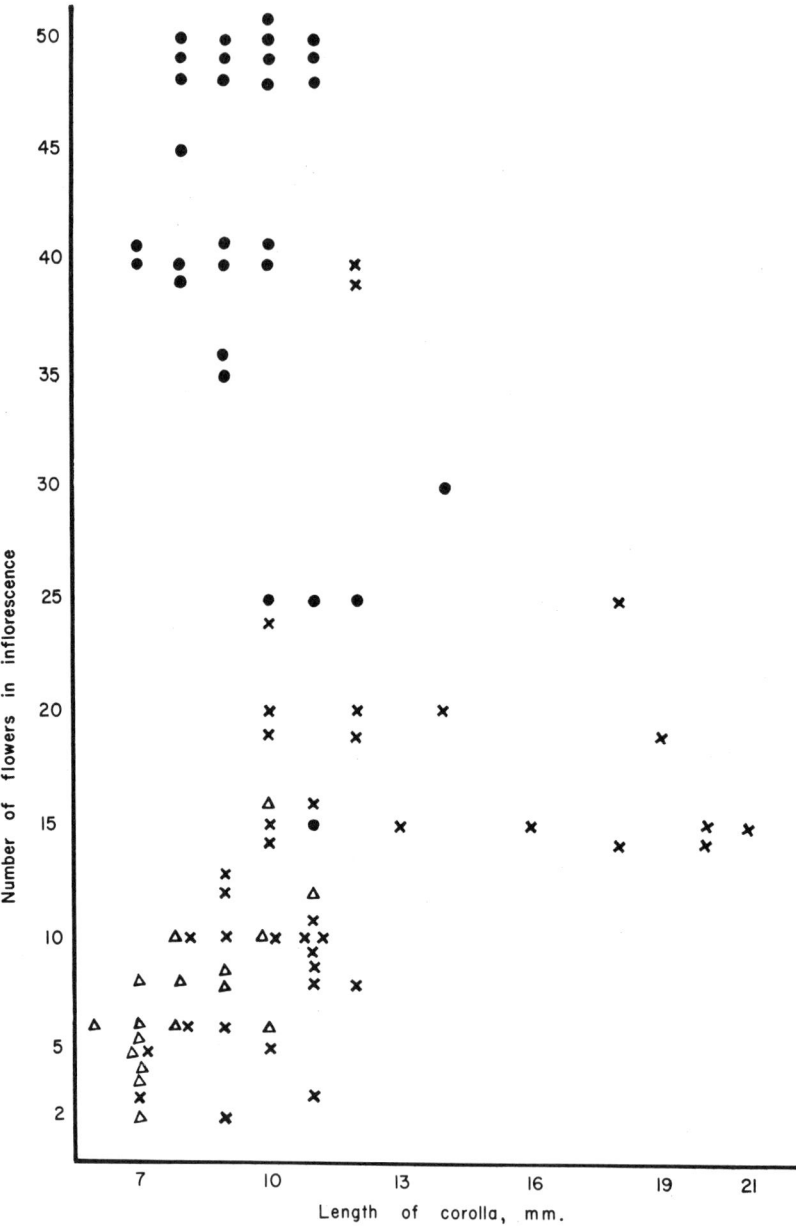

Fig. 1. Overlapping range of variation of three intersterile species of Leafy-stemmed *Gilia* with respect to two important diagnostic characters. Each dot represents the average characteristics of a single local population. ● *G. capitata abrotanifolia,* × *G. achilleaefolia,* △ *G. angelensis.* From *Evolution,* 7 (1953)

the ordinate the number of flowers in the inflorescence and the abscissa the length of the corolla. The absence of a discontinuity in the variation patterns of the different species is apparent. Similar results are obtained when other characters are used.

Relatively few biologists subscribe any longer to the old idea that species can be defined by the sterility of hybrids. There are too many cases of good species which can produce fertile hybrids under laboratory or garden conditions although they remain amply distinct in nature. Furthermore, all degrees of sterility exist so that the distinction between "intersterile" and "interfertile" is frequently an arbitrary one. The existence of intraspecific sterility phenomena also makes any general application of the sterility criterion of species hazardous. Nevertheless, there can be no doubt that sterility barriers comprise some of the most effective mechanisms which separate species, and their presence must be weighed carefully in any attempt to devise a classification of species which accords with the realities of nature.

The most essential property of a species is its ability to maintain its distinctive features, or rather the gene patterns which determine those features, from generation to generation in a community of organisms. Sympatry is the final test of species status. Two sympatric populations will normally be regarded as belonging to separate species. It by no means follows as an automatic rule, however, that related allopatric populations are necessarily conspecific.

We have several pairs of related allopatric species in *Gilia*. *Gilia tricolor* and *G. angelensis*, for example, belong to the same species group and occupy similar ecological habitats in contiguous geographical areas (Grant, 1952). It was at first assumed as a working hypothesis that they were members of the same polytypic species. This hypothesis had to be abandoned when it was learned that the two forms are completely intersterile in the experimental garden and probably possess different genomes as well. The intersterility of *G. tricolor* and *G. angelensis* stands in marked contrast to the absence of any known intraspecific sterility within either entity. The facts are very similar in the case of *G. millefoliata* and *G. laciniata; G. splendens* and *G. leptalea;* and

*G. sinuata* and *G. crassifolia*. Examples such as these should warn us against automatically combining related allopatric forms in the same species. The category of *superspecies*, proposed by Mayr (1942), is available for groups of geographically isolated but nevertheless valid species.

It is no more justifiable in certain cases to treat sympatric populations as distinct species than to reduce all allopatric populations automatically to the status of subspecies. We now know that the terminal races of a polytypic species may overlap in range in one part of the distribution area and yet be connected by more or less continuous intergradation in other parts of the area. When the population system is considered as a whole, it is commonly found impossible to split it naturally into more than one species.

How frequently the condition of sympatry arises within the limits of a single polytypic species is a moot point. Goldschmidt (1952) is "disappointed" that "only a few examples, quoted over and over again, exist to demonstrate what should be a most frequent occurrence." Why we should expect overlapping rings of races to be a commonplace phenomenon is not clear; but if fresh examples are desired, it may be recorded that three have turned up in the last few years in the family Polemoniaceae. Thus *Gilia capitata*, the largely allopatric *G. tenuiflora-latiflora* group, and *Ipomopsis aggregata* all exhibit terminal overlap of the extreme subspecies. The evolutionary significance of this situation cannot be considered here. As far as the species problem is concerned, it indicates that marginal sympatry is not, in itself alone, a reliable species criterion.

## Magnitude of the Species Problem in Plants

A biological species exhibits three characteristics (Mayr, 1948, 1949). It possesses its own distinctive morphological characters, which are separated from those of other species by a prominent hiatus in the variation pattern. It possesses its own particular ecological requirements, which reduce the competition with other sympatric forms. It has a combination of reproductive isolating mechanisms, which prevent or greatly inhibit gene exchange with other species.

These characteristics are clearly displayed in most groups of vertebrate animals. The same facile generalization can *not* be made for the higher plants. The circumstance that the biological species concept has been almost entirely a product of work in systematic zoology and preeminently in vertebrate systematics is a reflection of this fact. The only botanist who contributed in an important way to its development in the early period was Du Rietz (1930). Botanists as a whole have accepted the biological species concept much less generally than have the zoologists.

Zoologists are inclined to believe that the difficulties of the species concept in plants are confined to a few exceptional cases, i.e., Fisher (1954, p. 87). To the botanist, on the other hand, it frequently appears as though the clear-cut species is the exception and the taxonomically critical group is the rule. So characteristic is this impression that one plant taxonomist even felt obliged recently to apologize for publishing an excellent monograph of a large genus in which there were no outstanding difficulties of species delimitation. The true state of affairs in higher plants should be expressed in quantitative terms. This large task can only be commenced here.

In the California flora 32 of the families of seed plants have four or more genera and 25 or more species in all. Every one of these 32 large or middle-sized families contains critical groups in which the species are not clearly defined. In addition many smaller families in this flora also present real problems of species delimitation, i.e., the Typhaceae, Iridaceae, Juncaceae, Fagaceae, Salicaceae, Nyctaginaceae, Apocynaceae, Loasaceae, Lobeliaceae. The species concept thus runs into practical difficulties in all the large families, all the medium-sized families, and many of the small families in the California flora.

Similarly, there is a species problem of varying proportions in nearly all the large angiospermous genera of the north temperate zone. To list them all, from *Achillea* to *Zinnia*, would be a tiresome task. If by "good species" is understood those entities which are set apart from all their congeners by a prominent discontinuity in the variation pattern so that botanists have long been in agreement as to their validity, it is much easier to cite the few large genera in which all the species are good than the numerous genera in which at least some of the species are poorly behaved. Plant genera, like plant families, which have attained any con-

TABLE I. The Frequency of Good Species in Some Genera of Seed Plants

| Genus | Region | Family | No. species | No. good species | Good species, % |
|---|---|---|---|---|---|
| Trees | | | | | |
| *Pinus* | Pacific slope | Pinaceae | 18 | 12 | 67 |
| *Quercus* | Pacific slope | Fagaceae | 17 | 7 | 41 |
| Shrubs | | | | | |
| *Ceanothus* | California | Rhamnaceae | 43 ? | 0 | 0 |
| *Opuntia* | Southern California | Cactaceae | 17 | 6 | 35 |
| *Diplacus* | Pacific slope | Scrophulariaceae | 7 | 3 | 43 |
| Perennial herbs | | | | | |
| *Aquilegia* | Northern hemisphere | Ranunculaceae | 5 ? | 2 | 40 |
| *Asclepias* | North America | Asclepiadaceae | 108 | 108 | 100 |
| *Crepis* | Western North America | Compositae | 10 | 3 | 30 |
| Annual herbs | | | | | |
| *Clarkia* | Western hemisphere | Onagraceae | 33 | 4 | 12 |
| *Mentzelia* | California | Loasaceae | 13 | 9 | 69 |
| *Gilia* | Western hemisphere | Polemoniaceae | 52 | 21 | 40 |

siderable size and yet present a complete assemblage of well-defined species are exceptional.

It would be of interest to have some definite data on the relative frequency of good and poor species in various plant genera. Table I shows the proportion of good species in eleven genera representing as many plant families and all the major life forms. All these genera have been more or less carefully studied by modern methods. They were selected primarily for that reason and secondarily in order to illustrate a wide range of evolutionary patterns. The standard of reference is in each case the total number of species for a given region according to some recent revisional study or Munz' *A California Flora* (in press).

As Table I shows, the number and percentage frequency of discrete species varies from genus to genus within wide limits. Of the genera represented only *Asclepias* apparently is without problematical species for the area considered, and only *Ceanothus* is without any good species in the area considered; but there may be critical species of *Asclepias* in Mexico and Central America, and, conversely, *Ceanothus* seems to have about two good species in eastern North America. All the genera thus contain some good species as well as some poorly defined ones.

The frequency of clearly demarcated species varies from group to group within a single phylad. Thus within the family Polemoniaceae the two small genera, *Allophyllum* and *Langloisia*, both consisting of herbaceous annual plants, may be contrasted. There are four species of *Langloisia*, all of which are clearly marked; not one of the five species of *Allophyllum*, on the other hand, has well-defined boundaries. The 40 per cent of good species estimated for the genus *Gilia* in this same family is an average figure which covers extreme fluctuations from one infrageneric group to the next. Of the species composing the sections *Saltugilia* and *Giliastrum*, for example, 60 per cent are good, whereas the sections *Gilia* and *Arachnion* contain only 20 and 14 per cent good species.

Of interest to the zoologist will be the conclusion that critical species and species groups are the rule rather than the exception in higher plants. Of interest to those botanists who regard the species as a completely subjective concept is the demonstration that at least some discrete units, generally recognized as species, are to be found in almost every genus.

## Causes of the Species Problem

The difficulties encountered in the practical application of a biological definition of the species in many groups of higher plants may be referred to several causes.

One factor, which has long been acknowledged to contribute

to the imperfect delineation of many species, is the process of gradual speciation. We find species in all stages of divergence from the ancestral forms. Some examples of borderline cases between races and species, as manifested in superspecies composed of very different and discontinuous allopatric forms, in overlapping rings of races, and in species groups exhibiting marginal sympatry, have been discussed by Mayr (1942).

Whether to regard a pair of populations in transition from races to species as one or the other entity may at times require an arbitrary decision. Such arbitrary decisions have to be made here and there in most, if not all, of the genera listed in Table I. This difficulty of interpretation is, however, in these genera and elsewhere in higher plants, minor in comparison with the difficulties arising from other causes yet to be considered.

A second contributing factor to the species problem is the occurrence of sibling species. Species which are virtually indistinguishable morphologically are not uncommon in the higher plants. Every specialist can cite examples from his own group.

Cryptic species in plants are frequently associated with polyploidy. Two diploid species, AA and BB, combine to form a new allotetraploid species, AABB. The new population does not merely fill the morphological gap between the original diploids, but proceeds to segregate variations in the direction of each parental species. The result is that until cytotaxonomic and cytogenetic work is carried out the true biological species will not be recognized. Instead there will stand some artificial and admittedly unsatisfactory classification. In the hands of a conservative taxonomist the entire assemblage may be treated as one species, whereas a morphologically minded taxonomist is likely to recognize two species consisting respectively of AA plus certain variants of AABB, and BB plus other variants of AABB. As polyploidy continues to the higher levels of hexaploidy, octoploidy, and so on, the number of species involved in a morphologically indistinguishable complex of course increases.

Some of the most familiar plants comprise clusters of sibling species. The sagebrush or *Artemisia tridentata* group, for example, is a complex consisting of an unknown number of biological species. Being the dominant plant of the Great Basin and a characteristic subdominant in the mountains and high plateaus of western North America, the sagebrush has inevitably received much attention from stockmen, foresters, range specialists, botanists, and other naturalists. Nevertheless, despite this general attention, as well as two taxonomic revisions, one in 1923 based on field and herbarium studies, and the other in 1953 utilizing chromosome counts in addition to the classical methods, a classification of the sagebrushes which accurately represents the naturally occurring biological units still eludes us. The recent studies of Ward (1953) have at least defined the problem and pointed the way toward its eventual solution. There is reason to hope that the continuing investigations of Ward, Beetle, and others will result ultimately in the discovery and delimitation of all the constituent species in the *Artemisia tridentata* complex. The prospects that botanists will ever be able to determine correctly many of these species, without field observations, cytological observations, or both, are not, however, encouraging.

Among the genera listed in Table I, sibling species associated with polyploidy contribute to the taxonomic difficulties in *Gilia*, *Clarkia*, and *Mentzelia*.

A third cause of obscure species lines in plants is natural hybridization including introgression. If two or more species begin to hybridize, and if the derivative forms succeed in establishing themselves as natural populations, the original discontinuities between the species may become utterly blurred. This is the cause of the overlapping ranges of variation of three species of *Gilia* shown in Fig. 1. It is the most important cause of taxonomic difficulties in the Leafy-stemmed Gilias and the Cobwebby Gilias, the two most critical sections of the genus. The boundary lines between four species of Cobwebby Gilia are obscured by past hybridization of each species with two or more of the other four to such an extent that it is not apparent whether there are four species or one (Fig. 2). Interspecific hybridization is the cause of the fraction of poorly defined species in *Pinus, Quercus, Ceanothus, Diplacus,* and *Aquilegia,* listed in Table I, and in numerous other genera of higher plants.

The fourth difficulty of the species which will be mentioned here is the reversion in many plant groups to asexual reproduction. The biological species is a community of cross-fertilizing individuals linked together by bonds of mating and isolated reproductively from other species by barriers to mating. Both the internal organization and the external boundaries of species are defined in terms of gene exchange. When sexuality is absent, the biological species concept breaks down. Since the individual members of asexual assemblages do not share in a common gene pool, the group is not integrated into true species. The taxonomic difficulties in apomictic groups like *Crepis* sect. *Psilochaenia* and the clonally reproducing *Opuntia,* to cite only two examples listed in Table I, are due to the development of swarms of variants, including various hybrid types, which propagate themselves asexually.

Hermaphroditism, which prevails in the flowering plants, opens up the possibility of another deviation from sexuality and cross fertilization as it is known in the higher animals. Many hermaphroditic plants set seeds by a mixture of outcrossing and selfing, and many others, particularly among the annual herbs, are autogamous or automatically self-pollinating. It might be supposed that these deviations would contribute to partial or complete breakdown of species integrity in partially or completely self-pollinating plants. Only under certain circumstances (as in the autogamous but highly heterozygous *Oenothera biennis* group) does this appear to be true. Regular self-pollination, although it modifies the intraspecific variation pattern in important respects, does not prevent the integration of the separate populations into species units in most autogamous groups.

The explanation of the compatibility of self-pollination with species integration seems to be that autogamy is rarely if ever absolute. The self-pollination, although predominating for a while, is supplemented by enough rare outcrossing to tie the individuals together into populations and the populations into species.

To summarize, natural hybridization ranks high as a source of a species problem in the higher plants. Sibling species and asexual reproduction, but not self-pollination, are important contributing factors. A part of the species problems is due to the continuity of the speciation process. As far as the higher plants are concerned, however, the latter factor seems to be a minor source of difficulty in comparison with the other three causes mentioned.

## Biological Species Concept Reconsidered in the Light of Its Difficulties

Four factors which contribute to the difficulty in application of the biological species concept have been considered. These factors are: (1) sibling species; (2) the gradual formation of species from races; (3) hybridization; and (4) asexual reproduction. Each factor has a somewhat different bearing on the question of the objective reality of the species as a biological unit.

*Sibling Species.* The difficulties of the species concept arising from the existence of sibling species are practical, not theoretical, difficulties. The distinction between sibling species and normal species is relative. Most sibling species when studied long enough and hard enough are found to be separable morphologically. External diagnostic characters were eventually discovered even for the example par excellence of a pair of cryptic species, *Drosophila pseudoobscura* and *D. persimilis*.

The naming of sibling species in plants as a result of biosystematic studies has provoked a certain discontent among many herbarium curators and floristic taxonomists. The biosystematist does indeed have a responsibility to determine and annotate as many large herbarium collections as is feasible. He also has a responsibility to science not to suppress his findings merely in order to facilitate the task of herbarium filing. Two courses are open to the curator or general taxonomist who is confronted with undetermined material of sibling species. Let him determine it as to species group rather than as to species. Or let him scrutinize the material with more than average thoroughness.

If the taxonomic procedures in different phyla are compared, as Dobzhansky and Epling (1944) pointed out, the amount of effort considered necessary for making a determination is not a standard quantity. Dipterists are in the habit of killing and pinning a specimen before identifying it. Bacteriologists, on the other hand, are accustomed to carrying out physiological tests with live specimens. In the case of sibling species in *Drosophila* or *Anopheles*, etc., the dipterist may have to emulate the bacteriologist by employing physiological tests.

In a similar way phanerogamic taxonomists usually identify species by characters visible with the naked eye, a hand lens or a stereoscopic microscope, whereas cryptogamic botanists have long been habituated to the use of a compound microscope in taxonomic work. The size of pollen grains and guard cells in flowering plants is a microscopic character which varies proportionately with the level of ploidy in some (but not all) polyploid complexes. Where sibling species differ in ploidy level this character may prove diagnostic. Preliminary studies carried out by the author and an advanced botany class on herbarium material in both the *Artemisia tridentata* and *A. vulgaris* groups have revealed significant differences in pollen grain and stomatal size in each one of these polyploid complexes, some of which can be correlated with known chromosome numbers. Species recognition in these and other critical groups might be facilitated by measurements of cell size. Of course the chromosome number itself is a morphological character. Nothing but convention stands in the way of using characters which require a magnification greater than $25\times$ or $50\times$ in the keys to difficult genera of higher plants.

In the last analysis, if the search for workable macroscopic or microscopic characters has been unsuccessful, it may prove necessary to give up the attempt to distinguish the specimens of certain sibling species in the herbarium. There will be no objections on the part of the biosystematists who make the segregations in obscure groups to the filing of their indistinguishable species in the same folder. The species still stand, however, as objectively real biological units, whether the taxonomist can recognize them on visible characters or not. A system of classification of the biological species must be judged, not on the basis of convenience, but according to whether it represents accurately or inaccurately the realities in nature.

*Gradual Formation of Species.* Contrary to the belief held by

Darwin, Wallace, and their early followers, the objective existence of species is not compromised by the fact that they come into being by gradual processes of evolution. All the biological units are capable of orderly growth and development in space and time. The duplication of the chromosome strands appears to be achieved in stages, but no one has thought to disqualify the chromosome as an organizational unit because of its gradual mode of reproduction. Similarly, the origin of species in most cases is a process which seems to be completed by stages. During a transitional stage the diverging population is neither one nor two species but is an incompletely separated pair. The circumstance that the taxonomic hierarchy provides no category for taxa caught midway between races and species will inevitably obscure the biological process nomenclaturally, but should not be allowed to obscure it descriptively.

The above difficulty is probably in the majority of cases more theoretical than real. Although the evolution of species from races may be continuous, evolution crosses boundary lines. The boundary line between race and species, as defined by Dobzhansky (1935), is that stage of the evolutionary process when an actually or potentially interbreeding population becomes segregated into two or more reproductively isolated populations. Compared with the duration of life of an average species, the crossing of this boundary line is probably a fairly rapid process. Most populations at any given instant will consequently be found to exist either in the stage of races or in the stage of species, and not in an intermediate stage, in so far as their status is affected solely by the gradualness of speciation.

*Natural Hybridization.* The difficulties of the species concept which are brought about by natural interspecific hybridization are both theoretical and practical. The difficulty is not apparent and is tacitly overlooked by taxonomists when the hybridization is limited to the rare formation of a few hybrid individuals. It cannot be overlooked in either theory or practice when the hybrid derivatives establish themselves as an intermediate population in nature. A mere trickle of genes across an interspecific barrier, which is manifested in the production of an occasional hybrid, is a very different situation from a mass flow of genes between species by way of a bridging population of hybrid origin. In the one case the hybridizing entities are still good species in the taxonomic sense and only slightly contaminated in the biological sense; in the second case they have probably become geographical subspecies; and between these extreme conditions there is every intermediate stage in the merging of species.

If it should prove useful to designate the merging of separate species into a single species by a special term, the process might well be called *secondary speciation.* Secondary speciation is that stage of the evolutionary process at which two or more reproductively isolated populations become combined into one interbreeding population. The process is a gradual one and, as with primary speciation, presents a formal difficulty during the period of transition.

A concrete example of secondary speciation documented by fossil records is the progressive fusion of *Pinus muricata* and *P. remorata* since Pleistocene time (Mason, 1949). These two closed-cone pines were distinct and sympatric on the Santa Barbara mainland of California in the Pleistocene. Since that time a single variable population with the combined characteristics of both *P. muricata* and *P. remorata* has developed in this area. In Pleistocene time *Pinus remorata* was uniform and distinct on Santa Cruz Island thirty miles off the Santa Barbara coast. During this period *P. remorata* was probably the only species present on the island. Today both species exist on the island, *P. muricata* probably having arrived in the Late Pleistocene or Post-Pleistocene. Some of the recent populations of *P. remorata* are no longer uniform and sharply distinct from *P. muricata* but, on the contrary, are variable and introgressive with characteristics which indicate that they are merging with *P. muricata.*

If the merging of species proceeds at a rapid rate and goes to completion, the number of populations which cannot be assigned to the alternative categories of species or subspecies will be relatively small. Enough cases are actually known, however, of species groups in an intermediate stage of secondary speciation as to give rise to serious difficulties of species delimitation. Furthermore, so far as the botanist can judge from the study of numerous cases of natural hybridization, the process of secondary speciation does not necessarily always go to completion, but may remain more or less indefinitely in an undefined halfway stage.

*Asexual Reproduction.* The reversion to asexual reproduction, as numerous authors have pointed out, spells the end of species in the biological sense of the word. The smallest integrated unit above the individual is the clone or biotype; the smallest well-defined taxonomic unit may be a huge polymorphic complex, an agamic or clonal complex.

For purposes of classification the taxonomist must name and describe selected morphological types in asexual groups. Whether these aggregations of individuals should be called "species" or not is a matter of taste. One tradition favors a general and hence indiscriminate use of the category of species in all groups of organisms. Some botanists would qualify this usage in asexual groups by employing the terms *agamospecies* or *taxonomic species*. The term *agameon* has also been suggested (Camp and Gilly, 1943).

The present author prefers the neutral term, *binom*, suggested by Camp (1951). The grounds for this preference are that it saves the term species for the isolated population system and avoids the confusion of concepts inherent in the application of the same word to two basically dissimilar phenomena.

In summary, the existence of the species as an objective biological unit is not impaired by morphological indistinctness or by the continuity of the evolutionary process. The loss of sexuality, on the other hand, removes the very foundations on which the species exists as a type of breeding population. As a result, biological species do not exist in asexual groups.

The consequences of natural hybridization for the biological species concept cannot be stated so categorically. The discreteness of the species unit is a relative matter in a hybridizing complex. An attempt to evaluate the significance of natural hybridization for the objective reality of species will be made in the next section.

## Significance of Hybridization for the Species Concept

The bearing of natural hybridization on the species concept may be illustrated by three concrete examples selected from *Gilia*. All these examples involve diploid sexual organisms with a normal chromosome cycle and a rate of natural cross fertilization which varies from high to low but never reaches exclusive inbreeding. The integrity of the species as a biological unit is slightly impaired in the case of the first two examples but it breaks down in a more serious way in the third example. It is this third case that proves the most instructive for the question of the objective reality of species limits in a hybrid complex.

The first case is that of *Gilia capitata*, a widely distributed species on the Pacific slope of North America, and *Gilia millefoliata*, a species of the coastal sand dunes. The two species come into contact in a number of places along the Pacific coast. Morphologically, they are amply distinct, *G. capitata* being an erect plant with globose flowering heads and *G. millefoliata* a prostrate spreading plant with spotted flowers borne in small clusters. They are also well isolated by sterility barriers and can be crossed only with difficulty in the experimental garden. The hybrids then are highly sterile and are characterized by a low degree of chromosome pairing. On Cape Mendocino in northern California a local breakdown in the complex of reproductive isolating mechanisms has led to a limited amount of natural hybridization. The boundaries of the species are, however, quite clearly defined even here, which suggests that the populations are maintaining themselves as two distinct species.

The second case concerns *Gilia capitata* itself (Grant, 1950). This is a polytypic species composed of eight subspecies which align themselves morphologically and ecologically into three main groups. These major subdivisions of the species in all probability represent a trio of formerly distinct species. Vestiges of the original reproductive isolation between the ancestral species still persist in a few areas of sympatric overlap. The extreme populations are connected by a complete series of intergradations. Since the intermediate populations occupy geologically more recent habitats than the extremes, since they are highly variable from colony to colony whereas the extreme forms are quite uniform, and since the variability of the intergrades shows the character correlations which are expected when a hybrid segregates for two or more multifactorial characters, it is judged that the intermediates are historically more recent than the extremes and are of hybrid origin between them. If *Gilia capitata* can be regarded as a single species, therefore, it must be one of secondary origin from the union of previously isolated populations.

The most serious difficulties of classification are encountered in the *Gilia tenuiflora-latiflora* group (Grant and Grant, 1956). From the standpoint of intergradation and gene exchange the whole group, comprising *Gilia tenuiflora, G. latiflora, G. diegensis, G. leptantha,* and *G. cana,* could be regarded as a single species. That it is in reality more than this, however, is indicated by various facts.

The major constituents of the complex are more different from one another than any conspecific elements known in the genus. Each major constituent is comparable to other large species in *Gilia* and is in fact polytypic with varying numbers of well-marked subspecies. The reasoning involved in this argument does not presuppose a morphological species definition but rather a certain general correlation within a genus, or in this case even within a section, between morphological differentiation and biological species formation (Mayr, 1942; Rollins, 1952).

The treatment of *Gilia tenuiflora, G. latiflora,* and *G. cana* as a single species would obscure rather than accurately portray their relationships. Each species named has other genetically close relatives which are without question specifically distinct, whereas a pronounced reduction in fertility associated with a lowered degree of chromosome pairing is found in the hybrids between the major elements. Finally, a number of sympatric occurrences are known between the major elements. All these facts are difficult to reconcile with the idea that the major entities in the *Gilia tenuiflora* complex are the members of a single species in any normal sense of the word.

The existing evolutionary pattern in the *Gilia tenuiflora* group can be explained on the hypothesis that *G. tenuiflora, G. latiflora,* and *G. cana* went through a stage of barely completed primary speciation which was followed by extensive hybridization in various combinations (Fig. 2). The hybridization has gone far enough to prevent the original extreme populations from persisting as clear-cut species, but has not fully completed the process of secondary speciation. The major constituents thus possess some of the characteristics of subspecies in so far as they replace one

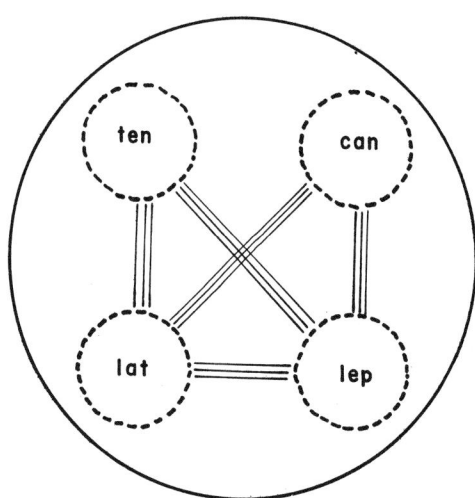

Fig. 2. The syngameon of *Gilia tenuiflora-latiflora,* consisting of four semispecies linked by natural hybridization in nearly every possible combination. The abbreviations stand for *Gilia tenuiflora, G. cana, G. latiflora,* and *G. leptantha.*

another geographically, intergrade in the zones of contact, and retain a certain degree of interfertility; they also possess some characteristics of species, as manifested in marginal sympatric contacts and in a degree of morphological and genetic divergence which is comparable to that of species elsewhere in the phylad.

The situation just described is not confined to *Gilia.* It is known also in *Pinus, Quercus,* and *Aquilegia,* to cite other examples from among those listed in Table I, and in numerous other genera.

Davidson (1954) has pointed out, "If we wish to argue that species have an objective reality, we may select examples such as Ginkgo. If we wish to argue that species do not exist in nature at all, we may select such examples as the willows." The species of willow are notoriously confused by natural hybridization, whereas this factor can scarcely affect *Ginkgo biloba*, which is the sole living member of its phylad.

The frequent occurrence of the willow pattern of evolution, as contrasted with the ginkgo pattern, is perhaps reflected in the species concept held by so many of the nineteenth century plant hybridizers. Their practical experience impressed upon many of these early students, Naudin, Nägeli, Darwin, Mendel, Kerner, De Vries, and others, as it has also on many modern botanists, an explicit disbelief in the notion that the species as a unit is qualitatively distinct from the race.

The failure of conformity of the evolutionary developments in certain groups of plants to the biological species concept in its present formulation suggests the need for a reexamination of our definitions. It sometimes proves necessary to split a biological unit into two or more units. As previously mentioned, the inconsistencies of the "concept of the individual" when applied to clones forced a recognition of both individuals and clones; and a splitting of the gene into functional and recombinational genes is apparently in the offing. Similarly, the evidence derived from evolutionary studies of higher plants indicates the existence of a unit of interbreeding higher than the species. That unit is the sum total of the species linked together by frequent or occasional hybridization.

Categories supplementary to the species have been proposed by several authors. The *coenospecies* of Turesson (1922) and the *commiscuum* and *comparium* of Danser (1929), being based on the fertility relationships of species in the experimental garden, are artificial categories of use only for the discussion of a special aspect of natural populations. The *coenogamodeme* and *syngamodeme* of Gilmour and Heslop-Harrison (1954) are synonymous with coenospecies and comparium respectively. The term *micton* of Camp and Gilly (1943), although corresponding to a natural

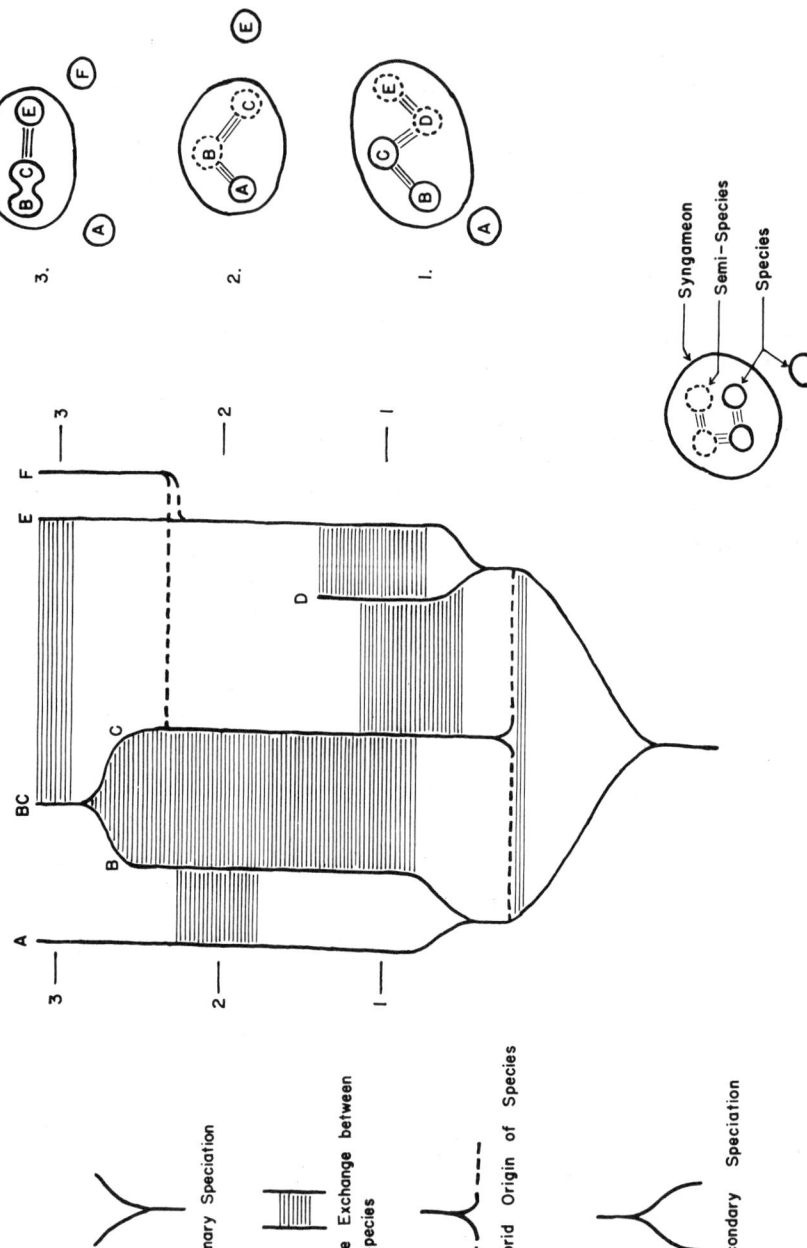

Fig. 3. Phylogenetic development of a homogamic complex showing its constituent syngameons at three different time levels. Hypothetical.

unit, is by definition a particular type of species, a species namely of hybrid origin, and is not applicable to an assemblage of hybridizing species. The commonly used term, *species group,* on the other hand, connotes an assemblage of related species but carries no implication of natural hybridization between them, and the same is true of the *superspecies* of Mayr (1942).

The *homogamic complex* proposed by Grant (1953) for a hybrid complex, the derivative forms of which are sexual, meiotically normal, and entirely or predominantly diploid, is not entirely suitable for our present purpose either, since this concept carries the connotation of past as well as present hybridization on an extensive scale. The homogamic complex is a type of phylogenetic reticulum which, at any given instant of time, may or may not correspond to the group of interbreeding species. The former is thus a more inclusive category than the latter, for it may contain species which were formerly united by hybrid connections but no longer actually hybridize.

In order to find a convenient means of designating the hybridizing species complex we must go back to the term *syngameon* of Lotsy (1925, 1931). The syngameon was defined by Lotsy as a "habitually interbreeding community" or "pairing community." Thus stated, the syngameon is nothing more than the breeding population or perhaps the biological species. The fact that Lotsy used the term species in a very narrow sense for groups of like individuals lends strength to the interpretation that his syngameon was equivalent to what we would now call species. By the context and the choice of illustrative examples, particularly the European *Betulas,* however, it is perfectly clear that Lotsy conceived the syngameon in a broader sense than the biological species. His syngameon, as applied to concrete cases, was a cluster of hybridizing species. We are justified, therefore, in taking up this old, generally familiar, and etymologically appropriate term for the interbreeding unit higher than the species.

The *syngameon,* as redefined according to present-day concepts, is the sum total of species or semispecies linked by frequent or occasional hybridization in nature; a hybridizing group of species; the most inclusive interbreeding population. A diagrammatic illustration of the changing composition of a syngameon within a homogamic complex is given in Fig. 3.

Reference to the fertility or sterility of the hybrid types is purposely omitted from the foregoing definition, contrary to earlier efforts in the era of Lotsy, Turesson, Danser, etc., as not being of major importance for the development of a syngameon.

In a syngameon the original biological species either may persist as discrete entities, as exemplified by *Gilia capitata* and *G. millefoliata* in the syngameon to which they belong, or come to occupy a definitely subordinate position to the total complex. In the latter case the species as a biological unit is found in varying degrees of dissolution. When species-like entities are discernible on the basis of partial reproductive isolation externally contrasting with free gene exchange internally, these constituent units can be designated, following Sibley (1954), as *semispecies*. (This term was used earlier by Mayr (1940) in a somewhat different sense.) The syngameon of *Gilia tenuiflora-latiflora,* for example, consists of four semispecies as shown in Fig. 2. In some syngameons, finally, the original species have become more or less completely swamped by hybridization, as has happened with *staminea* and *capitata* in the development of the present *Gilia capitata.*

The species, semispecies, or remnants of species are organized into a higher unit by frequent or infrequent gene exchange across presently or formerly existing reproductive barriers. The most inclusive interbreeding population is, consequently, not the species but the syngameon. Depending on the degree of integration attained in a syngameon, its constituent species may be preserved as such or partially or completely lost. The syngameon is in any case a biological unit in its own right.

## The Botanical versus the Zoological Species Problem

If comparisons are made between taxonomically well studied groups of seed plants and vertebrate animals, the boundaries of species are found to be, on the average, less well marked in the plants than in the animals. The causal factors of the species problem can accordingly be expected to exhibit a contrasting degree of influence in plants and animals. A comparative approach to the species problem may thus throw some light on its causes in higher plants. Such an approach may even shed light on the ultimate causes behind the immediate causes.

Since the factor of gradual speciation is common to both animals and plants, it cannot be cited as a cause of the proportionately greater species problem in plants unless certain additional assumptions are made. If the rate of speciation is higher in plants than in animals, and if this speciation is still continuing in plants but has tapered off or ceased in many animal groups, the proportion of poorly defined species due to incomplete divergence will be greater in plants.

The first assumption can be safely granted. There are about six times as many species of angiosperms as vertebrates in the world (ca. 250,000 to 40,000), even though the angiosperms originated later in geological history than the vertebrates. Whether this speciation is continuing into the recent period more actively in plants than in animals, however, is difficult to judge with the evidence available. The Age of Man has probably retarded evolution in the larger land mammals and speeded evolution up in weedy and cultivated plants, but the contrast between higher plants and animals with respect to the clarity of species is not confined to these particular groups. All that we can safely conclude is that the process of gradual species formation will result in a larger number, but not necessarily a higher proportion, of poorly defined species in higher plants as compared with vertebrate animals.

Since, so far as we can judge, incomplete speciation is not a major contributing factor to the species problem in plants, any differences that do exist between plants and animals with regard to the relative frequency of incompletely formed species can probably be ignored for the purpose of the present analysis.

A second cause of the species problem, the occurrence of sibling species, does show marked differences in frequency in plants and animals. Sibling species are of course well known in animals, being fairly common in the Diptera and occurring also in the vertebrates. In the latter they are apparently not common. In the higher plants, by contrast, cryptic species are not at all uncommon; examples can be found in nearly every large family and in a majority of large genera. When we recognize the fact that cryptic species are a common byproduct of polyploidy, and that polyploidy is a commonplace mode of speciation in plants but an exceptional phenomenon in animals, we can readily account for the observed differences between plants and animals as to the frequency of sibling species.

Similarly, asexual reproduction, which is extremely rare in the vertebrates and uncommon in the insects, is a normal means of propagation in higher plants. Vast polytypic assemblages of asexual microspecies, such as the agamic complex in *Crepis* sect. *Psilochaenia* or the clonal complex of the *Opuntia phaeacantha* group, have no known counterpart in the higher animals, although they represent a standard pattern of evolution in plants.

A prominent difference in the mode of reproduction between higher plants and higher animals is the almost universal dioecism of the latter and the very frequent hermaphroditism of the former. Many hermaphroditic plants, moreover, reproduce partially or predominantly by self-pollination. As already noted, however, the species problem does not seem to be materially increased in plants over that in animals by this difference in mode of reproduction alone. Apparently a very low rate of natural cross pollination in many autogamous plants is consistent with an integration of the members into species units.

Natural hybridization seems to play a rather different role in the evolution of plant and animal species. Natural hybrids have of course been recorded in various animal groups, as exemplified by Cockrum's (1952) check list for birds. Attention has also been paid to the effects of natural hybridization on the variation pattern and on speciation in several groups of animals, as for instance birds (Mayr and Gilliard, 1952; Sibley, 1954; Miller, 1955), mammals (McCarley, 1954; Rudd, 1955), amphibia (Blair, 1941, 1955; Volpe, 1952), fishes (Hubbs, 1955), butterflies (Hovanitz, 1949), and snails (Hubendick, 1951), to cite only a few representative studies. In the fishes this hybridization may be a fairly common and significant process. The consensus among zoologists, however, as expressed by Mayr (1942), Blair (1951), Hubendick (1951), Dobzhansky (1953), Fisher (1954), and other authors, is that hybridization is a rare and abnormal event which usually ends with the production of a few interesting but inconsequential freaks.

No similar categorical conclusion with regard to the plant kingdom can be seriously considered by botanists. That natural hybridization has evolutionary consequences of considerable importance was suggested in an early period by Naudin (1863), Kerner (1891), Lotsy (1916), and others. Botanical studies, in numerous groups of higher plants, utilizing more refined techniques developed especially by Anderson, have amply confirmed and extended the conclusion reached by the pioneer hybridologists that hybridization has profoundly affected the course of plant evolution (cf. Anderson, 1953; Grant, 1953; Anderson and Stebbins, 1954). This conclusion seems to be as true for the lower vascular plants as for the seed plants (Manton, 1950; Wagner, 1954).

A consideration of both the zoological evidence and the botanical evidence leads us to a tentative general conclusion. The zoological evidence now available, accepted at its face value, suggests that natural interspecific hybridization may have been a small or even a negligible factor in the evolution of many higher animals. The botanical evidence proves that natural interspecific hybridization has been an important evolutionary process in the plant kingdom.

The direct effects of natural hybridization on the species problem in plants are plain enough. These effects range in magnitude from a slight blurring of species distinctions to the development of a huge intricate syngameon in which the original species are more or less lost as distinct entities. It remains to point out that natural hybridization is indirectly involved in two other sources of a species problem in plants.

Sibling species, as previously noted, are a byproduct of polyploidy; polyploids outside the experimental garden are in the great majority of cases allopolyploids; and allopolyploids are the polyploid derivatives of interspecific hybrids. Asexual reproduction, similarly, affects the species problem most markedly where it is associated with hybridity, as is the case in both the agamic complex and the clonal complex. Even self-pollination, which under normal conditions is compatible with good species divisions, can if associated with hybridity and other peculiarities of the genetic system cause a disintegration of true species, as illustrated by the heterogamic complex of the *Oenothera hookeri-biennis* group.

Natural hybridization is thus directly or indirectly involved in nearly all the major sources of a species problem in plants. The conclusion is inescapable that this process is the most important single factor in the breakdown and indistinctness of plant species. The conclusion that plant species are more affected by hybridization than animal species agrees well with the observed differences in the average clarity of species boundaries in the two kingdoms.

### Physio-Genetic Nature of the Species

Among evolutionists there has been some speculation as to the reasons for the generally higher level of reproductive isolation in animals than in plants. Ethological isolating mechanisms probably account for a part of the difference. Specific behavior patterns which stimulate conspecific matings but inhibit interspecific matings constitute an important element in the complement of species-separating barriers in most animals. In higher plants such isolating mechanisms, while not entirely wanting, are certainly represented in a much weakened form. The sterility barriers also seem to be more nearly absolute in many animal groups than in plants. This may be a consequence of the apparent prevalence of genic sterility in animals and of chromosomal sterility in plants (Dobzhansky, 1951), as well as of the greater absolute fecundity of long-lived plants (Stebbins, 1950).

Explanations such as these, however, merely touch the surface of the problem. If animal species are in general less contaminated by hybridization than plant species, the ultimate cause must be sought not in the strength of the isolating mechanisms but in the evolutionary forces which build up those isolating mechanisms. A possible explanation may be that natural selection favors a certain amount of transgression of species barriers in plants, whereas in animals a more rigid isolation of species is of selective value. This hypothesis is consistent with other underlying differences between plants and animals.

The most basic difference between plants and animals is in the degree of complexity of the individual organism. The development of the relatively simple plant body is commensurate with an open system of growth by which individual parts are built up in series. The animal body is a far more complex and delicately balanced system which must develop as a whole without disruption of the internal organization.

Although very little is actually known about this question, it is probable that the genetic determinants of growth and development correspond in complexity to the degree of integration of the individual organism. Thus the factorial basis of individual character differences between races and species of plants is not infinitely complex but can often be resolved into a fairly small number of genes (Clausen, 1951; Baker, 1951; Burton, 1951; Grant, 1950). The more simple characters in animals, such as coat color or blood types in mammals (Castle, 1954; Race, 1950), can likewise be explained in terms of a few genes. By contrast, such fundamental characters of animals as the sex differences are determined by exceedingly complex systems of genes which defy an exact factorial analysis.

The genic complement of any species is the product of a long period of natural selection. The injurious effects of an unrestricted influx of foreign genes from other species are the probable basis of *ad hoc* isolating mechanisms (Dobzhansky, 1951). A species possessing a highly integrated and finely balanced gene system will, moreover, suffer relatively worse effects from hybridization than one with a simpler and more loosely integrated set of genetic determinants. In the case of the latter type of species, the disadvantageous effects of interspecific gene exchange may even be outweighed by the advantages resulting from an increase in variability.

The ability of a species to tolerate the presence of foreign genes is thus a function of the degree of complexity and integration attained by its system of genic determinants. The chances of obtaining a favorable recombination of the genotypes of two species will vary from nil to high, depending upon the amount of internal coordination in each specific genotype. A perfectly functioning bicycle can be built up out of the parts of several different makes of bicycle, but a watch containing the cogwheels of two or more different makes will probably not run accurately and may not run at all.

It is highly unlikely that the partial to complete breakdown of the species as a biological unit could continue on a wide scale in many plant genera without the sanction of natural selection. In order to explain this result we must show first that the relaxation of isolating barriers is not necessarily disadvantageous to a population, and secondly that the resulting gene flow may confer positive advantages. The hypothesis that a simple physiological-morphological organization in plants, reflecting a correspondingly simple genic pattern, permits a certain freedom in the exchange of genes between species which is not enjoyed by the vastly more complex animals, satisfies the first condition. The increase in variability resulting from the pooling of the mutations of not one, but two or several species populations is the advantageous aspect of hybridization demanded by the second condition. Together these conditions are capable of explaining an important difference between plants and animals as to the nature and behavior of their species.

## Conclusions

According to the biological species concept the species is a population set apart from the rest of the living world by reproductive isolating mechanisms. Its boundaries, therefore, should be marked by a prominent gap in the variation pattern. Good discrete species are in fact found in all major plant groups. These well-defined species, however, constitute only a fraction of the populations in their respective phylads. The botanist finds problematical species intermixed in varying proportions with good species in most of the families and genera. This is in marked contrast to the situation which the zoologist finds in higher animals in which clearly circumscribed species are the rule and poorly defined ones the exception.

A survey of some typical plant genera in which systematic difficulties are encountered shows that the species problem is due to a variety of causes. (1) Where species originate from races by a continuous and gradual process of evolution, some populations at any given instant will be in a halfway stage between race and species. (2) Some populations which must be acknowledged as species on the basis of reproductive isolation do not happen to possess easily visible distinguishing characters. (3) The isolation of species may not be complete with the result that some intermediate populations of hybrid origin may arise. (4) Plants often revert to asexual means of propagation in which case the very basis of the organization of populations into species, namely interbreeding and exchange of genes internally and reproductive isolation externally, is lost. The most important single cause of a species problem in plants is natural hybridization.

How does the biological species concept stand up in the light of the various difficulties of application listed above? The objective reality of the species as a biological unit is not impaired by the gradual formation of species or by the existence of cryptic species. The loss of sexuality, on the other hand, simply means that true biological species do not exist in certain plant groups. The effects of natural hybridization on the objective reality of the species, finally, cannot be summarized in categorical terms, since the discreteness of the species unit is a relative matter in a hybridizing complex.

Inasmuch as the species is defined as an interbreeding population isolated from other populations by reproductive barriers, natural hybridization is a factor opposed to the integrity of the species. If the hybridization is very limited in extent, its effects may be overlooked in taxonomic practice. But if the hybridization continues on an extensive enough scale, the result may be the breakdown of the original species. Between these extreme conditions there is a wide range of intermediate stages in species disintegration due to hybridization. Where the process of hybridization runs the full course to secondary speciation, the original extreme populations undergo a progressive change in status from species through semispecies to subspecies.

Many existing plant groups, as exemplified by certain sections of the willows, oaks, birches, columbines, *Ceanothus,* and *Gilia,* are aggregations of hybridizing semispecies. The major elements composing these vast polymorphic assemblages are not good species in the usual sense because they interbreed more or less freely with one another. It is equally inappropriate to regard them as merely races and the whole assemblage as a single species, because the extreme forms show partial reproductive isolation from one another and have attained a degree of morphological differentiation equivalent to that of good species in other branches of the phylad. The solution suggested for this difficulty of the species concept is to recognize the whole assemblage as a unit of interbreeding higher than the species and to designate it by the term *syngameon* of Lotsy. The syngameon is defined as a group of hybridizing species or semispecies.

The "splitting of the species" is not without precedents in biology. The conceptual difficulties of defining the individual in plant groups which propagate asexually forced a recognition of two classes of units: individual, as defined on the basis of physiological autonomy, and clone, the sum total of the individuals descended from a single zygote. Physiological geneticists are currently splitting the gene into two distinct units, having found that the unit of biochemical activity in the chromosome is not the same as the unit of crossing over. Similarly in higher plants it is necessary to recognize the existence of both the species and the syngameon or complex of species linked together by hybridization.

The final question as to why the syngameon has developed as a common evolutionary unit in higher plants and not in higher animals is tantamount to the question why plant species have been so much more affected by hybridization than animal species. A comparison of higher plants and animals with respect to the degree of organization of the genotype may help to explain these differences. The great complexity and internal coordination of the animal body probably reflects an equally complex and delicately balanced ensemble of polygenes, whereas the far more simply organized plant body is the resultant of a simpler and more loosely integrated gene system. Foreign genes will consequently be much less likely to fit harmoniously into the highly integrated genotype of an animal than into the more loosely organized genotype of a plant. Natural selection will accordingly eliminate the products of interspecific hybridization, or prevent the hybridization from occurring by the erection of breeding barriers, to a greater extent in animals than in plants.

*Acknowledgments.* The manuscript was critically reviewed by Dr. Jay M. Savage, of the University of Southern California, to whom the author is indebted for numerous helpful suggestions. Mr. Howard Latimer of the Rancho Santa Ana Botanic Garden also made constructive suggestions with regard to one section of the manuscript. The problems of classification in certain difficult western American genera of plants were discussed with Dr. Philip A. Munz of the Rancho Santa Ana Botanic Garden. The interest and advice of these colleagues is gratefully acknowledged. Finally, the deliberations in Atlanta brought out several new points, which were taken into consideration in the preparation of the final manuscript.

The work on the Cobwebby Gilias, incorporated into the discussion of the effects of hybridization on species distinctness, was accomplished with the aid of a research grant from the National Science Foundation, which is likewise thankfully acknowledged.

## REFERENCES

Anderson, E. 1953. Introgressive hybridization. *Biol. Revs. Cambridge Phil. Soc.*, **28**, 280-307.

Anderson, E., and G. L. Stebbins, Jr. 1954. Hybridization as an evolutionary stimulus. *Evolution*, **8**, 378-88.

Baker, H. G. 1951. Hybridization and natural gene-flow between higher plants. *Biol. Revs. Cambridge Phil. Soc.*, **26**, 302-37.

Blair, A. P. 1941. Variation, isolation mechanisms, and hybridization in certain toads. *Genetics*, **26**, 398-417.

Blair, A. P. 1955. Distribution, variation, and hybridization in a relict toad (*Bufo microscaphus*) in southwestern Utah. *Am. Museum Novitates*, **No. 1722**, 1-38.

Blair, W. F. 1951. Interbreeding of natural populations of vertebrates. *Am. Naturalist*, **85**, 9-30.

Burma, B. H. 1954. Reality, existence, and classification: a discussion of the species problem. *Madroño*, **12**, 193-209.

Burton, G. W. 1951. Quantitative inheritance in pearl millet (*Pennisetum glaucum*). *Agron. J.*, **43**, 409-17.

Camp, W. H. 1951. Biosystematy. *Brittonia*, **7**, 113-27.

Camp, W. H., and C. L. Gilly. 1943. The structure and origin of species. *Brittonia*, **4**, 323-85.

Castle, W. E. 1954. Coat color inheritance in horses and in other mammals. *Genetics*, **39**, 35-44.

Clausen, J. 1951. *Stages in the Evolution of Plant Species*. Cornell University Press, Ithaca, New York.

Cockrum, E. L. 1952. A check-list and bibliography of hybrid birds in North America north of Mexico. *Wilson Bull.*, **64**, 140-59.

Danser, B. H. 1929. Über die Begriffe Komparium, Kommiskuum und Konvivium und über die Entstehungsweise der Konvivien. *Genetica*, **11**, 399-450.

Darlington, C. D. 1940. Taxonomic species and genetic systems. In *The New Systematics*. Oxford University Press, Oxford, England.

Davidson, J. F. 1954. A dephlogisticated species concept. *Madroño*, **12**, 246-51.

Dobzhansky, T. 1935. A critique of the species concept in biology. *Philosophy of Science*, **2**, 344-55.

Dobzhansky, T. 1951. *Genetics and the Origin of Species*, 3rd ed. Columbia University Press, New York, N.Y.

Dobzhansky, T. 1953. Natural hybrids of two species of *Arctostaphylos* in the Yosemite region of California. *Heredity*, **7**, 73-79.

Dobzhansky, T., and C. Epling. 1944. Contributions to the genetics, taxonomy, and ecology of *Drosophila pseudoobscura* and its relatives. *Carnegie Inst. Wash. Publ.* **No. 544**.

Du Rietz, G. E. 1930. The fundamental units of biological taxonomy. *Svensk Botan. Tidskr.*, **24**, 333-428.

Fisher, R. A. 1954. Retrospect of the criticisms of the theory of natural selection. In *Evolution as a Process,* Allen and Unwin, London.

Gilmour, J. S. L., and J. Heslop-Harrison. 1954. The deme terminology and the units of micro-evolutionary change. *Genetica,* 27, 147-61.

Goldschmidt, R. B. 1952. Evolution, as viewed by one geneticist. *Am. Scientist,* 40, 84-98.

Grant, A., and V. Grant. 1956. Genetic and taxonomic studies in *Gilia.* VIII. The Cobwebby Gilias. *El Aliso,* 3, 203-87.

Grant, V. 1950. Genetic and taxonomic studies in *Gilia.* I. *Gilia capitata. El Aliso,* 2, 239-316.

Grant, V. 1952. Genetic and taxonomic studies in *Gilia.* III. The *Gilia tricolor* complex. *El Aliso,* 2, 375-88.

Grant, V. 1953. The role of hybridization in the evolution of the Leafy-stemmed Gilias. *Evolution,* 7, 51-64.

Hovanitz, W. 1949. Increased variability in populations following natural hybridization. In *Genetics, Paleontology, and Evolution.* Princeton University Press, Princeton, New Jersey.

Hubbs, C. L. 1955. Natural hybridization in fishes. *Systematic Zool.,* 4, 1-20.

Hubendick, B. 1951. Recent Lymnaeidae. Their variation, morphology, taxonomy, nomenclature, and distribution. *Kgl. Svenska Vetenskapsakad. Handl.,* 3, 1-223.

Kerner, A. 1891. *Pflanzenleben.* 2 vols. Leipzig and Vienna.

Lotsy, J. P. 1916. *Evolution by Means of Hybridization.* The Hague.

Lotsy, J. P. 1925. Species or Linneon. *Genetica,* 7, 487-506.

Lotsy, J. P. 1931. On the species of the taxonomist in its relation to evolution. *Genetica,* 13, 1-16.

Manton, I. 1950. *Problems of Cytology and Evolution in the Pteridophyta.* Cambridge University Press, Cambridge, England.

Mason, H. L. 1949. Evidence for the genetic submergence of *Pinus remorata.* In *Genetics, Paleontology, and Evolution.* Princeton University Press, Princeton, New Jersey.

Mason, H. L. 1950. Taxonomy, systematic botany and biosystematics. *Madroño,* 10, 193-208.

Mayr, E. 1940. Speciation phenomena in birds. *Am. Naturalist,* 74, 249-78.

Mayr, E. 1942. *Systematics and the Origin of Species.* Columbia University Press, New York City, N.Y.

Mayr, E. 1948. The bearing of the new systematics on genetical problems. The nature of species. *Adv. in Genetics,* 2, 205-37.

Mayr, E. 1949. Speciation and selection. *Proc. Am. Phil. Soc.,* **93,** 514-19.

Mayr, E., and E. T. Gilliard. 1952. Altitudinal hybridization in New Guinea honeyeaters. *Condor,* **54,** 325-37.

McCarley, W. H. 1954. Natural hybridization in the *Peromyscus leucopus* species group of mice. *Evolution,* **8,** 314-23.

Miller, A. H. 1955. A hybrid woodpecker and its significance in speciation in the genus *Dendrocopos. Evolution,* **9,** 317-21.

Naudin, Ch. 1863. Nouvelles recherches sur l'hybridité dans les végétaux. *Ann. sci. nat.* (Paris) (Botan.), **19,** 180-203.

Race, R. R. 1950. The eight blood group systems and their inheritance. *Cold Spring Harbor Symposia Quant. Biol.,* **15,** 207-20.

Rollins, R. 1952. Taxonomy today and tomorrow. *Rhodora,* **54,** 1-19.

Rudd, R. L. 1955. Variation and hybridization in shrews. *Systematic Zool.,* **4,** 21-34.

Sibley, C. G. 1954. Hybridization in the red-eyed towhees of Mexico. *Evolution,* **8,** 252-90.

Stebbins, G. L., Jr. 1950. *Variation and Evolution in Plants.* Columbia University Press, New York City, N.Y.

Turesson, G. 1922. The genotypical response of the plant species to its habitat. *Hereditas,* **3,** 211-350.

Volpe, E. P. 1952. Physiological evidence for natural hybridization of *Bufo americanus* and *Bufo fowleri. Evolution,* **6,** 393-406.

Wagner, W. H. 1954. Reticulate evolution in Appalachian Aspleniums. *Evolution,* **8,** 103-18.

Ward, G. H. 1953. *Artemisia,* section *Seriphidium,* in North America. A Cytotaxonomic study. *Contribs. Dudley Herbarium,* **4,** 155-205.

# THE SPECIES PROBLEM IN FRESHWATER ANIMALS

JOHN LANGDON BROOKS: YALE UNIVERSITY, NEW HAVEN, CONNECTICUT

Most of the problems encountered in the systematics of freshwater animals are of the same nature as those met in the systematics of terrestrial or marine animals. The only justification for considering these problems in freshwater animals is the possibility that some of the complicating factors may operate with a peculiar intensity or in unusual combinations in the freshwater fauna. Whereas a comprehensive survey of the systematics of the diverse animal stocks that comprise this freshwater fauna might reveal the information we seek, such a survey would be lengthy and our labors might not find a proportionate reward. What I propose to do in this paper is to select from biologically different, but equally successful, freshwater stocks two genera the systematics of which have proved difficult to comprehend. By analyzing the "species problem" in each, it will be possible to see whether the sources of the systematic difficulty in the two groups are at all similar. If any similarities are evident, it can tentatively be concluded that these represent evolutionary processes working with peculiar intensity in fresh waters.

Generalizing from selected, extreme examples must be done with proper caution both in the selection of the cases and in the force and extent of the generalizations. The two genera selected are *Daphnia*, small planktonic crustaceans, and *Coregonus*, the true whitefishes. These are two of the roughly half-dozen systematically most troublesome genera of common freshwater animals. *Daphnia* was chosen because that is the group with which I am most familiar, and for this reason the greater part of our attention will be devoted to it. Much of the information about this genus is based upon investigations described more fully in a monograph on the systematics of North American *Daphnia* (Brooks, 1956). *Coregonus* was selected as the second group because it represents a very different kind of organism (one especially different from *Daphnia* in its reproductive behavior), its systematics has proved particularly thorny, and, possibly most important, it has been the subject of a penetrating study by Gunnar Svärdson (1949-1953), of the Institute of Freshwater Research at Drottningholm, Sweden. Whereas the genera chosen represent a reasonable biological diversity, they are both groups which flourish especially in the glaciated areas of North America and Eurasia. Thus both groups are alike in having been subjected to the disruptive influences of the Pleistocene glaciations. However, as most of the bodies of standing fresh water are now in areas which were glaciated, the rich fauna which has developed in these waters will have, in greater or lesser degree, shared the same history. Any assessment of the role of these Pleistocene events in the evolution of even these two genera is well outside the scope of this study.

### The Species Problem in *Daphnia*

*Daphnia* are small (½ to 4 mm. long), planktonic Cladocera common in lakes and small ponds. They feed primarily on small algae which they filter from the water by means of combs on some of their thoracic appendages, which are enclosed within a bivalve carapace. The genus is worldwide in distribution (as many genera of Cladocera tend to be). The ranges of some species are rather limited, embracing but a portion of a continent, whereas those of others are broad, extending over several continents.

A recent study (Brooks, 1956) indicates that there are fifteen species living in North America, out of a total of about thirty well-characterized species. However, it is probable that there are about fifty extant species. While the distributions even of these thirty known species are not so completely known as one might wish, a rough estimate, on the assumed basis of forty to fifty species, for the other major land areas would be: South

America, 7; Eurasia, 25; Africa, 15; Australia, 3. Although the subsequent discussion of the systematic problems in *Daphnia* is based primarily on a study of the genus as it occurs in North America, it is almost certain that Eurasia and Africa present problems of equal or greater magnitude, and probably of a similar nature.

An indication of the extent of the difficulties of *Daphnia* systematics can be found in the varied taxonomic treatment that the members of this genus have received. During the past seventy years several hundred names have been proposed for *Daphnia* populations, and these have been trinomial and quadrinomial combinations, as well as the usual binomials. For example, in Birge's widely used key to North American Cladocera (in Ward and Whipple, 1918), six binomials, seven trinomials, and four quadrinomials are used to indicate the fourteen entities that he recognized. However, these polynomials, although usually based on collections from one or a few localities, were usually considered to designate "morphological variants" which might occur anywhere, rather than geographically limited variants. Thus the trinomials, which are common in the literature on *Daphnia* systematics, seldom referred to a geographical subspecies, in the manner now common in the systematics of, for example, birds.

When Wagler, about 1936, attempted a revision of the genus he was faced with this accumulated taxonomic chaos. His solution, seemingly a reaction against the prevailing system, was to classify all *Daphnia* into eleven species all indicated by binomials. The simplicity of this system is in large part due to the fact that several of these "species" are aggregates of what, in reality, are clearly distinct species. This, however, was not an example of the lumping as opposed to the splitting of groups of closely related species, because many of the aggregated species are quite dissimilar.

Two of these aggregates are *Daphnia longispina* and *Daphnia pulex*. Since many phenotypic characteristics of each are variable, both seasonally and geographically, all the species could be put into "present" or "absent" categories on the basis of only one relatively constant morphological feature. This was the middle pecten of the postabdominal claw. Either the teeth of this pecten were much larger than those of the proximal and distal pectens, the "*pulex*" criterion, or they were the same size as those of the other pectens, the "*longispina*" criterion (see Fig. 8). However, classifying species on the basis of a single character, while often convenient, frequently, if not usually, fails to indicate the phylogenetic relationships between these species. It has become increasingly apparent that Wagler's simplified classification of *Daphnia* does not avoid this pitfall.

We can possibly appreciate the results of these taxonomic philosophies, which we might call the "polynomial" and the "simplified," where they have both been used to classify the same assemblage of forms. In 1950 Kiser applied the polynomial philosophy to the North American *Daphnia*. His classification named 51 entities. These comprised four species, 14 "subspecies," and 33 "forms." His *Daphnia pulex* consisted of seven subspecies, three of which have one, three, and five forms, respectively. His *Daphnia longispina* is composed of six subspecies, four of which have four, six, six, and eight forms, respectively. Pennak (1953) divides the North American *Daphnia* into the same four species. He considers that *Daphnia pulex* and *Daphnia longispina* are each highly variable species, but that little is gained by naming the more distinctive forms of each. However, the systems of classification of both Pennak and Kiser are basically the same, because each places primary emphasis on variation in a single character, the relative size of the teeth in the middle comb of the postabdominal claw. Detailed study has revealed that there are six North American species that would be considered *Daphnia pulex* in either of the previously mentioned classifications (*D. pulex* Leydig, *D. middendorffiana* Fischer, *D. schødleri* Sars, *D. catawba* Coker, *D. parvula* Fordyce, *D. retrocurva* Forbes); and seven that would have been considered *D. longispina* (*D. ambigua* Scourfield, *D. longiremis* Sars, *D. rosea* Sars, *D. laevis* Birge emend. Brooks, *D. dubia* Herrick emend. Brooks, *D. galeata mendotae* Birge, *D. thorata* Forbes). Interestingly, *D. longispina* O. F. Müller, originally described in Europe, does not occur in North America. Examination of Figs. 8 and 9 will reveal some of the distinctive characters of ten of the thirteen species by which the subgenus *Daphnia* is represented in North America.

Our main task is to discover the reasons why the characterization and identification of these common freshwater animals has presented so much difficulty. But before proceeding to a consideration of these sources of difficulty, we should consider some of the basic biological peculiarities of the animals being studied.

The average life span of a *Daphnia* is between a few weeks and several months, depending on the temperature. At the temperature (about 20° C.) common in ponds and the upper waters of lakes in summer, the female of most species lays her first brood of eggs about a week after she is born. Subsequent broods of eggs are laid at two- or three-day intervals. Although it is difficult to assess the life spans under natural conditions, probably two or three weeks would be a reasonable estimate of the most frequent life span at these temperatures (Brooks, 1946). These same species grow much more slowly at the low temperatures (4° C. or less) characteristic of temperate lakes (*dimictic* lakes in the terminology of Hutchinson and Löffler, 1956) during winter. A few of the individuals born in the fall survive until the waters warm in the spring. Then they produce a few broods of eggs and die. Although it is clear that the few females which do survive have a life span of three or four months, it is difficult to assess the average life span at these temperature under natural conditions.

A *Daphnia* population living in a large permanent body of water may have as many as ten or fifteen generations per year (see below for the basis of this estimate). At the other extreme are the temporary-pond populations, which may have only one or two generations per year. The species maintains itself for the remainder of the year in the form of dormant early embryos (so-called resting eggs) within a resistant case, the ephippium.

The populations of *Daphnia* consist almost entirely of females (the exceptional presence of males is discussed later) that are able to maintain themselves because the eggs produced under propitious environmental conditions are diploid, and they begin cleavage soon after they leave the oviduct. The diploidy of this succession of individuals is apparently maintained by a failure of the second reduction division during maturation of the oöcyte nucleus. The eggs so produced are often referred to as "parthenogenetic" eggs, and although in some exceptional situations (Edmondson, 1955) this designation loses its utility, it is sufficient for present purposes. This derivation of one generation of females

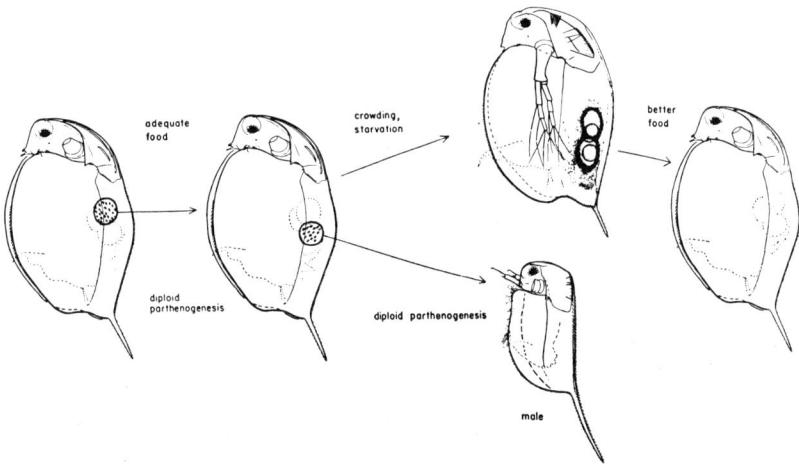

Fig. 1. Diagrammatic representation of the modes of reproduction in *Daphnia*. Usually reproduction is uniparental and all the individuals are females. Occasionally, however, the females produce eggs requiring fertilization, and more or less simultaneously some of the parthenogenetic eggs develop into males instead of females. Females bearing ephippia and "sexual" eggs, and males usually appear just after population maxima. Eggs (actually early embryos) shed in the ephippium are in a dormant state and they can often withstand drying and freezing. If conditions improve, the female can resume parthenogenetic egg production after she has shed an ephippium. Males are short-lived.

from the previous by diploid parthenogenesis is represented diagrammatically in Fig. 1 by the two females in the right half of the diagram. As long as the food supply remains adequate, the female generations will succeed one another parthenogenetically for an apparently indefinite period (Banta, 1939). Shortly after each molt, the parthenogenetic eggs are extruded into the space between the body proper and the dorsal part of the bivalve carapace. They are prevented from falling out of this space, referred to as the brood pouch or brood chamber, by the finger-like projections on the posterior part of the abdomen (see Fig. 1). The number of eggs is influenced by the size and age of the female, but for females of a given size, the level of the food supply is the principal determinant of clutch size. While very large individuals may produce twenty, or thirty, or even up to fifty eggs per brood, the smaller adults under the "usual" range of conditions may have from none to five eggs per brood (Slobodkin, 1954; Brooks, 1946). These eggs develop directly into free-swimming individuals which escape from the brood chamber during the next molting of the parent. It should probably be noted here that *Daphnia* embryos develop at the expense of the material stored in the egg; there is no nourishment otherwise derived from the mother while the embryos are in the brood chamber.

Although this diploid parthenogenesis is repeated as long as nutritive conditions remain adequate, a sudden drop in food intake can, in some of the females, provoke a change in the reproductive physiology. This change involves the cytoplasm of the developing oöcytes as well as the meiotic events of the nucleus and the structure of the developing carapace. The cytoplasm of the oöcytes becomes dark brown and opaque, as opposed to the light-colored (green-gray-blue), hyaline appearance characteristic of parthogenetic eggs. The nucleus undergoes both reduction divisions so that the resulting egg is haploid. These eggs must be fertilized before they will develop. The cells of the dorsal part of the carapace change from flattened and colorless to columnar and pigmented, and form the ephippium (the saddle-shaped area apparent on one female in Fig. 1). The ephippium clasps the two fertilized eggs which have been extruded into the brood chamber. The eggs undergo partial development while in the brood chamber, but soon reach a quiescent stage. In this dormant state the eggs can withstand freezing or drying. At the time of the next molt the ephippium, clasping its eggs, is shed with the rest of the exoskeleton. It may sink to the bottom or be caught in the surface film. Its ability to stay in the surface film is a consequence of the hydrofuge nature of its cuticle.

After the female has shed her old exoskeleton with its ephippium, she may, if nutritive conditions continue to be poor, have no eggs in the brood chamber (no oöcytes having matured while she was carrying the ephippium). If conditions should have improved, however, it is possible for her to resume quickly the production of parthenogenetic eggs.

The males necessary for the fertilization of the haploid ephippial eggs develop from diploid parthenogenetic eggs. Males usually appear both in nature and in laboratory culture shortly after a population maximum. The exact nature of the environmental stimulus for the development of males instead of females is not clear, nor is the mechanism by which this is accomplished (see Banta, 1939, for a summary of investigations into this problem). Shortly before the time that a population maximum is reached, the amount of food per individual usually falls rapidly. The apparent environmental determinant of ephippial egg formation (diminution of food supply) thus initiates the slower process of ephippial egg formation just before males begin to develop (at the time of population maxima).

There can be little doubt of the selective value of uniparental reproduction to Cladocera, and especially to *Daphnia* in which it is especially prominent. When food is adequate, every feeding member of the population is an egg-producer—making a most efficient transformation of food into eggs. This enormous reproductive potential, however, carries with it the danger of overshooting the population size and density that the food supply can support. Possibly the chief significance of the episodes of sexuality from the point of view of population dynamics is that they quickly reduce both the reproductive rate and the feeding rate of the population. The diversion of some diploid parthenogenetic eggs into males reduces the number of egg producers developing from oöcytes already formed. The oöcytes beginning to form in other females follow a path of development which, if fertilization ensues, produces an embryo with a dormant period of varying duration. Thus a part of the population will exist for a time as nonfeeding, nonreproducing individuals. These mechanisms permit a temporary reduction in the size of the feeding population, so that there is less chance that death by starvation will reduce it more drastically.

The general reproductive pattern of a population determines its genetic system even as the pattern is genetically controlled. The range of genetic variation among the offspring produced by diploid parthenogenesis (of the type with which we are concerned) will very nearly be coextensive with that of the parents. Although mutation and crossing over will probably produce some increased variation, this increase in the amount and kind of variation can be expected to be considerably less than that which would occur after sexual reproduction. (See Banta, 1939, for a summary of work by Banta and co-workers on variation following the two types of reproduction.)

The relative frequency of uniparental and biparental inheritance differs according to the environment which the population inhabits. By and large *Daphnia* are most successful in two types of environments: either large, permanent bodies of water, or small ponds, which become regularly (or irregularly) uninhabitable because of the disappearance of the water or the formation of ice. A species is usually better adapted to one type or the other, although some can maintain themselves in both. In many lake-dwelling populations, reproduction is almost exclusively uniparental. Sexual forms occur only at infrequent intervals and then comprise only a very small percentage of the population. The offspring produced sexually probably have, in most cases, little effect upon the genetic composition of the population because of the relative unhatchability of ephippial eggs after long periods of uniparental reproduction. Banta (1939) believes that this reduced viability of sexually produced offspring is due to the deleterious recessive mutations that had accumulated during the periods when the population was reproducing by diploid parthenogenesis. However, it is possible that a single recombinant might be more successful than many of the other clonal genotypes which constitute the population. Such an individual could give rise by parthenogenesis to a clone of individuals with similar constitutions. This clone, if truly superior in that particular body of water, might slowly replace the other clones which comprise the population.

In a population of a pond-inhabiting species the ability to produce a relatively large number of viable resting eggs (embryos) would have considerable selective value. For most species sexual reproduction is a necessary prelude to the formation of the resting stage. [It is of interest to note in passing, however, that the mechanism for producing resting eggs parthenogenetically has become firmly established at least once, and it has probably arisen several times (Edmondson, 1955; Brooks, 1956).] In these pond populations it is to be expected that natural selection will favor the development and retention of high sensitivity to the environmental conditions which trigger the processes leading to the formation of males and to the production of haploid eggs. A rapid and extensive conversion of the population from uniparental to biparental reproduction would thus reduce the demands on a dwindling food supply as well as produce the (usually) necessary resting stages.

After this consideration of the reproductive peculiarities of *Daphnia* we are now in a position to appreciate more fully the factors that appear to be responsible for the difficulties encountered in the systematics of this genus. These are:

1. The wide range of phenotypic variation due to the sensitivity of the developmental processes to environmental conditions.

2. The frequent coexistence of flourishing populations of several species in the same body of water.

3. The introgression of genetic material between these coexisting species.

*Phenotypic Variation.* The phenotypic variation in *Daphnia* is readily discernible as it changes the body shape. Figure 2 indicates some of the variants to be found in a pond-dwelling species (*Daphnia schødleri* Sars). The form of any crustacean is structurally determined by the form of its exoskeleton. The differences between A, B, and C of Fig. 2 are due to the different extent to which the exoskeleton of head, carapace, and shell spine have grown. In the species illustrated here, as well as in all the other members of the genus with which we are concerned in this

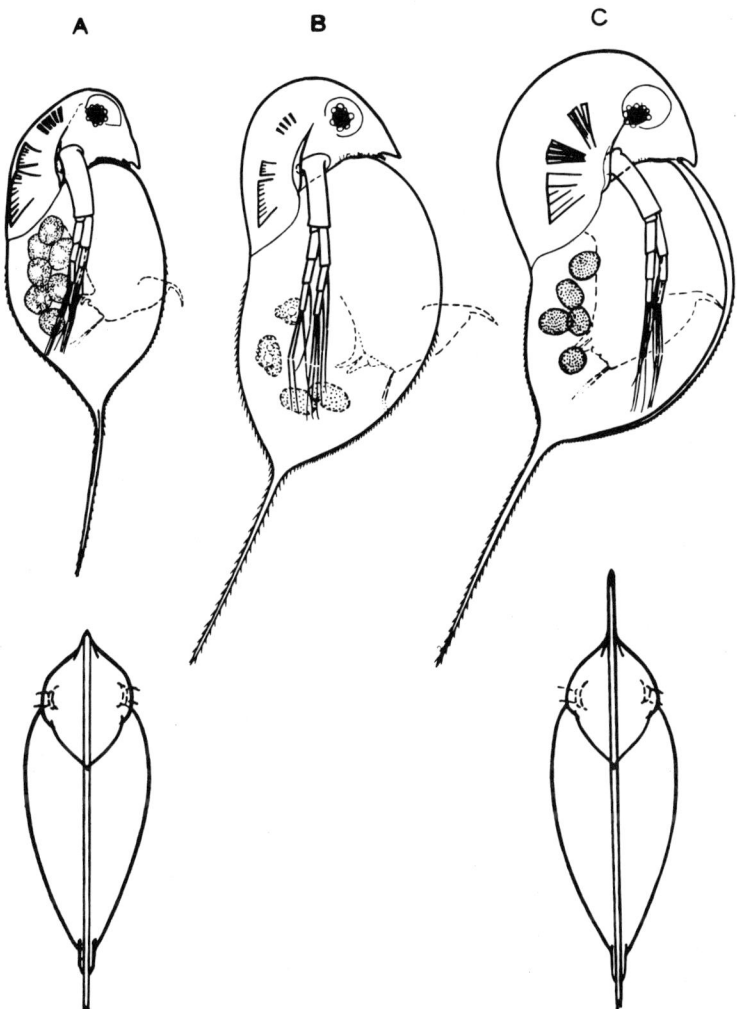

Fig. 2. Phenotypic variation in *Daphnia*. These three specimens of *Daphnia schødleri* Sars exemplify the variation in body form that characterizes most species of *Daphnia*. The principal differences are in the length of the shell spine and the extension of the exoskeleton along the head margins. In lateral view the extent of the crest, as this lamellar extension is called, can be judged by the distance between the head margin and such structures as the optic vesicle and the origins of the antennal muscles. Dorsal views of specimens A and C indicate lamellar nature of the crest.

paper, the differences in the relative growth of the exoskeleton of these body parts are especially noticeable because they alter the outline of the body when the animal is viewed laterally, as it usually is. When the extreme forms (A and C) are viewed in the dorsal aspect, the chief difference is in the extension of a very thin lamella which develops over the midline of the head and carapace. It is this thin lamella, composed of two closely oppressed sheets of exoskeleton separated by a thin film of haemolymph, which is responsible for the most striking shape differences between the variant individuals. The mid-dorsal extension of the head margin is called a "crest" when it covers a relatively broad section of the anterior and dorsal portions of the head margin, and a "helmet" (see Figs. 3, 4) if narrower at the base and taller. The bases of the antennule muscles are included in the drawings to indicate the position of the body margin (in lateral view) if the crest or helmet were not present. As might be expected, the relatively excessive exoskeletal growth of B and C as compared with A is also expressed in the larger and more numerous spinules along the dorsal line of the carapace and on the ventral margins of the valves, and in the relatively greater length of the shell spine. Although in the species depicted here the most exuberant form (C) is larger than the most conservative (A), this is usually not true; the individuals with the most exaggerated shapes often are smaller. It is therefore not a question of increasing disproportion with increasing size.

This wide range of phenotypic variation occurs in generations that have developed parthenogenetically. Various laboratory experiments on *Daphnia* (see especially Brooks, 1946, 1948) indicate that a clone formed from any individual picked at random from a natural population can express the same range of phenotypic variation that the natural population exhibits under similar environmental conditions. Although various genetic mechanisms are known in diverse organisms by which recombinations are possible even when the orthodox meiotic process is absent, there is no evidence that they play a role in the determination of the phenotypic variation in *Daphnia*. This does not mean to imply that some such mechanism may not be effective in increasing the genetic diversity during the periods when the population is reproducing by diploid eggs.

As the determinants of this phenotypic variation are not genetic, they must be environmental. The simplest hypothesis is that the size of the crest or helmet is dependent on the environmental conditions during the growth of the individual concerned. In order to study this relationship it is desirable to have a population showing an extreme range of phenotypic variation in a natural situation in which the population's environment is as homogeneous at any one time as is possible.

A fortunate combination of these desiderata occurred in Bantam Lake, Connecticut. This is a large, shallow lake with three basins, the largest and deepest of which is about a mile in diameter and seven meters deep. In such a body of water, the water is likely to be nearly the same temperature from top to bottom because the winds are able to mix the water throughout most of the season when there is no ice cover. As there was evidence to indicate that temperature plays a major role as a determinant of the phenotype, such a lake is to be preferred over one in which the warmer upper strata are underlain in midsummer by strata of cool or cold water. Although the behavior of most of the species that can develop helmets is such that they remain, day and night, in the upper strata, the possibility that a portion of any population might have penetrated into the cold, lower waters would complicate the analysis of the environmental effects. In the year with which we are concerned the principal species of *Daphnia* was *D. retrocurva* Forbes. This species exhibits probably the most extreme variation of any North American species and one of the most striking in the genus and, indeed, in the Cladocera.

The population of *D. retrocurva* in Bantam Lake during 1945 was sampled at approximately two-week intervals from April 1, one week after the ice melted, until August 10. By the middle of August the population had so dwindled in size that sampling yielded only a few specimens. (However, it must be remembered that even a population so sparse as to yield but a few specimens to this standard sampling procedure could amount to a total size

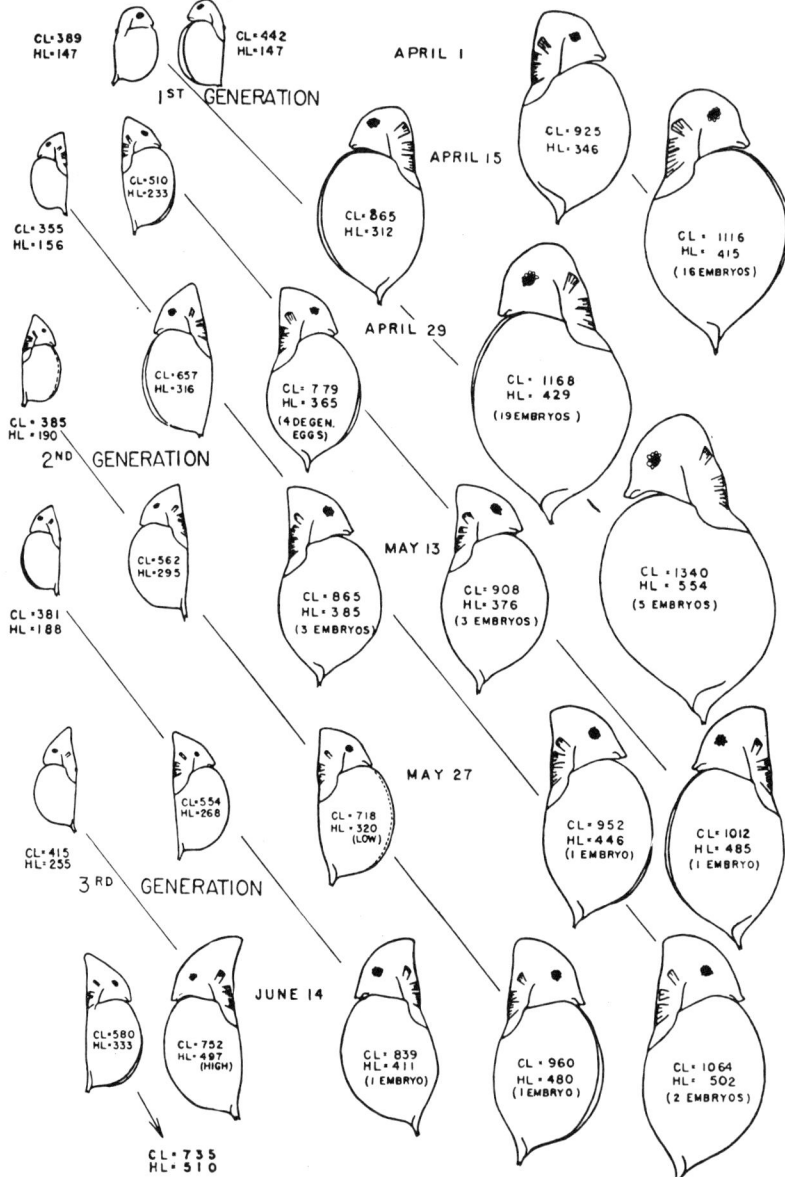

Fig. 3. See legend with Fig. 4.

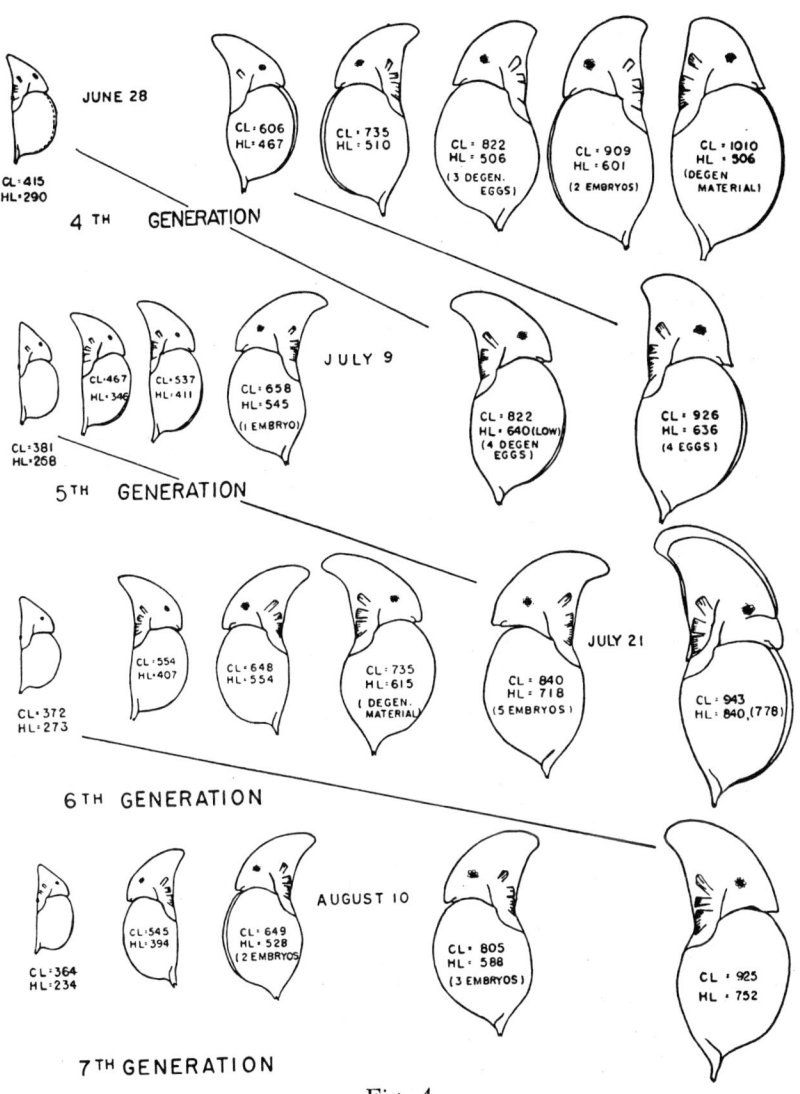

Fig. 4.

Figs. 3 and 4. Seasonal phenotypic variation in population of *Daphnia retrocurva* Forbes in Bantam Lake, Conn., 1945. Series of camera lucida drawings in each horizontal row depict characteristic shape of each size group on date indicated. Head length (HL) and carpace length (CL) of each given in microns. Diagonal lines trace development of newborn individuals of each sampling date. All have same magnification.

of something of the order of a million individuals.) The sampling was done so that all depths of the lake were equally represented at all dates and so that each series of samples was quantitatively equivalent. In equivalent columns of water there were three specimens of *retrocurva* on April 1. By the end of that month there were over three thousand individuals, and the number remained between three and five thousand until the middle of June. By the end of June this had dropped to about one thousand, and on July 21 and August 10 there were less than two hundred *retrocurva* in the standard column of water. These population densities are of interest in relation to the phenotypic variation.

The appearance of the individuals that occurred at the dates of sampling is indicated in Figs. 3 and 4. The proportions of all undistorted specimens in a sample (up to a maximum of about one hundred) were measured, and the figures drawn in each horizontal row depict the average proportions (of head vs. carapace) for each size grouping that occurred. By studying the distribution of size and proportions in this temporal series of population samples, it was possible to estimate whether or not the majority of the newborn in the population sample represented a new generation or merely successive broods of the same generation. For example, on April 15 the newborn individuals were almost certainly the offspring of the over-wintering females (seen as the largest females on both April 1 and April 15). The female with a carapace length of 865 microns ($CL = 865$) represents the most numerous size group. These are the individuals which were the first-born of the year, and which were approaching maturity. However, the old, large over-wintering females were the only ones producing eggs, and these in large numbers. It was not until the end of April that the second generation predominated among the newborn. Although such an analysis becomes less clear-cut later in the season, it is reasonable to estimate that there were seven generations during the four-and-a-half-month period of observation. On this basis one can make the even rougher estimate that there are probably about ten or twelve generations of *Daphnia retrocurva* per year in this lake.

The diagonal lines in Figs. 3 and 4 trace the history of the newborn of each sample.

With this brief description of the population in mind, we can proceed with a consideration of the correlation between the characteristics of the phenotypic variants and the environmental conditions under which these variants developed. Measurements of the relative size of head and carapace of the individuals of the population provide insight into the relative developmental rate of the helmet as compared with that of the rest of the body, of which carapace length is taken as a characteristic dimension. Furthermore, the ratio of head length to carapace length in the newborn provides an index to the rate of relative growth prior to birth. Field and laboratory studies indicated that the relative length of the head at birth is controlled by the temperature during development. However, the rate of relative growth of the helmet after birth when the animal is free-swimming is not directly correlated with the temperature. It appears that both high temperature (above 18-20° C. for many species) and turbulence of the water are necessary conditions for the development of helmets of the maximum size of which a population is capable (Brooks, 1946, 1947). The details of the interrelation of these major determinants of the relative growth rate of the helmet have not been worked out, but a promising hypothesis is that both factors (temperature and turbulence) have their effect by raising the metabolic rate.

Although laboratory experiments to determine the importance of turbulence in species other than *Daphnia galeata* have not as yet been done, there is a considerable body of observational evidence indicating that large helmets develop in those lake-dwelling species that live in the upper waters of a lake. In the majority of lakes in the temperate zone this will mean that these populations during the winter months will live in water that is less than 4° C., often near 0° C. However, in midsummer these water strata are commonly above 18-20° C. except in very large lakes, in which, incidentally, large helmets usually are not found. These animals thus have to swim, feed, grow, and reproduce in a very wide temperature range. The exuberant form which develops at high

98 FRESHWATER ANIMALS

temperatures is an almost inevitable result of the action of a genotype which has to be able to maintain activities at slightly above 0° C.

In the past it has usually been assumed that the ability to produce these extreme helmets has been retained in the genetic makeup because of selective advantage conferred by the possession of such a bizarre body shape. Although this is not the place to develop the inadequacies of this group of hypotheses,

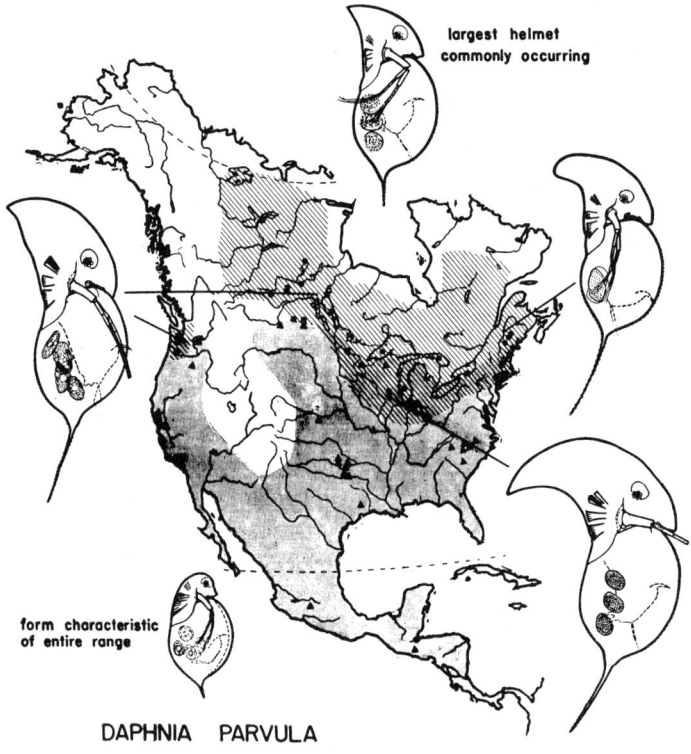

Fig. 5. Geographical variation in the shape of the helmets of maximum size in *Daphnia retrocurva* Forbes. The three specimens with large helmets typify the helmet shapes developed in the three regions indicated. The specimen at the top is a reminder that the vast majority of specimens throughout the range of the species have helmets smaller than the ones depicted. For comparison the distribution in North America and the characteristic invariant shape of *D. parvula* Fordyce are indicated. *D. parvula* is the only species closely related to *D. retrocurva*.

several kinds of information should possibly be considered here that are not consistent with the hypothesis of advantage conferred by the helmet. The first such evidence is presented in Fig. 5, which indicates the range of *Daphnia retrocurva* and some of the forms of the helmet. The animal at the top is a reminder that most of the individuals in any population have a small helmet, smaller than the one depicted. In Bantam Lake, for example, the population size when the helmets were large was less than one-tenth that when the helmets were unspectacular (cf. Figs. 3 and 4). The other three specimens of *retrocurva* indicate the geographical variation in the extreme form of the helmet. One of the elaborate hypotheses (Woltereck, 1930) for the functional significance of the helmet maintains that a retrocurved helmet, of which the specimen from Lake Mendota is probably the most extreme in the genus, has a different effect from one which is elongated in the body axis. Many specimens of *D. retrocurva* have tall helmets that are not retrocurved. The specimen drawn as characteristic of the Canadian and West Coast areas has a more retrocurved helmet than many in these areas. The special pleading necessary to account for this cline and for the fact that most of the helmeted forms are intermediate in shape between the two extremes, with (supposedly) quite different effects, tends to cast doubt on the various hypotheses of this type so far proposed.

However, rather different and more telling evidence is available which makes it improbable that the ability to produce a helmet is maintained in the genotype because of selective advantage conferred by the helmet itself. This comes from the consideration of species such as *Daphnia longiremis*, which usually has no more than the slight crest indicated on the topmost of the three specimens drawn in Fig. 6. Throughout the indicated range in North America, and in northern Eurasia as well, *longiremis* occurs in populations of such round-headed individuals. On rare occasions, however, populations occur in which helmeted individuals are common. The places where such populations have been found in North America are indicated on the map. Two types of helmets occur, and the differences are probably geneti-

100  FRESHWATER ANIMALS

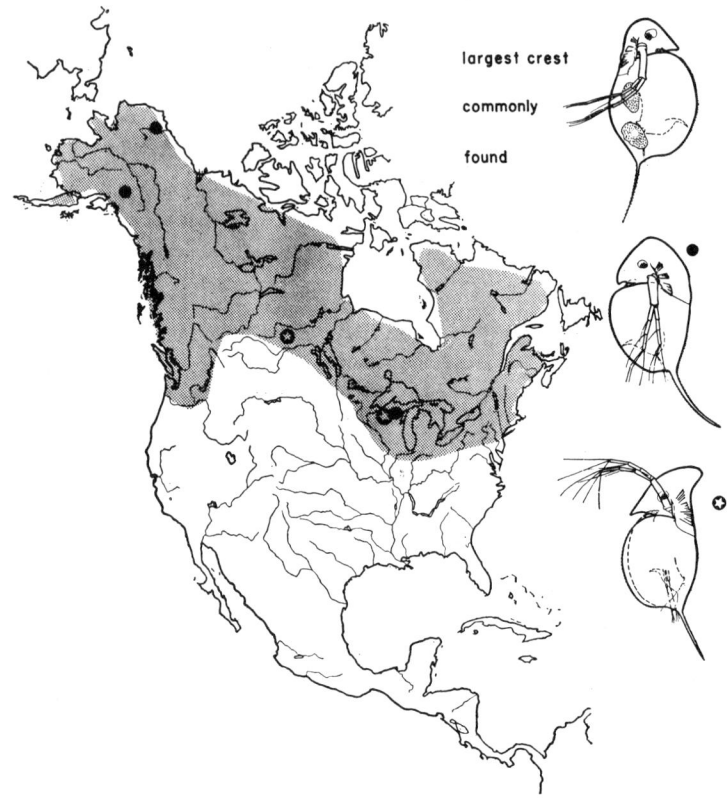

Fig. 6. Phenotypic variation in *Daphnia longiremis* Sars. *D. longiremis* is distributed throughout northern North America and northern Eurasia. Throughout this range most populations are composed of individuals resembling the topmost specimen. However, a few scattered North American populations are known in which individuals with large helmets have appeared. These helmets are of the two forms indicated. The environmental circumstances in which the two populations (one in Wisconsin, one in Saskatchewan) produced retrocurved helmets are indicated diagrammatically in Fig. 7.

cally determined. These extreme helmets occur in widely scattered localities, yet the broad type from northern Alaska is very like the broad type from Wisconsin. Furthermore, we know that the population in Saskatchewan, which during one particular season contained individuals with retrocurved helmets, in other years consisted of individuals with nothing but a small crest

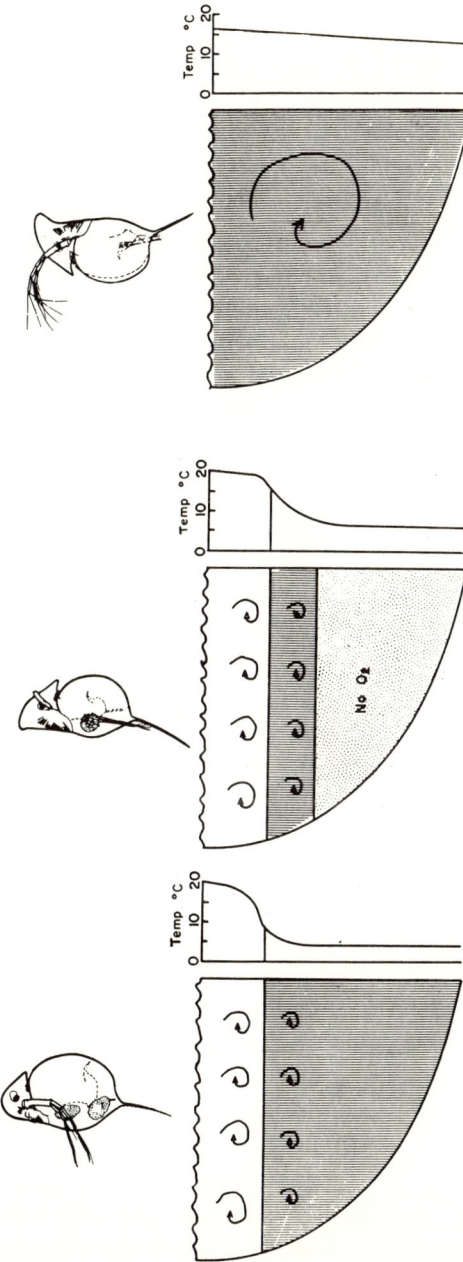

Fig. 7. Environmental correlates of helmet development in *Daphnia longiremis*. The three specimens indicate the head shape characteristic of the vast majority of *D. longiremis* populations (left), the retrocurved helmets that developed in Muskellunge Lake, Wisconsin, in the summer of 1931 and (right) those that developed in Lake Waskesiu, Saskatchewan, in midsummer of 1932. Below each specimen is a diagrammatic vertical half-section of the lake in which each phenotype developed. Vertical hatching indicates the water strata in which the *longiremis* populations lived. Temperature and estimated amount of turbulence are also indicated. Diagram for Muskellunge Lake based on data in Woltereck (1932) and for Lake Waskesiu in Rawson (1936).

(Brooks, 1956). It is difficult to believe that a character manifest in such sporadic fashion could be of such adaptive significance that the ability to produce this character would be carried latent in the germplasm of a species with an extensive distribution over two continents.

We fortunately know something of the physical properties of the water in two of the lakes in which populations with retrocurved helmets occurred. We are thus in a position to compare the environmental characteristics at the time of helmet development with the more usual conditions under which helmets are not produced. This comparison is made in diagrammatic fashion in Fig. 7. Under a specimen representing the helmet characteristic of each population is depicted half of a vertical section through the lake in which the population lives, and a graph showing the temperature at different depths. The section at the left is characteristic of the majority of lakes of the temperate zone in midsummer. The graph indicates that an upper stratum of water is warm, about 20° C.; below this there is a region where the temperature decreases rapidly with increasing depth. Most of the bottom waters are slightly above the temperature of maximum density (4° C.). The vertical hatching on the lower part of the section indicates that *D. longiremis* usually lives in the cool bottom waters. *D. retrocurva*, on the other hand, would live in the warm, turbulent upper layers. Such populations, living in cool waters in summer as well as in winter, show little seasonal variation, and at most develop the small crest indicated.

The center section represents the situation in Muskellunge Lake, Wisconsin, on August 26, 1931. This was reported, but differently interpreted, by Woltereck (1932). Here *Daphnia longiremis* with tall, slightly retrocurved helmets occurred, and as we might expect, the population was living in warmer (and more turbulent) water than the species usually does. The bottom waters were devoid of oxygen, and the population could maintain itself only in the middle strata of water. The diagram at the right represents the physical conditions in Lake Waskesiu, Saskatchewan, during midsummer of 1932 and the helmeted *longiremis* that developed therein. The thermal conditions in

Waskesiu in this and other years, has been extensively studied by Rawson (1936). The diagrammatic representation is based upon his data and the specimens examined were collected by him at this time. Waskesiu is a large, relatively shallow lake which usually becomes stratified during midsummer, although the bottom temperatures are usually higher than those considered typical in the diagram at the left. In 1932, however, exceptionally strong winds caused the stratification to break down and mixed the surface waters to greater and greater depths, until finally the water temperature at all depths was very nearly the same (around 18° C.). Under these conditions *longiremis* could live throughout the entire depth of the lake. Specimens were found living in the upper five meters! They bore retrocurved helmets, as drawn.

Examination of these two instances of helmeted *longiremis* indicates that this extreme body shape was evoked when the populations were forced to live in warm (and turbulent) water. It also makes it seem unlikely that the genetic ability to develop this phenotype under rare and abnormal circumstances is preserved in the genotype of a widespread population because of any advantage conferred by this extreme phenotype during these exceptional circumstances.

It seems rather more likely that the variations in body shape in *D. longiremis,* as in the rest of the genus, are a consequence of maintaining a genetic constitution that will allow any individual to develop and function under any of the large range of environmental conditions which the species can tolerate. The morphogenetic processes must permit development at temperatures slightly above 0° C. Morphogenesis controlled by the same genotype must also produce an effective organism at a temperature twenty degrees higher. That there are differences in the shape of the organisms produced by these same processes functioning at such different temperatures is not surprising.

Although the morphological changes in *Daphnia* are especially pronounced and have considerably complicated the problem of species determination, such extreme variation occurs in many freshwater organisms. From our investigation of the nature of this phenotypic variation, we should expect to find it whenever short-lived animals have to develop and remain active under a wide range of thermal conditions. The plankters of freshwater lakes of the temperate zone might be expected to exhibit this variation to a pronounced degree, and they do. Short-lived planktonic Cladocera (*Bosmina*, *Daphnia*), rotifers, and copepods, all show striking morphologic variation, whereas related species and genera which, for example, are benthic or longer-lived show much less pronounced variation.

*Species Associations.* The confusion which the wide intraspecific range of body shapes causes for the investigator seeking to delimit the species is confounded by the frequent coexistence of large populations of several species of *Daphnia* in the same body of water. Although the associations in lakes tend to be larger, even small, temporary ponds commonly have two species living at the same time. In addition to the fact that several species may occur in a lake at the same time, different species may comprise the dominant associations in different years. Various evidence indicates that such shifts are possible because a species may exist in a lake in a population so small as to be undetected except with very extensive sampling and careful examination of the material collected (Brooks, MS). However, under favorable conditions this population may attain sufficient size to be readily collected.

Figure 8 indicates the array of species that can be found living in New England lakes. In some parts of North America even larger arrays of species can be found, but this will suffice as an illustration of the problem. Collections in the upper waters of any New England lake may disclose populations of any one or more of the first five species living in the upper waters. In the bottom waters in midsummer usually either *D. ambigua* or *D. longiremis* will be found, with *D. ambigua* more common in the lakes of southern New England. There is no indication of preferential associations. In a survey of 107 lakes in southern and central Maine, it was found that the seven species occurred with the approximate percentages indicated in Fig. 8 (Brooks, 1956, MS). The most frequent associations were between the species most successful in the area; that is, chance appeared largely to determine the species to be found living together. The largest number of species yet found simultaneously in a lake in North America (or the world) are the six that lived in Aziscoos Lake, Maine. The one species, *parvula,* which was not found in Aziscoos at that time was one of the least common in the Maine lakes studied.

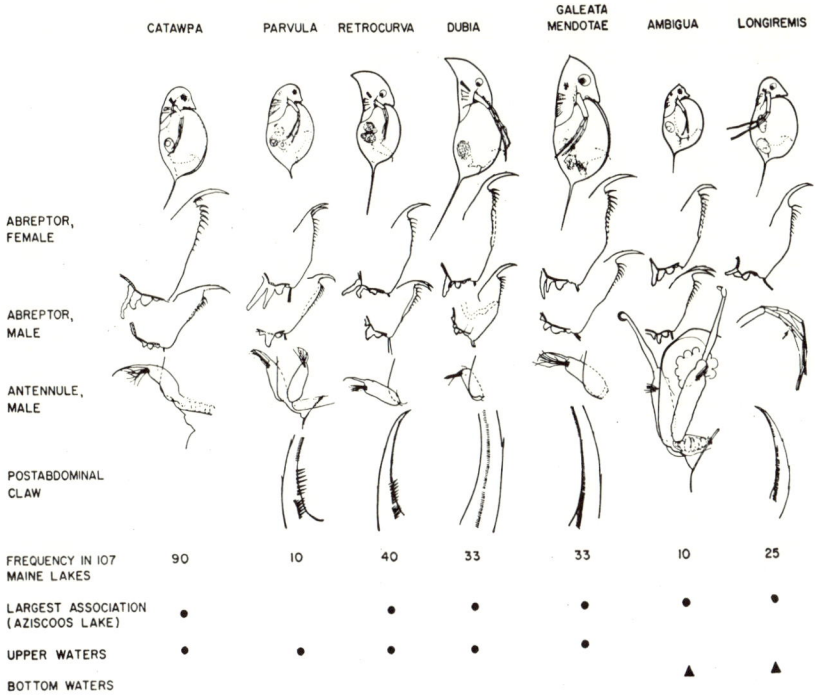

Fig. 8. Species of *Daphnia* commonly occurring in New England lakes. Most lakes have more than one species common at any time. The largest association found was in Aziscoos Lake, Maine, which had the six species indicated. Characteristic morphological and ecological traits of each species are indicated.

Although the coexistence of this many closely related species raises many interesting theoretical questions (which will be discussed elsewhere), our immediate interest is in the effect on the problems of species recognition. The confusion which the presence of as many as four or six species in a lake presents for the unwary investigator is great, because each population may exhibit considerable variation. Furthermore, in the late spring and early summer, when the *Daphnia* are likely to be most numerous, their shapes tend to be least distinctive (cf. Figs. 3, 4).

There is, however, a further possible complication to species recognition which may be the result of the continued coexistence of several species. This is the oft-noted tendency for two populations of different species inhabiting a lake to show some resemblance to each other (see, for example, some of Woltereck's papers; Kiser, 1950). Thus, each of the two common species in Moosehead Lake, Maine, *Daphnia dubia* and *Daphnia galeata mendotae*, tend to differ slightly in body shape from their characteristic forms (cf. Fig. 8)—the helmet of *galeata* may be more slender, that of *dubia* broader usual. Other structures may show some intermediacy, yet the specific identity of each is not in doubt. This phenomenon has been interpreted by Woltereck as being a direct effect of the peculiar features of the water of the lake in question on the species living in it. A more reasonable hypothesis is that the genes of the two species are somehow admixed in the individuals of such populations. One possibility is that these populations have developed from a chance hybrid between the two species, at one of the infrequent periods of sexual reproduction. A hybrid individual genetically very like one of its parents might be able to compete successfully with its parents. Of course, only one viable female hybrid is necessary to start a clonal population. If its genotype were superior in the peculiar environment of that lake, especially at some critical time, it could conceivably replace the parental species most like itself.

*Hybridization.* Thus we are introduced to the third aspect of the species problem in *Daphnia*. This is the occurrence of populations partaking of the characteristics of two species. The following facts about these intermediate populations are noteworthy.

1. They occur more commonly in pond-dwelling species than in lake species. This is consistent with the fact that periods of sexual reproduction are more frequent in the pond species and

affect a greater percentage of the population. Thus the chances for forming hybrids is considerably greater.

2. The hybrid populations occur in regions where both putative parent species live.

3. The hybrids themselves are usually chimaera-like in that in some features they are quite like one parent, in some quite like the other. In this they are very like the only experimentally produced hybrids between species of *Daphnia*, those obtained from

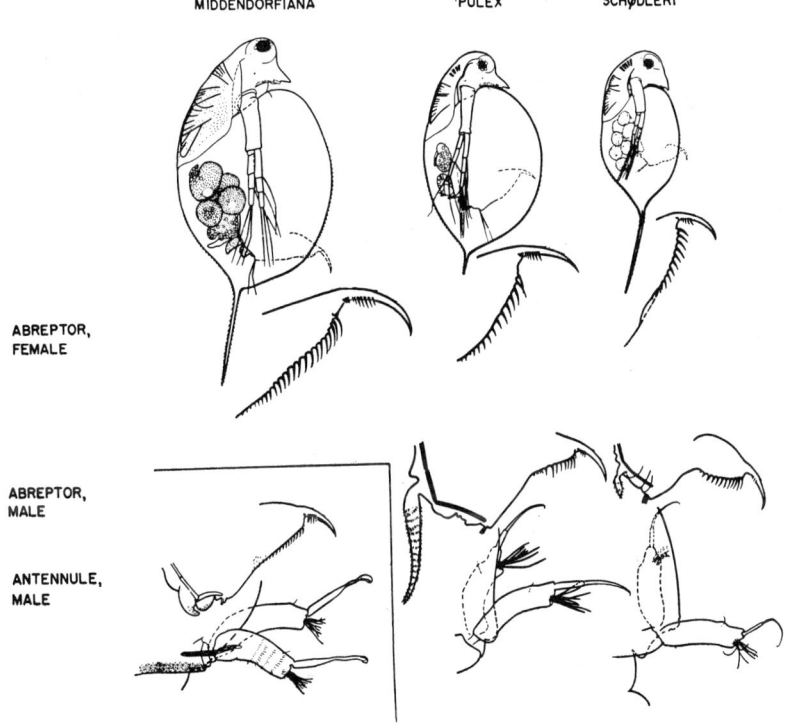

Fig. 9. Three similar species of pond-dwelling *Daphnia* common in northern and western North America. Although the postabdominal claws (on abreptor) of females are similar (all of *pulex* type), other morphological differences in female are apparent (head shape, length of shell spine, shape of carapace). Structure of males indicates distinctiveness of each species more clearly than that of female. Male of *middendorffiana* very rare. Many *Daphnia* populations in regions where these species coexist appear to be hybrids between various pairs of these species. (See Fig. 10.)

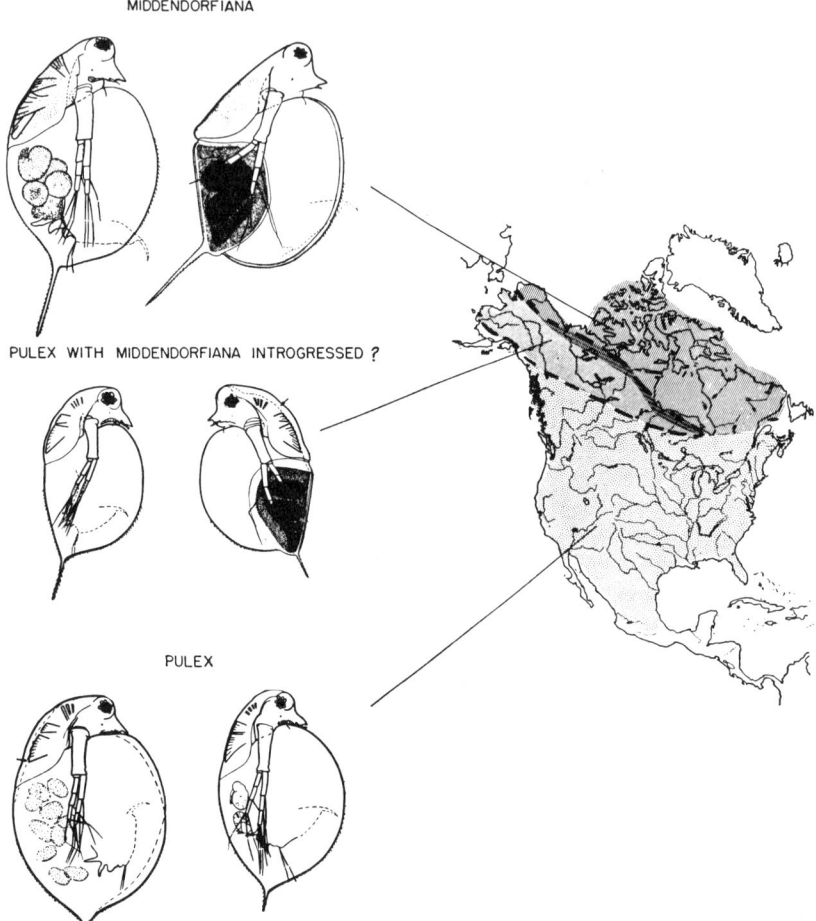

Fig. 10. Structure and distribution of *Daphnia middendorffiana* Fischer and *D. pulex* Leydig and of populations that appear to be hybrids between them. In a broad belt between the northern region where *middendorffiana* is one of the few species of *Daphnia* and the southern region where *pulex* is common, neither of these species is found in its "pure" form. In this belt there are many populations, the individuals of which have some characteristics of both *middendorffiana* and *pulex*.

a cross between European *Daphnia pulex* and *Daphnia obtusa* performed by Agar and reported by Scourfield (1940).

The species most frequently involved in these presumably hybrid populations, called, for the sake of brevity, simply "hybrid," are three pond-dwelling species of the northern and western parts of North America. These are *Daphnia middendorffiana* Fischer, *Daphnia pulex* Leydig, and *Daphnia schødleri* Sars. Although each of these is variable, the females drawn in Fig. 9 can be considered characteristic. The postabdominal claw in each is similar and variable, so that it is not difficult to understand that all three have in the past usually been assigned to *Daphnia pulex* when the possession of such a claw was taken as the sole species criterion. The abreptor and antennule of the male are distinct in all three, and when males occur, the populations can be readily determined. Both *pulex* and *schødleri* range widely over the western parts of this continent (*pulex* actually over most of it), and hybrid populations between them occur scattered throughout the large area of overlap. However, the best illustration of such hybrid populations concerns those between *pulex* and *middendorffiana*. As can be seen in Fig. 10, *Daphnia middendorffiana* is most common in the very northern parts of North America, and is one of the chief zooplankters in the numerous tundra ponds and lakes of that region. *Daphnia pulex* extends fairly far to the north but apparently cannot compete successfully with *middendorffiana* in the extreme north. Between the areas in which these two species are common there is a broad belt, from central Alaska eastward at least to Hudson Bay, where most of the *pulex*-like *Daphnia* exhibit a combination of *pulex* and *middendorffiana* characteristics. These have been labeled "*pulex* with *middendorffiana* introgressed" in Fig. 10, merely to indicate a guess as to their probable nature and origin. Although it is not indicated on Fig. 10, *Daphnia middendorffiana* occurs in scattered populations south as far as central California and Wyoming. Scattered populations in this region of western United States resemble the hybrids common to the region indicated on the map.

This complex of *middendorffiana, pulex,* and *schødleri,* the females of which are sometimes difficult to distinguish, together with scattered populations combining characteristics of any two in somewhat chimaera-like fashion, has posed a considerable systematic problem, probably the most difficult one encountered in the North American assemblage.

Populations intermediate between other species also occur, but they are largely confined to regions such as the Pacific Northwest where northern and southern species assemblages have come into contact (Brooks, 1956). The distribution and properties of these populations bear the same resemblance to the putative parents as did those in the case just discussed.

On the basis of this kind of evidence it seems justified to propose that introgressive hybridization is a factor that has complicated the systematics of *Daphnia*. As yet, however, neither experimental crosses nor analyses of the sort that Anderson (1949) has developed for plants have been done with these *Daphnia* populations. Because hybridization has been viewed with much skepticism, often justified, comparable data on its importance in the systematics of most other groups of animals are wanting. Hubbs (1955) has provided an excellent summary of hybridization in North American fishes, which are the most adequately investigated group of animals from that point of view. We shall again mention the possibility of introgressive hybridization in the discussion of the species problem in *Coregonus*.

## The Species Problem in *Coregonus*

*Coregonus*, the genus of the true whitefishes, is one of the three genera that comprise the whitefishes. These genera are considered by Berg (1940) to constitute one of the two subfamilies of the Salmonidae. Members of the genus *Coregonus* live in the rivers (many species tolerate brackish water) and lakes of northern Eurasia and North America. These whitefish are exceedingly numerous in the fresh waters of the very northern parts of these two continents.

The classification of these fish has aroused interest, possibly because some provide excellent food. Aside from the inherent difficulties in their classification, considered below, it has been difficult to grasp the problem in its entirety because the bulk of the whitefish populations live north of the regions heavily populated by man. The numerous and diverse attempts following the standard practices of ichthyology have resulted in cumbersome systems employing many tri- and quadrinomials (see, for example, Berg, 1948, and Hubbs and Lagler, 1947). The chief source of difficulty has been the diversity of the morphological and physiological characteristics of these *Coregonus* populations. Not only does each major drainage system have its peculiar forms, but also several forms may occur in any one lake. It was not until the investigations by Gunnar Svärdson into the biology of the populations living in various Swedish waters that we have had any insight into the nature of the systematic relationships within this genus. The remainder of the discussion will be concerned with the factors contributing to the difficulty of recognizing species of *Coregonus*, as these factors have been uncovered by Svärdson in a series of papers (1949-1953).

Unfortunately Svärdson has not yet attempted to name the species that his investigations have disclosed, nor has he attempted to unravel the chaotic taxonomy of the genus. His general feeling, however, is that the genus comprises two major species groups (superspecies in the sense of Mayr). Svärdson calls these two superspecies the *lavaretus* group (after *Coregonus lavaretus* Linn.) and the *albula* group (after *C. albula* Linn.). Not only are the species within each group nearly indistinguishable but even the two superspecies are also very similar. The best character distinguishing the superspecies apparently is that the lower jaw in species of the *albula* group is always longer than the upper. The *lavaretus* group, on which all of Svärdson's work was done, has its greatest number of species in Europe and western Russia, where the *albula* group is represented by a much smaller number of species. In eastern Asia and in North America, however, it appears that the *albula* group has formed many species, whereas there are relatively few *lavaretus*-like species. Walters (1955) pointed out that the glaciated portions of Eurasia and North America have far more *Coregonus* species than do the unglaciated portions. However, too little is known of the species and their distribution to permit meaningful speculation about the historical zoogeography of this genus.

Svärdson has found that the chief source of difficulty in the recognition of *Coregonus* species is the extraordinary phenotypic variability of each genetic stock (species). The coexistence of several species in a lake (see Fig. 11A), a circumstance which often complicates the problem for the systematist, has actually provided Svärdson with the major clues to the identity of the species. In these ways both *Coregonus* and *Daphnia* present somewhat similar problems and, as in *Daphnia,* there are evidence of the formation of hybrids and some resemblances between the species of major drainage systems which suggest introgressive hybridization.

*Phenotypic Variation.* The primary reason for the failure of orthodox taxonomic methods in *Coregonus* systematics has been that the phenotypic characteristics are not constant for, and therefore not indicative of, a given genotype. Quite the contrary is true; the phenotype is highly variable, depending upon the environmental conditions under which the genetic stock developed. These plastic characteristics include meristic ones (number of vertebrae, fin rays, scales), size and proportions, and also physiological ones (growth rate, onset of sexual maturity, length of life). The allometric growth relationships are complex because the proportions within a stock differ not only in fish of different size (length) but also in fish of the same length but of different age (Svärdson, 1949, 1950). The number of gill rakers is the only phenotypic characteristic that is apparently unaffected by the environmental circumstances during development. This is the only phenotypic manifestation indicative of the genotype, and, therefore, of taxonomic utility. However, the number of gill rakers is under close selective control and may vary somewhat between allopatric populations of a species. Svärdson was able to modify the gill raker count in a population introduced into a formerly *Coregonus*-free lake by imposing an artificial selection for this factor at the time of introduction. It is little wonder that the

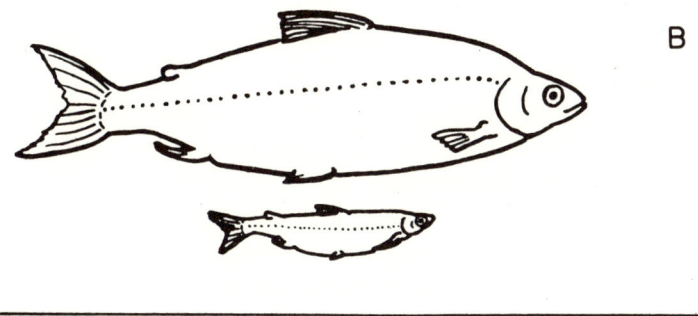

Fig. 11. Mature whitefish of various species of the *Coregonus lavaretus* species group inhabiting Swedish lakes. A, spawners of two different species inhabiting the same lake. B and C indicate the phenotypic differences between populations of the same genotype that have developed in different environments. Lower drawing in B indicates phenotype of spawning fish of a certain population. When fry from this population are transplanted to a different lake they develop into mature fish resembling the upper specimen. Upper figure in C represents a mature fish of a different population. When *mature* fish of this population were transplanted to another lake, they grew to size of the lower specimen after two summers in the different environment. All figures from Svärdson (1949); B after Olofsson (1934); C after Runnström (1944).

systematics of this large number of very similar species has been confused!

Svärdson (1952) was able to demonstrate the environmentally determined plasticity of all these phenotypic characteristics by transplanting stocks with well-known characteristics to lakes that lacked *Coregonus* populations. Different and unrelated aspects of the environment may affect the phenotypic expression of certain of the above-noted plastic features. For example, the number of scales along the body, a feature that has frequently been used to distinguish subspecies and even species, was shown by Svärdson to be affected both by temperature during development and by the state of nutrition of the female parent. The eggs produced by well-fed populations are larger than those of poorly fed ones, and more scales are initially laid down in the larger fry that develop from the larger eggs.

The extraordinary plasticity of these fish can be seen in comparisons of specimens from such transplanted stocks with corresponding ones from the original populations. In Fig. 11B the lower specimen is the normal size of a spawning whitefish from Lake Lomsjön. When this stock was transplanted to Lake Oxvattnet, the normal size at spawning was that of the upper specimen. Some stocks retain this plasticity into maturity, as is demonstrated in Figure 11C. Spawning whitefish of Lake Näckten attain the size indicated in the upper specimen. When adult fish from this stock were transplanted to Lake Algsjö, they had, in two summers, achieved the size of the lower specimen. Further differences in growth, size, and length of life between populations of the same species in different lakes can be seen by comparing the graphs of Fig. 12.

The studies on the environmental plasticity of these fish are summarized by Svärdson (1953, p. 165) with these words:

. . . species of whitefish, which as far as can be judged must be the same form from lake to lake, can have different forms in one lake or another. The same species can have a long life or die young, grow rapidly or have extremely poor growth, live in large or small populations, spawn in running or still water at different depths and at extremely varying times. Some variation is already found in the habits of the species in one and the same lake, but when the same species is compared from different localities the variation is very considerable.

The intra-lacustrine variation mentioned in the last sentence of the quotation refers primarily to differences between fish of different year classes.

*Species Associations.* Many lakes are known to have several populations of *Coregonus* in them, and Svärdson records two lakes each of which has five species of the *lavaretus* group. The populations that we now observe coexisting in some lakes are known to have done so for long periods of time (of the order of a century or two) because fisheries records of this age mention the local names for the different kinds of whitefish. It seems reasonable to assume that many, if not most, of the *Coregonus* species associations are of equal, and probably greater, stability. However, the stability of such associations can be destroyed, as has been indicated (Svärdson, 1949) by an intriguing case, the details of which are but sketchily known. In this instance a Lapland lake was known to have had two species of *Coregonus* which maintained their integrity by having temporally separated spawning periods. When a third stock was introduced (apparently in an attempt to improve the fishing), the result was the creation of the large, panmictic, highly variable population that exists there today. It is supposed that the introduced stock had a spawning season that overlapped the hitherto separated seasons of the other two.

In his 1953 paper Svärdson presents data concerning the species of *Coregonus* found associated in three lakes of as many different river systems in central Sweden. These rivers all flow southeastward into the Baltic Sea from the mountains that run down the middle of the Scandinavian peninsula. From south to north the lakes are Lake Idsjön, Lake Storsjön, and the Hornavan-Uddjaur series of lakes. Figure 12 indicates the species of *Coregonus* in each lake and some of the characteristics of each population. Svärdson considers the populations indicated in these graphs to be species, but he did not assign them scientific names. Their common names are therefore used. As indicated in the previous section, the phenotypic characteristics of different year

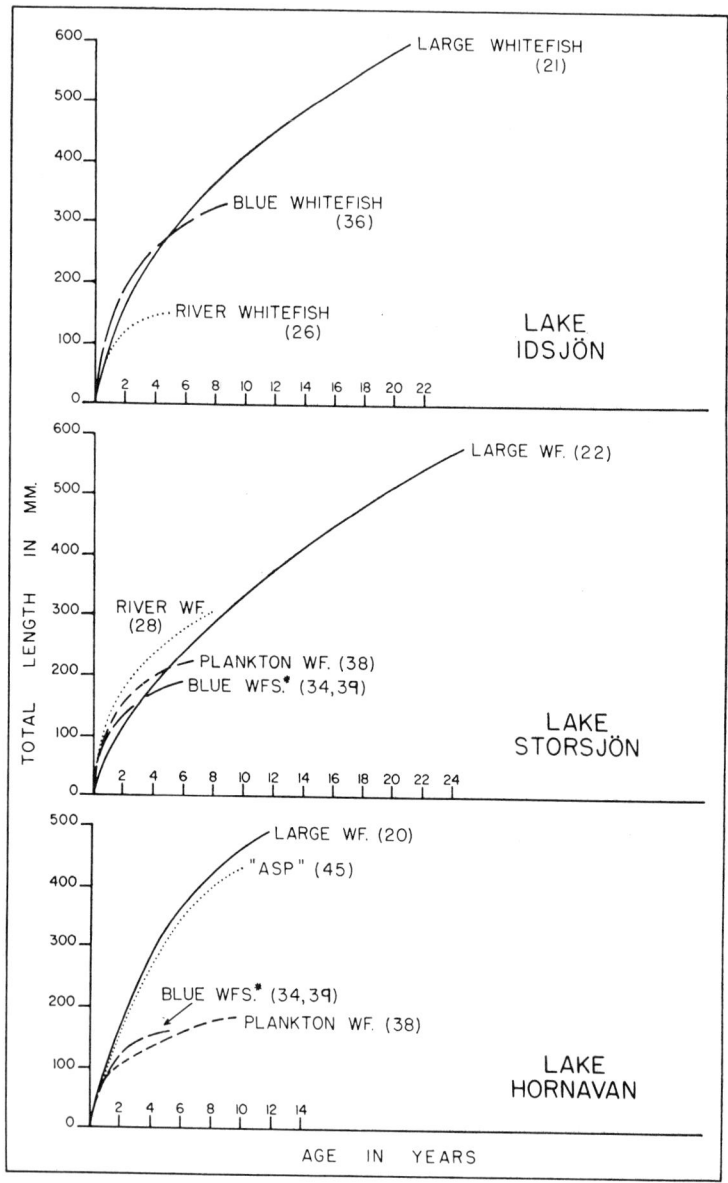

Fig. 12.

classes of a population are somewhat different, and the data given here can be considered normal for these stocks. It can be seen from Fig. 12 that Lake Idsjön has three species of *Coregonus,* whereas each of the other two lakes has five, although the graphs for the last two present the data for only four species each. The reason for this discrepancy is that subsequent to the publication of his 1953 paper Svärdson discovered (reported in Runnström, 1953) that the "blue whitefish" of several Swedish lakes were in reality two closely similar species. Lakes Storsjön and Hornavan are two of the lakes discovered to have two species of "blue whitefish." Populations of these two can be distinguished by their gill raker counts; one has an average of 33-35 (considered 34 in Fig. 12) and the other 38-39 (considered 39).

For populations of closely similar biparental organisms to maintain themselves within the same ecosystem it is necessary (1) that each be able to find sufficient food in the face of actual or potential competition from the others, and (2) that each be able to maintain its genetic integrity in the face of possible admixture with the other population. As the *Coregonus* populations of these lakes have been able to maintain themselves, it is clear that they must be able to meet the above conditions. Svärdson's conclusion that each such population belongs to a species different from the others with which it continues to coexist is entirely reasonable.

The manner in which these populations have been able to meet the two conditions for continued coexistence is not known in detail, although we know more of the breeding habits than of the food requirements. We can appreciate the ways in which

Fig. 12. Characteristics of sympatric *Coregonus* populations of three Swedish lakes. Common names are used because scientific ones for these populations have not been established. The only certain species criterion is the number of gill rakers. The average number of rakers on the first left gill arch of each population is given in parentheses. The two numbers and the asterisk after the blue whitefish of Storsjön and Hornavan indicate that two very similar species of "blue whitefish" occur in these lakes. Comparisons of characteristics of same species in different lakes indicate lability of such phenotypic characteristics as size at given age and longevity. Slight differences in average number of gill rakers (in part may be due to sampling errors) are in part, apparently, genetically determined.

the coexisting species maintain their genetic integrity by considering the spawning behavior of the species of the Hornavan lake system (Svärdson, 1953). The pertinent information is summarized in Table I. These differences in time of spawning and types of spawning ground apparently suffice to permit the populations to reproduce with relatively little intermixing.

However, the reproductive isolation between the populations is by no means complete. Ripe "asp" have been taken on the spawning grounds of the great whitefish in December, even though their normal breeding time is a month and a half earlier.

TABLE I. Spawning Behavior of *Coregonus* Species of Hornavan Lake System

| Species (local name) | Spawning period | Spawning place |
| --- | --- | --- |
| "Asp" | November | In running water of rivers between lakes and in mouths of influent rivers |
| Large whitefish | December | Shallow water above stony bottom; widely distributed in lakes |
| Blue whitefish[a] (two species) | December January | Deeper water over stony bottom; widely distributed except in northern part of Lake Hornavan |
| Plankton whitefish | February | Restricted in feeding and spawning to deep waters of Lake Hornavan; spawns on bottom at relatively great depth (30–100 m.) |

[a] The spawning habits of the two species known by this name have not been differentiated.

The local fishermen have long recognized a special kind of whitefish that they call "albask," which Svärdson thinks are most likely hybrids between the "asp" and the great whitefish. Hybridization between late spawners among the blue whitefish and early plankton whitefish in areas where their spawning grounds meet is to be expected, and Svärdson finds some evidence that would seem to substantiate this possibility.

*Intermediates and Hybridization.* Individuals that appear intermediate between various of the species (like the "albask" just mentioned) occur with a low frequency. However, only a portion of these can be established as hybrids; the only hybrids readily detectible are those between species with widely divergent gillraker counts. Many of these aberrant individuals owe their intermediate appearance to the phenomenon of "shoal-trapping." This term refers to the circumstance in which an individual becomes separated from the shoal of similar-sized fish of its own species, with which it would normally swim, and joins a shoal of a different species. A shoal-trapped fish is thus exposed to the range of environmental conditions proper to the alien species comprising the shoal. Because of the extreme sensitivity of the developmental processes in the whitefish to environmental conditions, the shoal-trapped individual develops a phenotype determined genetically by one species, but so modified by the environment characteristic of the alien species that it strongly resembles the species that it has inadvertently adopted. It is only when the species involved have very different numbers of gillrakers that the genetic stock of the shoal-trapped individual can be identified. Furthermore, it is only when the trapping has occurred fairly late in the life history that the foreign fish is sufficiently different from its shoal-mates to attract the attention of an observer. If the trapping occurs in the fry stage, the phenotype of the trapped fish will be little different from that of its adopted species, so little different that it will very likely escape the attention even of a trained observer.

Not only are shoal-trapped fish difficult to distinguish from hybrids, but they are likely to increase the production of hybrids because such a trapped fish is likely to spawn along with the rest of the shoal. The resultant mixture of gametes is likely to produce hybrids, if we can judge from the high yield in cross fertilizations between species of the *Coregonus lavaretus* group. Although $F_1$ hybrids are readily produced in artificial crosses, it has not yet been established whether $F_2$ hybrids and backcrosses are produced with proportionate facility. The detected natural hybrids are rare and could be almost entirely of the first generation. The progeny from a backcross would probably be extremely difficult to detect in nature. However, if they did occur, even with a very low frequency, they might account for the introgression of genetic material from one species to another. Svärdson remarks that the various species of a river system often exhibit similarities which, aside from mutual introgressions, are difficult to understand. Further investigations on hybridization and introgression apparently are in progress at Drottningholm.

**Summary**

1. The systematics of two "difficult" genera of freshwater animals, the crustacean *Daphnia* and the whitefish *Coregonus*, have been examined in an attempt to discover any common sources of the species problem in these groups that might be considered as peculiar, at least in their intensity, to freshwater animals.

2. The extraordinarily wide range of phenotypic variants that may develop from the same genotype emerges as the chief source of the species problem in both *Daphnia* and *Coregonus*. For the systematist this means that the phenotypic characteristics that are taken as indicators of the genotype, i.e., the species, must be chosen with great care. From an evolutionary point of view, this wide range of phenotypic variation is a reflection of the ability of the genotype to prosecute successful development over the wide range of environmental conditions occurring in the fresh waters that these animals inhabit. It is probably not fortuitous that animals with this ability are among the more successful inhabitants of these inland waters (both *Daphnia* and *Coregonus* and the respectively related groups, *Bosmina* and the chars, for example). For *Daphnia* the extreme environmental conditions succeed each other cyclically, with the seasons, so that the entire range of phenotypic diversity is seen in the annual succession of generations of these short-lived animals. In *Coregonus* the extraordinary phenotypic diversity, while in part expressed in different year classes of the same population, is principally expressed by allopatric populations. The conditions of life are sufficiently different from one lake to another so that nearly all the phenotypic characteristics of a species are different from lake to lake.

3. A second source of difficulty in the recognition of species in both genera has been that in many bodies of fresh water several highly variable species may coexist. In North America as many

as six species of *Daphnia* are known to occur together in the same lake, while five species of *Coregonus* of the same species group (superspecies) have been reported from Sweden (the area studied by Svärdson).

4. The coexistence of several closely related species in the same body of water provides the possibility for hybrid formation and introgression. There is evidence for both of these phenomena in both *Daphnia* and *Coregonus*, evidence consistent with the different reproductive biology of the two groups. The hybrid individuals in *Coregonus* and hybrid populations in *Daphnia* are an additional source of taxonomic difficulty. In neither group has the phenomenon of introgression been carefully investigated, although in both groups it stands as the most likely explanation for certain general similarities between the species of a given region.

5. These factors, extreme phenotypic variation, coexistence of several closely related species, and occasional hybridization (and introgression?), although by no means peculiar to freshwater organisms, do appear to be largely responsible for the species problem in two very different groups of freshwater animals. Whether the sources of difficulty common to the systematics of *Daphnia* and *Coregonus* will prove to be common to other "difficult" groups of freshwater animals will only be revealed by further investigation.

## REFERENCES

Anderson, E. 1949. *Introgressive Hybridization.* John Wiley & Sons, New York, N.Y.

Banta, A. M. 1939. Studies on the physiology, genetics, and evolution of some Cladocera. Paper No. 39. Dept. of Genetics. *Carnegie Inst. Wash. Publ.* No. 513.

Berg, L. S. 1940. Classification of fishes, both recent and fossil. [In Russian.] *Trav. inst. zool. acad. sci. U.R.S.S.*, 5. No. 2. Reproduced, with English translation by Edwards Bros. Inc., Ann Arbor, Mich., 1947.

Berg, L. S. 1948. The freshwater fishes of U.S.S.R. and the surrounding countries. [In Russian.] Part I. Moscow.

Birge, E. A. 1918. The water fleas (*Cladocera*). In *Freshwater Biology*. John Wiley & Sons, New York, N.Y.

Brooks, J. L. 1946. Cyclomorphosis in *Daphnia*. I. An analysis of *D. retrocurva* and *D. galeata*. *Ecol. Monographs*, **16** (4), 410-47.

Brooks, J. L. 1947. Turbulence as an environmental determinant of relative growth in *Daphnia*. *Proc. Natl. Acad. Sci. U.S.*, **33** (5), 141-48.

Brooks, J. L. 1956. The systematics of North American *Daphnia*. *Mem. Conn. Acad. Arts Sci.*

Brooks, J. L. (MS) The vertical distribution and associations of *Daphnia* species in New England lakes.

Edmondson, W. T. 1955. The seasonal life history of *Daphnia* in an arctic lake. *Ecology*, **36**, 439-55.

Hubbs, C. L. 1955. Hybridization between fish species in nature. *Systematic Zool.*, **4** (1), 1-20.

Hubbs, C. L., and K. F. Lagler. 1947. *Fishes of the Great Lakes Region*. The Cranbrook Press, Mich.

Hutchinson, G. E., and H. Löffler. 1956. The thermal classification of lakes. *Proc. Natl. Acad. Sci. U.S.*, **42** (2), 84-86.

Kiser, R. W. 1950. A revision of the North American species of the Cladoceran genus *Daphnia*. Edwards Bros., Ann Arbor, Mich.

Pennak, R. W. 1953. *Fresh-water Invertebrates of the United States*. Ronald Press, New York, N.Y.

Rawson, D. S. 1936. Physical and chemical studies in lakes of the Prince Albert Park, Saskatchewan. *J. Biol. Board Can.*, **2** (3), 227-84.

Scourfield, D. J. 1942. The "Pulex" forms of *Daphnia* and their separation into two distinct series, represented by *D. pulex* (De Geer) and *D. obtusa* Kurz. *Ann. and Mag. Nat. Hist.*, **9** (ser. 11), No. 51, 202-18.

Slobodkin, L. B. 1954. Population dynamics in *Daphnia obtusa*. Kurz. *Ecol. Monographs*, **24** (1), 69-88.

Svärdson, G. 1949. The coregonid problem. I. Some general aspects of the problem. *Rept. Inst. Freshwater Research Drottningholm*, **29**, 89-101.

Svärdson, G. 1950. The coregonid problem. II. Morphology of two coregonid species in different environments. *Rept. Inst. Freshwater Research Drottningholm*, **31**, 151-62.

Svärdson, G. 1951. The coregonid problem. III. Whitefish from the

Baltic successfully introduced into fresh waters in north of Sweden. *Rept. Inst. Freshwater Research Drottningholm*, **32**, 79-25.

Svärdson, G. 1952. The coregonid problem. IV. The significance of scales and gillrakers. *Rept. Inst. Freshwater Research Drottningholm*, **33**, 204-32.

Svärdson, G. 1953. The coregonid problem. V. Sympatric whitefish species of the Lakes Idsjön, Storsjon and Hornavan. *Rept. Inst. Freshwater Research Drottningholm*, **34**, 141-66.

Wagler, E. 1936. Die Systematik und geographische Verbreitung des Genus *Daphnia* O. F. Müller mit besonderer. Berücksichtigung der südafrikanischen Arten. *Arch. Hydrobiol.*, **30**, 550-56.

Walters, V. 1955. Fishes of western arctic America and eastern arctic Siberia. *Bull. Am. Museum Nat. Hist.*, **106**, Article 5.

Woltereck, R. 1930. Alte und neue Beobachtungen über die geographische und die zonare Verteilung der helmlosen und helmtragenden Biotypen von Daphnia. *Intern. Rev. Hydrobiol.*, **24**, 358-80.

Woltereck, R. 1932. Races, associations and stratification of pelagic daphnids in some lakes of Wisconsin and other regions of the United States and Canada. *Trans. Wisconsin Acad. Sci.*, **27**, 487-522.

# THE SPECIES PROBLEM WITH FOSSIL ANIMALS

JOHN IMBRIE: COLUMBIA UNIVERSITY, NEW YORK, N.Y.

In spite of the extended attention this problem has received, the nature of fossil species remains one of the most controversial topics in paleontology. That this should be so in a decade that has witnessed the publication of numerous synoptic studies on the origin of species may surprise some students unfamiliar with the materials and problems of paleontology. But two key questions are still being asked: What is a fossil species? How can fossil species be recognized?

In defining fossil species it would be possible to ignore living organisms completely and to frame definitions strictly in terms of fossils. In fact, such an unbiological approach to paleontological taxonomy is almost, but not quite, forced upon us in dealing with some groups such as the conodonts whose biological functions are obscure. But most attempts at a definition of fossil species begin with a review of species concepts held by students of living, sexual organisms, and from this foundation construct a theoretical model of fossil species. The concept of fossil species held by most paleontologists is largely an inference, an inference based both on the observed structure of living species and on a theoretical model of the evolutionary mechanism.

Paleontologists by no means agree on what a fossil species is. Burma (1954), for example, has given thoughtful expression to the thesis that species do not have an objective reality, a view that is rejected by Simpson (1951) and others. In a symposium on paleontological species, Eagar (1956) holds that for certain groups of fossil clams a workable species concept must be typological. Other students, including Newell (1956) and Sylvester-Bradley (1956), maintain that the concept of interbreeding populations is a necessary prerequisite for the definition and delineation of fossil species. Disagreement exists not only on the nature of species but also on the inevitable question of their proper scope. Thus Burma (1948) would distinguish as species the smallest statistically recognizable grouping of populations, whereas Imbrie (1956) argues that consistent application of such a criterion would lead to useless multiplication of species names.

Since all the problems alluded to in the preceding paragraph have their exact counterparts in neontology, it would be pointless to make of this paper a compendium of paleontological disagreement. Instead, we shall focus on those aspects of the species problem which are unique, or are at least uniquely developed, in dealing with fossil materials.

## Species Concepts

In dealing with sexual organisms, whether fossil or living, two fundamentally different species concepts can be employed. The *typological* concept defines species as a group of individuals essentially indistinguishable from some specimen selected as a standard of reference. The *biological* species concept, on the other hand, considers the species to be made up of one or more variable, interbreeding populations. Both of these concepts serve as theoretical bases for taxonomic work in paleontology today.

Different criteria may be emphasized in delimiting biological species. For many students the criterion of interbreeding is decisive, and species so defined may be referred to as *genetic* species. In order to emphasize the fact that most modern and all fossil species are distinguished more on the basis of morphology than on breeding habits, some taxonomists recognize a *morphological* species concept as distinct from the genetical concept. But this is really a trivial distinction. Morphological data are never really considered by themselves; their interpretation is always colored to some degree by prevailing theories on population structure. Moreover, descriptions of species, like descriptions of molecules and genes, are inferences drawn from various sorts of data, including observations on geographic distribution and ecology as well as morphology and genetics.

It is widely recognized that many anomalies arise when the genetical definition of species as actually or potentially interbreeding groups of organisms is strictly applied. Asexually reproducing organisms are left out of account, for example, as are the many instances on record in which local or temporary breakdowns occur in the genetical barrier between sympatric populations which on other evidence are classed as good species. Simpson (1951) argues that such inconsistencies occur only because the interbreeding criterion is taken as the final test of specific identity. He proposed that an *evolutionary* criterion be substituted for the genetical so that the species is defined as a segment of a phyletic lineage "evolving independently of others, with its own separate and unitary evolutionary role and tendencies." Clearly, this leaves wide latitude for judgment on the part of the taxonomist delineating a species; and when taken out of context Simpson's definition appears to be a different verbalization of the famous dictum that a species is "a community or number of related communities whose distinctive morphological characters are in the opinion of a competent systematist sufficiently definite to entitle it or them to a specific name" (Regan, 1926). Taken in context, however, Simpson's definition differs by insisting that a combination of morphological, biogeographical, associational, ecological, *and* genetical data be used to assess the evolutionary discreteness of a population or group of populations under study. By the nature of the evolutionary process we cannot eliminate arbitrary taxonomic judgments. Species-making will remain a practical art as well as a scientific discipline.

*Typological Species in Paleontology.* The typological concept of species is employed today either implicitly or explicitly by a considerable number of paleontologists. To illustrate this point we shall turn our attention to the work of a group of students who have contributed a great deal to our knowledge of British non-marine Carboniferous clams, notably Trueman, Weir, Leitch, and Eagar (for a good summary see Weir, 1950). Like their modern relatives the unionids, the shells described by these workers (*Anthraconaia, Carbonicola,* etc.) display an extraordinary amount of intrapopulation variation (Fig. 1). Now the mere

128    THE SPECIES PROBLEM WITH FOSSIL ANIMALS

fact of variation, however excessive, would not of course be used as a justification for a typological approach. But the difficulty in this case is enormously compounded by a combination of stratigraphic, biologic, and practical economic circumstances. In the first place, abundant collections of these shells have been obtained from widely distributed localities and very numerous

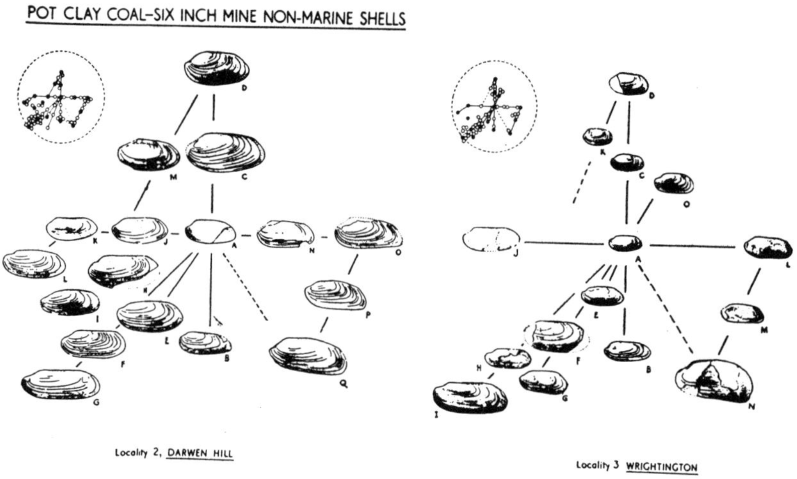

Fig. 1. Variation diagrams of two local populations of *Carbonicola* ? from a thin shell bed immediately above the Pot Clay and Six-Inch Mine Coals of Yorkshire and Lancashire. These shells are referred to *C. ? lenisulcata* Trueman and *C. ?* aff. *bellula* Bolton. Distribution diagrams, inset in the upper left-hand corner of each pictograph, show the numerical strength of variation trends. Black circles distinguish the figured variants, and white circles show the disposition of the remaining shells, the position of each one being controlled by its resemblance to one or more of the figured variants (Eagar, 1952b).

stratigraphic horizons in the British coal measures. The result is an embarrassment of riches. To make the problem more complex, many pairs of unit taxonomic characters vary independently, or at least show low total correlation. From this it follows that statistical characterization of populations by means of simple univariate or even bivariate distribution clusters will usually be unsatisfactory. By means of multivariate regression analyses Leitch (1940) has shown that it is possible objectively to identify variation norms and to place taxonomic discrimination on a quantitative basis. But such computations are far too laborious to enable the large number of population samples at hand to be processed. Another complicating factor is the labyrinthine course of evolution followed by these British clams: with one exception, rectilinear evolutionary trends have not been identified.

Faced with these complexities, Trueman and his co-workers have evolved an ingenious and workable taxonomic procedure. Each population sample is analyzed into variants which are illustrated and arranged in a systematic (and subjective) manner as in Fig. 1. Subsequent analyses of similar populations can then be abbreviated by means of appropriately placed dots on distribution diagrams. One or more modal or characteristic variants in any population may then be described as a species. As a consequence, an assemblage of shells from a single horizon and locality may be described as a group of species, even though there is every evidence that variation between the named species is continuous.

The students who employ the system described above recognize that their use of the term species violates the modern biological species concept, but the claim is made that this system of nomenclature has proved to be the only practical one for their peculiar biostratigraphic problems (Eagar, 1952a). Other workers have argued that the same practical results could be obtained within the framework of the biological species concept (Sylvester-Bradley, 1952) and without burdening international nomenclature with names of purely local application (Newell, 1956).

*Biological Species in Paleontology.* A majority of articulate paleontologists working today employ a biological species concept. Although this consensus reflects prevailing neontological views, the paleontologist's concept of species, which must take into account important segments of geological time, is inevitably more complex than the corresponding concept of zoologists. In order to clarify this point it will be helpful to examine the simplest possible phylogenetic model which includes the time dimension. Such a model, representing a small fragment of a phylogenetic tree, is shown in Fig. 2. The branches of this tree represent ancestral-descendent population sequences replacing one another through time as they undergo morphological and genetic divergence. For the sake of simplicity, it is assumed that the organisms involved are sexual, biparental, and that complications due to partial genetic barriers and *Rassenkreise* do not exist

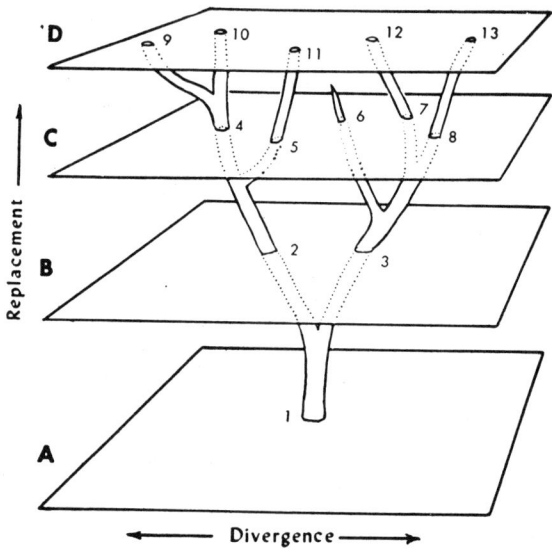

Fig. 2. A simplified theoretical phylogeny. At the indicated stratigraphic horizons (A, B, C, D) transient species (1-13) are discrete evolutionary units with no morphological or genetic overlap. In rare instances, where the record of a line of descent is nearly continuous, segments of evolving lineages may be designated as successional species. (After Newell, 1956.)

at levels A, B, C, and D. All the individuals living at time A are capable of interbreeding. By time B speciation has occurred so that populations 2 and 3 are no longer capable of genetic interchange. Five similar branchings are recorded in the diagram.

Given the simplified conditions of this model, contemporaneous organisms are always distributed among discrete, reproductively isolated groups. For reasons discussed below, paleontological samples almost always consist of the remains of individuals who lived during a fraction of geologic time so short that evolutionary morphological change within the sampled time interval cannot be demonstrated. Thus species recognized on time planes A, B, and C by the paleontologist represent stages in an evolutionary continuum and are thus analogous to neontological species on plane D. In some rather rare instances the paleontologist can assemble material representing time planes so closely spaced that modal shifts in an evolving succession of populations can be demonstrated. Such circumstances are relatively rare, however, and the paleontologist usually deals with widely spaced cross sections of evolving lineages. The term *transient species* will be used in this paper to identify such cross sections, i.e., reproductively isolated groups of individuals living during a single instant of geological time. Normally the paleontologist's cross section is limited geographically, and when this is the case, the transient species of the paleontologist corresponds to the nondimensional species of the neontologist (Mayr et al., 1953). In an increasing number of instances, however, paleontologists have been able to document taxonomically important infraspecific geographic differences, i.e., polytypic transient species.

For practical reasons the neontologist can rarely undertake the field sampling and breeding experiments required to document statistically significant seasonal shifts in gene frequencies. Hence the paleontologist of necessity and the neontologist by default describe transient species in the majority of cases now on record.

In favorable circumstances the paleontologist can obtain materials which enable him to document gradual morphological changes in a succession of fossil populations. In Fig. 2, for example, collections taken at a number of horizons might define the course of evolution between transient species 4 and 9. Quite naturally, paleontologists use the term species for such segments of phyletic lineages. Lineage segments of this sort can be called *successional species* to distinguish them from the static groupings referred to above as transient species. A number of theoretical taxonomic and nomenclatural problems which arise in defining the scope of successional species will be considered below in connection with a discussion of *Micraster*.

132 THE SPECIES PROBLEM WITH FOSSIL ANIMALS

|  | TYPOLOGICAL SPECIES | BIOLOGICAL SPECIES | | |
|---|---|---|---|---|
| BASIS | MORPHOTYPE | LIVING POPULATION | FOSSIL POPULATION | |
| DISTRIBUTION IN TIME | VARIOUS | CONTEMPORANEOUS | ESSENTIALLY CONTEMPORANEOUS | NON-CONTEMPORANEOUS |
| TYPE OF SPECIES | TYPOLOGICAL SPECIES | TRANSIENT SPECIES | TRANSIENT SPECIES | SUCCESSIONAL SPECIES |
| CRITERIA — GENETICAL | — | X | — | — |
| CRITERIA — MORPHOLOGICAL | X | X | X | X |
| CRITERIA — ASSOCIATIONAL | — | X | X | X |
| CRITERIA — ECOLOGICAL | — | X | X | X |
| CRITERIA — GEOGRAPHICAL | — | X | X | X |
| CRITERIA — BIOSTRATIGRAPHICAL | — | — | — | X |

Fig. 3. Tabular summary of some characteristics of typological, biological, transient, and successional species. Operational taxonomic criteria associated with each of these concepts are indicated.

Figure 3 summarizes aspects of the species concept which have been emphasized in the preceding discussion.

### Problems in Applying Biological Species Concepts to Fossils

*Transient Species.* In the foregoing discussion attention has been focused on the theoretical model of the biological species in use today by a majority of paleontologists. We shall now examine some of the practical problems which arise when this model is applied to real paleontological data. Most of these problems arise from three sources: the inadequacy of morphological data, the prevalence of biased frequency distributions, and the incompleteness of the available fossil record.

1. *Inadequacy of morphological data.* The most obvious shortcoming of paleontological data is that fossils normally preserve only hard skeletal parts. To a student of soft-bodied living animals (say *Euglena*) this difficulty might appear insuperable. Indeed, some taxonomists assume for this reason alone that clas-

sifications of living and extinct species are on an entirely different level. For the vast majority of genera commonly found as fossils, however, this judgment is exaggerated or untrue. There can be, of course, no conflict between the classification of living and extinct *Euglena:* like hosts of other soft-bodied forms this

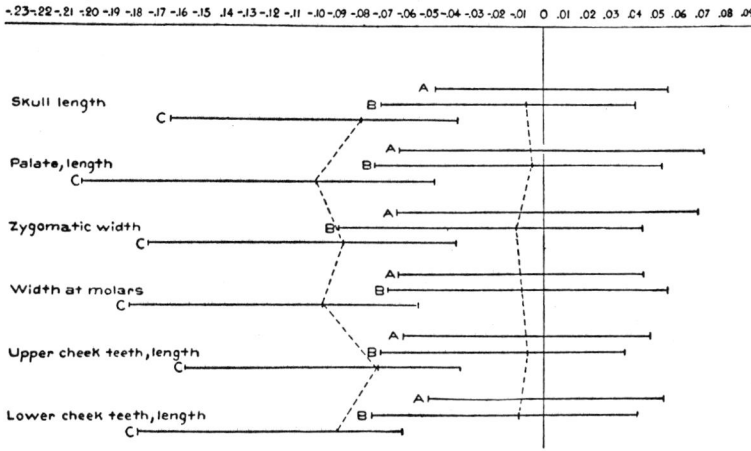

Fig. 4. Ratio diagram showing comparisons of certain osteological characters in two species of the modern marten *Mustela* and two modern subspecies of *Mustela sibirica*. The latter show a strong overlap of measured characters, whereas the two species show little overlap in the same characters. A, *Mustela sibirica fontanierii;* B, *M. s. davidiana;* C, *M. altaica kathiah*. Abcissal scale represents for each character the difference between the logarithm of any given value and the logarithm of the mean value of the population selected as reference standard (in this case, population A). Thus the horizontal distance between any two points represents the ratio of either one to the other. Three points connected by a horizontal line are plotted for a given character in each sample to represent the greatest, least, and mean dimensions. The several means for each character in a population are connected by dashed or solid lines. For discussion of ratio diagrams, see Simpson (1941). Mammal data from Allen (1938) (Colbert and Hooijer, 1953).

genus leaves essentially no fossil record. Furthermore, it is generally true that specific discrimination in groups of animals commonly found as fossils is (or can be) based on skeletal morphology. Consider, for example, the shell-bearing foraminifera, radiolarians, corals, ectoproct bryozoans, brachiopods, snails, clams,

134    THE SPECIES PROBLEM WITH FOSSIL ANIMALS

echinoids, and mammals. For such groups differences in the classification of fossil and living organisms at the species level are rarely attributable to the inadequacy of morphological data.

Students of some of the groups just listed may argue that many important specific taxonomic characters are not preserved in fossils—pelage characteristics in mammals, for example. This view, however, violates one of the basic tenets of taxonomy: that the *nature* of a unit biological character is not so important in

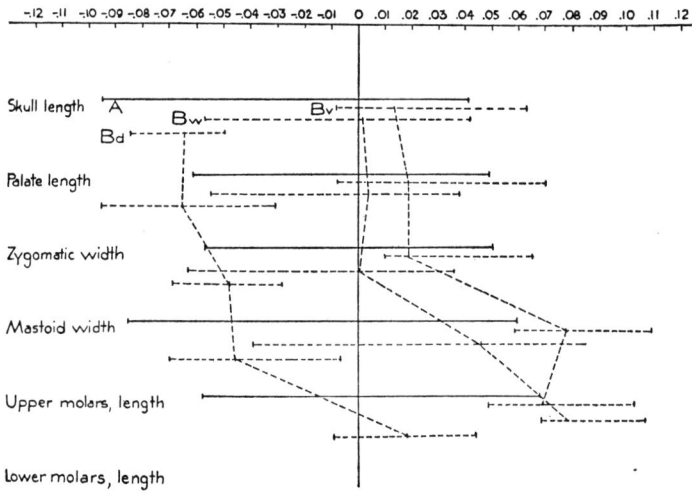

Fig. 5. Ratio diagram comparing range of size variation in the fossil bamboo rat (*Rhizomys sinensis troglodytes*) and three modern subspecies of *R. sinensis*. A, *Rhizomys sinensis troglodytes*; Bd, *R. s. davidi*; Bv, *R. s. vestitus*; Bw, *R. s. wardi*. Diagram constructed in the same manner as Fig. 4 (Colbert and Hooijer, 1953).

taxonomy as its statistical *distribution* within and between populations (Simpson, 1943). This point is so fundamental to an understanding of the paleontologist's approach to classification on the species level that it will be illustrated with three examples drawn from diverse sorts of animals.

In a study of Chinese Pleistocene mammals, Colbert and Hooijer (1953) analyzed skeletal data on several species of modern Asiatic mammals and made comparisons with similar data on fossils. A ratio diagram (Fig. 4) based on skeletal measurements of two species of the modern marten *Mustela* indicated that the amount of morphological overlap between two subspecies is much greater than that between two species, showing that osteological data alone reflect a classification (Allen, 1938) based on geographic, ecologic, and pelage data. With this in

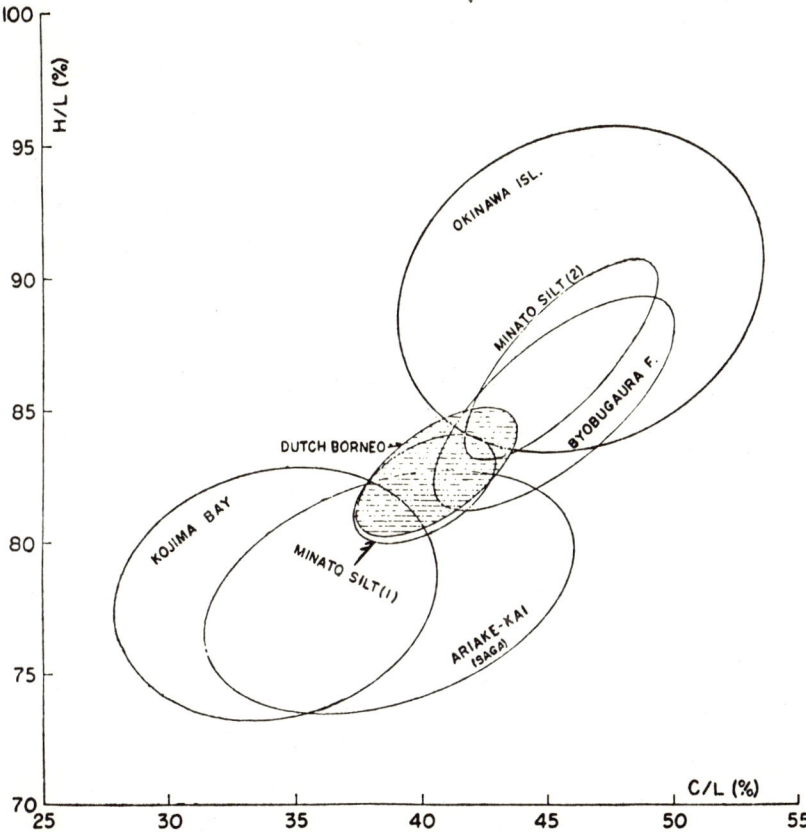

Fig. 6. Diagram showing estimated morphological ranges of seven populations of the arcid pelecypod *Anadara*. H, maximum distance between two lines parallel to the hinge; L, shell length; C, maximum depth of one valve. Ellipses are statistical estimates of ranges to include 95% of the total population ranges. Populations from Kojima Bay, Ariake Bay, and Dutch Borneo are recent. Other populations are fossil. Samples from Dutch Borneo and Minato silt (1) are assigned to *Anadara granosa granosa;* from Kojima and Ariake Bays to *Anadara granosa bisenensis;* and those from the remaining areas to *Andara obessa* (Kotaka, 1953).

136 THE SPECIES PROBLEM WITH FOSSIL ANIMALS

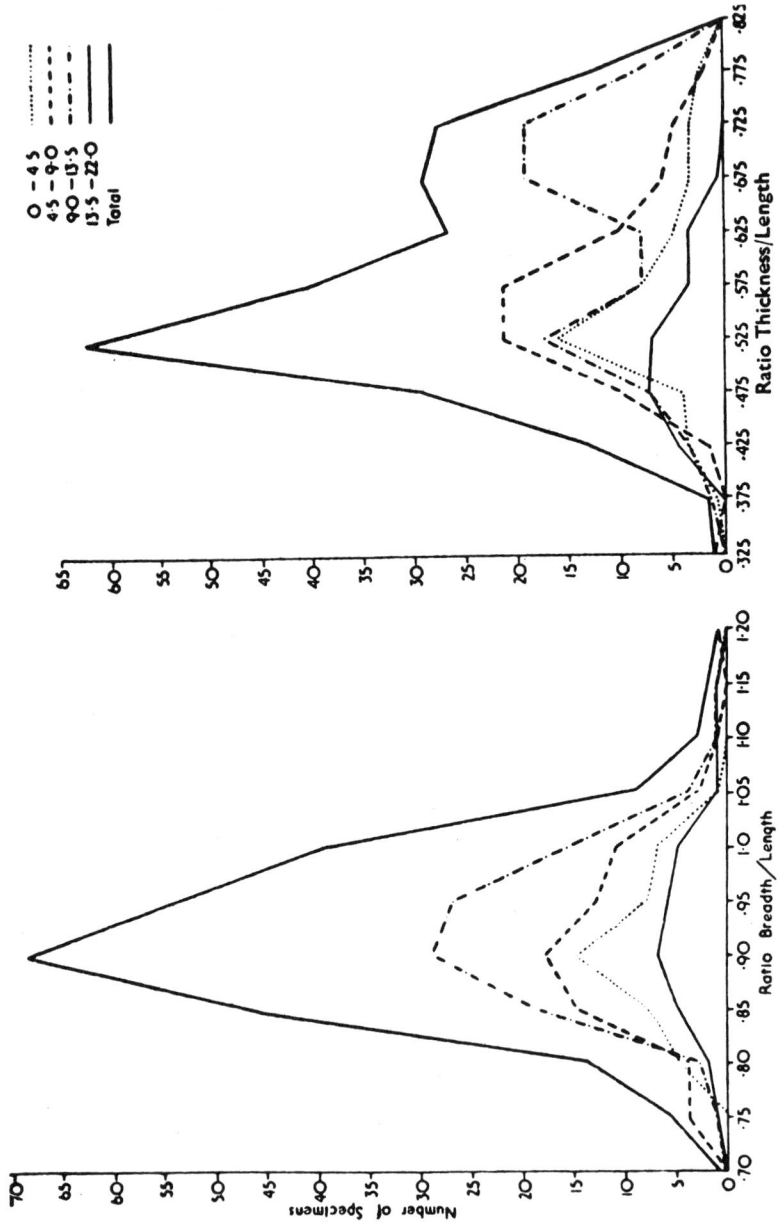

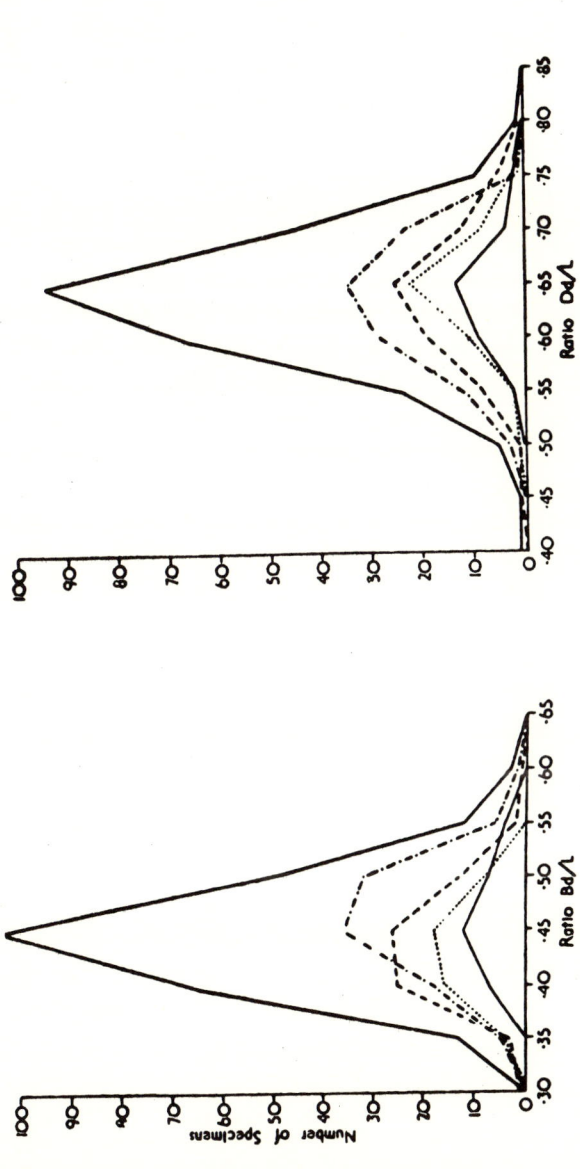

Fig. 7. Frequency polygons of four ratios measured in a collection of the Permian terebratuloid brachiopod *Dielasma elongata*. Bd, distance of maximum breadth and thickness respectively from posterior end of shell; L, total length of shell. Data analyzed into size groups based on total length (Westoll, 1950).

mind, data on a collection of the Pleistocene bamboo rat *Rhizomys* were compared with data on three subspecies of the modern *Rhizomys sinensis* (Fig. 5). The extensive mutual overlap among the four samples in the six measured skeletal features was then used as a basis for considering the fossil a subspecies, *R. sinensis troglodytes*.

A similar approach by Kotaka (1953) on recent and fossil populations of the arcid pelecypod *Anadara* is illustrated in Fig. 6. Having concluded that the population in question could best be distinguished on the basis of two shell indices (H/L and C/L), Kotaka calculated and graphed the 0.95 elliptical contour of each bivariate density distribution. Note that although four fossil and three recent populations are included in this study, each receives the same taxonomic treatment; and there is every reason to suppose that Kotaka's conclusions apply equally to fossil and modern forms.

An illustration of the taxonomic importance of intrapopulation variation patterns is furnished by Westoll's (1950) study of a large sample of the Permian terebratuloid brachiopod *Dielasma elongata* (Fig. 7). Frequency polygons of the thickness-length ratio reveal a distinct bimodality in larger size classes, although in all other respects shell characters display unimodal distribution patterns. This bimodality is interpreted as sexual dimorphism. Westoll's brachiopods and Kotaka's clams illustrate the same point: sound taxonomic inferences can be made solely on the basis of the distribution pattern displayed by shell features whose functional significance is incompletely known.

One troublesome limitation of paleontological data lies in the difficulty (usually the impossibility) of distinguishing phenotypic from genotypic variation. This problem is particularly evident to the taxonomist analyzing statistically significant morphological differences among a small number of samples whose stratigraphic relationships are poorly known. If suitable collections are available for study, however, the taxonomic (but not the genetic) problem may disappear. A case in point is McKerrow's (1953) detailed study of contemporaneous brachiopod communities sampled from many localities along the outcrop of a thin stratigraphic unit known as the Fuller's Earth Rock. Although the genus *Ornithella* displays a wide range of morphological variation at each locality and a considerable overlap among geographically separated populations, careful study reveals that certain modal variants tend to characterize different localities, as indicated

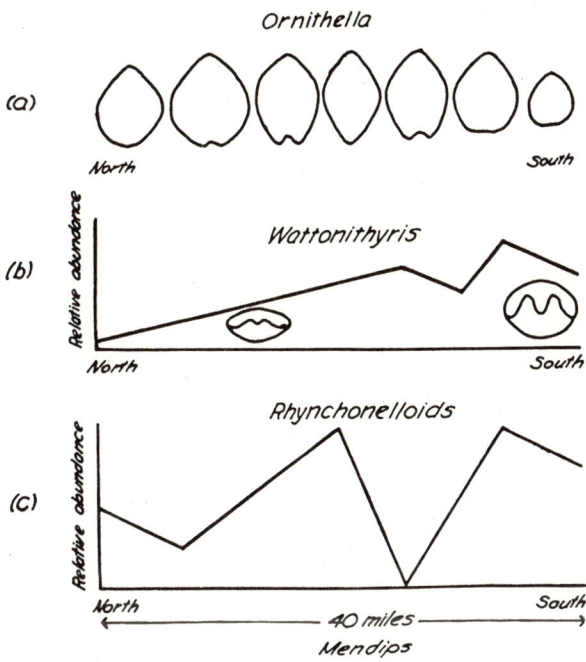

Fig. 8. Diagram of some lateral changes in brachiopod communities in the Jurassic Fuller's Earth Rock. (a) Changes in the outline of *Ornithella*, simplified and somewhat exaggerated; (b) changes in the relative abundance and in the anterior commissure of *Wattonithyris*; (c) changes in the relative abundance of the rhynchonelloids (McKerrow, 1953).

schematically in Fig. 8. Some morphological features in fact tend to be symmetrically disposed about Mendips. Are we to interpret these differences as the result primarily of genotypic or phenotypic variation? Even though this question must remain unanswered, the taxonomic expression of these data is in no way affected. In either case the entire assemblage will be termed a (transient) species.

Now suppose that McKerrow had had available only two geographically isolated but distinctive local populations. In this case, taxonomic disposition of these collections would be far more subjective. This serves to emphasize the point that the chief taxonomic difficulties which arise in dealing with paleontological

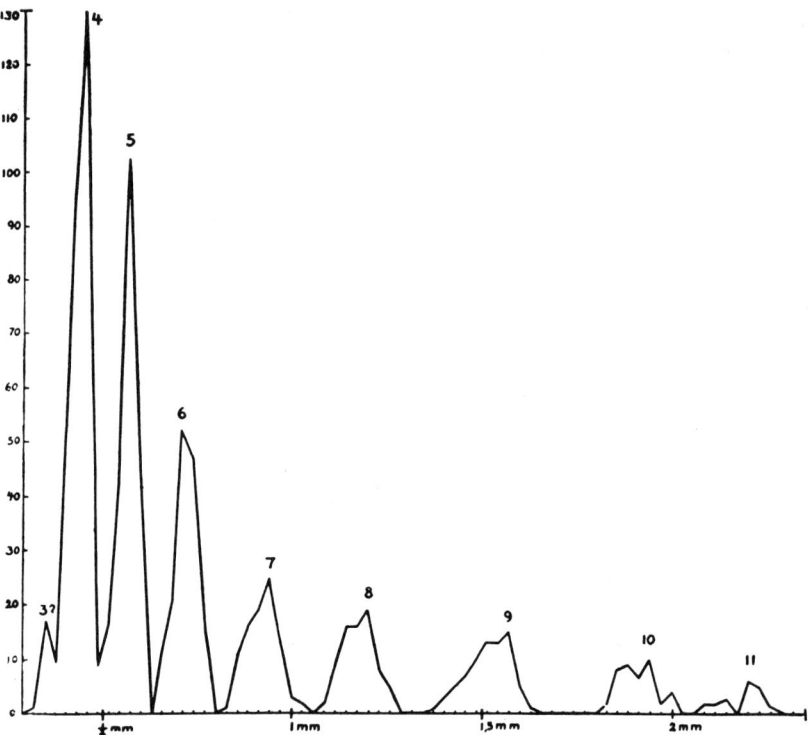

Fig. 9. Graph showing size and frequency distribution of growth stages represented in a population of the Silurian ostracod *Beyrichia jonesi* from Gotland (Spjeldnaes, 1951).

materials are due not so much to the limitations inherent in postmortem examination of skeletal anatomy, but rather to the incompleteness of the available fossil record.

2. *Biased frequency distribution.* The size-frequency distribution of a few carefully studied paleontological samples seems to reflect fairly accurately the size distribution of the original living

population. For example, in a large collection of the Silurian ostracod *Beyrichia jonesi* analyzed by Spjeldnaes (1951) there is a progressive depletion in frequency at least from the fourth to the eleventh instars (Fig. 9). This pattern has been taken as evi-

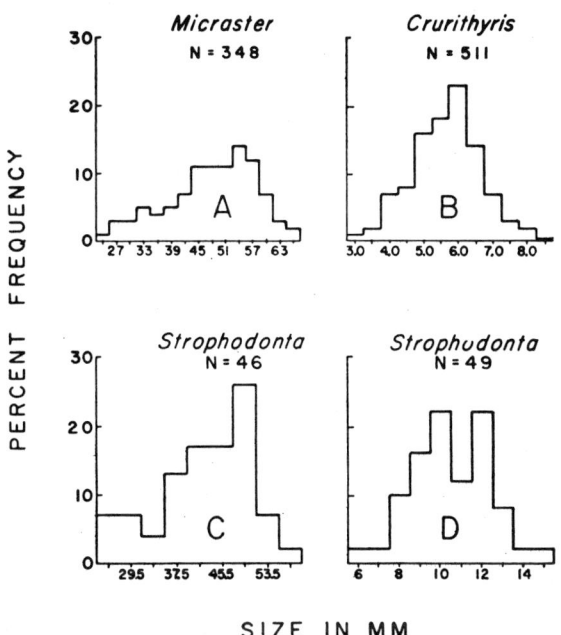

Fig. 10. Size-frequency distributions of four typical samples of fossil invertebrates. N, number of measured specimens; A, measurements of greatest width in a sample of the echinoid *Micraster coranguinum* from the Cretaceous of Northfleet, England. Data from Kermack (1954). B, measurements of maximum width in a sample of the brachiopod *Crurithyris planoconvexa* from the Dry Shale of Kansas. Data from Tasch (1953). C, measurements of maximum width of the brachiopod *Strophodonta* sp. from a single locality in the Gravel Point formation, Michigan. Original data. D, measurements of maximum length in a sample of *Strophodonta extenuata ferronensis* from a locality in the upper Ferron Point formation, Michigan. Original data.

dence that the collection approximates a random sample of the original population, and on this assumption Kurtén (1953) has calculated life tables giving mortality rates at each growth stage as well as other standard parameters of population dynamics.

Samples of the sort just described are by no means common, however. Four size-frequency distributions typical of fossil marine invertebrates are given in Fig. 10. Such distributions tend to be unimodal and either symmetrical or slightly skewed; many can be approximated by the normal distribution. In most instances, the observed distributions are strongly biased and reflect primarily a complex of ecologic, geologic, and operational factors which at best bear only obliquely on the taxonomic problem (Boucot, 1953; Kermack, 1954; Imbrie, 1955).

In dealing with groups of animals (mammals, for example) where practical and objective morphological criteria are available for the identification of definite growth stages, the existence of biased size distributions may cause little concern to the taxonomist. In dealing with fossil groups lacking criteria of this sort, however, particularly in groups displaying strongly allometric growth patterns, taxonomic judgments may be considerably affected. Once recognized, this problem can be quite easily solved by characterizing sampled populations in terms of growth patterns rather than growth stages (Kermack, 1954; Parkinson, 1954; Imbrie, 1956).

3. *Incompleteness of the available fossil record.* The most serious limitation of paleontological data is the sparsity of fossils. It is of course true that the total number of collecting localities which have yielded good fossils from every system younger than Precambrian is very large, and it is also true that future work will bring forth an unknown but certainly very large amount of new material from localities and horizons now unrepresented in existing collections. Nevertheless, from general theoretical considerations on the nature of sedimentation and diagenesis, and from practical experience in portions of the geological column which have been thoroughly examined for fossils, most paleontologists and stratigraphers would predict that no amount of future field work will ever fill a majority of existing phyletic gaps between transient species.

At least five factors are important in accounting for the incompleteness of the available fossil record, not including lack of adequate field work: nondeposition, erosion, migration, nonpreservation, and inaccessibility. Together, these factors account for the prevalence of transient species.

Since the publication of the classic paper by Barrell (1917), stratigraphers have generally admitted that very few local sedi-

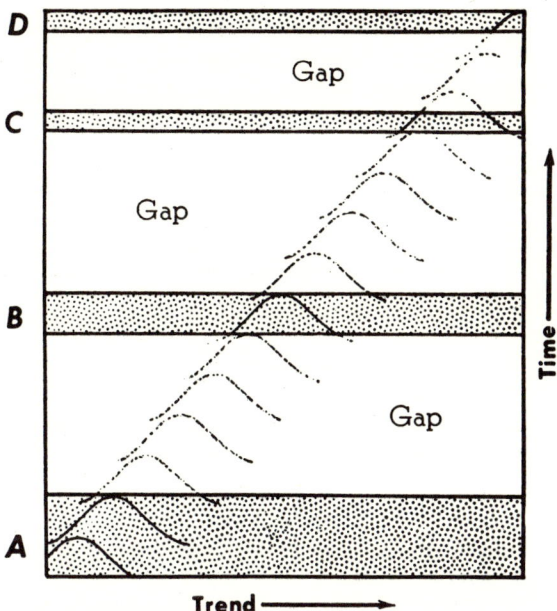

Fig. 11. Diagram to illustrate how gaps in the stratigraphic record result in a meager fossil record of a continuously evolving lineage. Gradual evolution of a lineage is indicated by a series of shifting normal frequency curves. Stippled areas represent preserved portions of the stratigraphic record. Solid frequency curves represent four transient species collected from layers A-D. Gaps in the fossil record, indicated by dashed curves, cannot be eliminated by collecting in this area. (After Newell, 1956.)

mentary sequences contain a direct, continuous record of the total time span represented in a given stratigraphic sequence. This is because many fossiliferous strata were formed in water so shallow that the permanent accumulation of a sedimentary layer (with entombed fossils) was a rare event, possible only when the surface of sedimentation was below the depth to which waves and currents were effective in moving sediment. Figure 11 illustrates the formation of sedimentary and paleontological hiatuses according to this principle.

Mere deposition of sedimentary layers does not safeguard their preservation since large volumes of such rocks are destroyed by uplift and erosion. Even if the sediments escape this fate, it does not guarantee that even a small sample of skeleton-bearing organisms present in the original living community will become collectable fossils. In order to be preserved a skeleton must escape

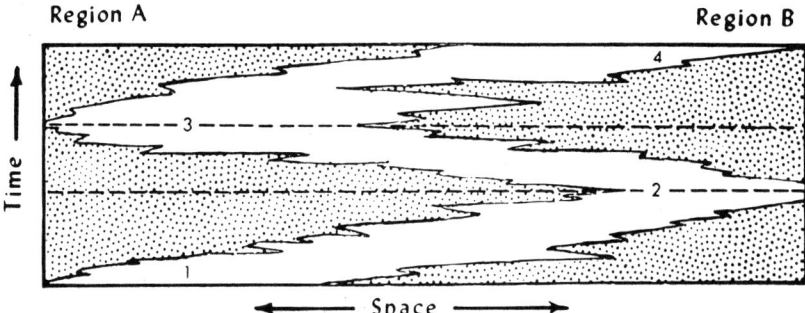

Fig. 12. The effect of facies migration on the fossil record of evolving lineages. Well-defined transient species occur at horizons 1 and 3 in region A. Different transient species occur at horizons 2 and 4 in region B. Even if deposition and preservation have been continuous in this area, inaccessibility of the rocks lying between regions A and B makes the available record of this lineage incomplete. (After Newell, 1956.)

chemical, biochemical, and mechanical destruction by agencies acting before, during, and after burial.

Figure 12 illustrates the combined effect of migration and inaccessibility in reducing the number of collectable fossils. Stippled and unstippled areas on this diagram represent three contrasting environments (as well as the geologic record of those environments, or facies) which have continuously occupied the area. Three environments, each with its associated organic community, are represented at any given instant. Under normal circumstances geological conditions responsible for the localization of physical environments change, and the associated communities shift accordingly. Now consider the effect of this migration on the fossil record of an evolving succession of populations. Four transient species belonging to one lineage are symbolized by numbers 1 through 4. Under ideal conditions, fossils representing the entire lineage could be studied in outcrops of the appropriate facies. In practice, however, the geological column can be studied only in isolated, accessible areas. Hence a paleontologist examining fossils from region A records two transient species 1 and 3, while in region B his observations are limited to transients 2 and 4. If

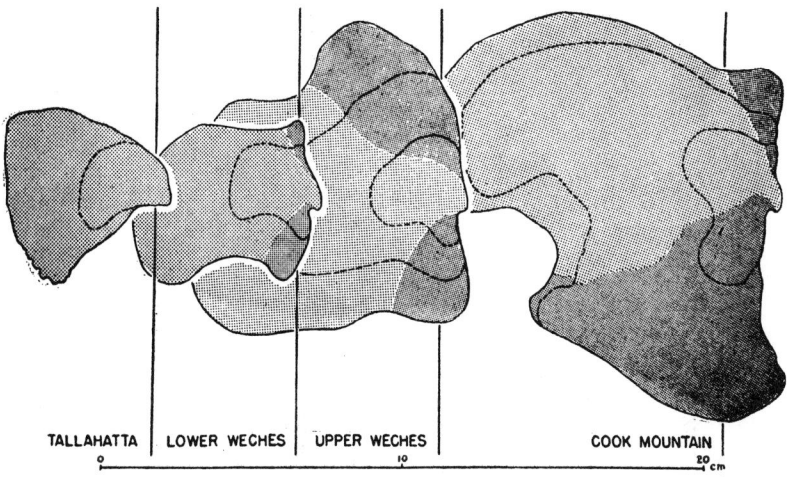

Fig. 13. Some of the characteristic features of four species of the Eocene oyster *Cubitostrea*. Heavy stippling represents auricles. Formation names indicate the stratigraphic position of individual species, as follows: Tallahatta, *C. perplicata;* Lower Weches, *C. lisbonensis;* Upper Weches, *C. smithvillensis;* Cook Mountain, *C. sellaeformis* (Stenzel, 1949).

morphological gaps between studied transients are too large, sound phylogenetic interpretations may be impossible.

One of the best documented examples of transient species recording portions of an evolving lineage is Stenzel's (1949) study of the oyster *Cubitostrea*. Four clear-cut species have been recognized, each restricted to a different stratigraphic level in the Eocene of the Gulf Coast. *C. perplicata*, the oldest known species in this line, is characterized by small size, strong ribbing, thin shell, absence of auricles, lack of arching, as well as other characters

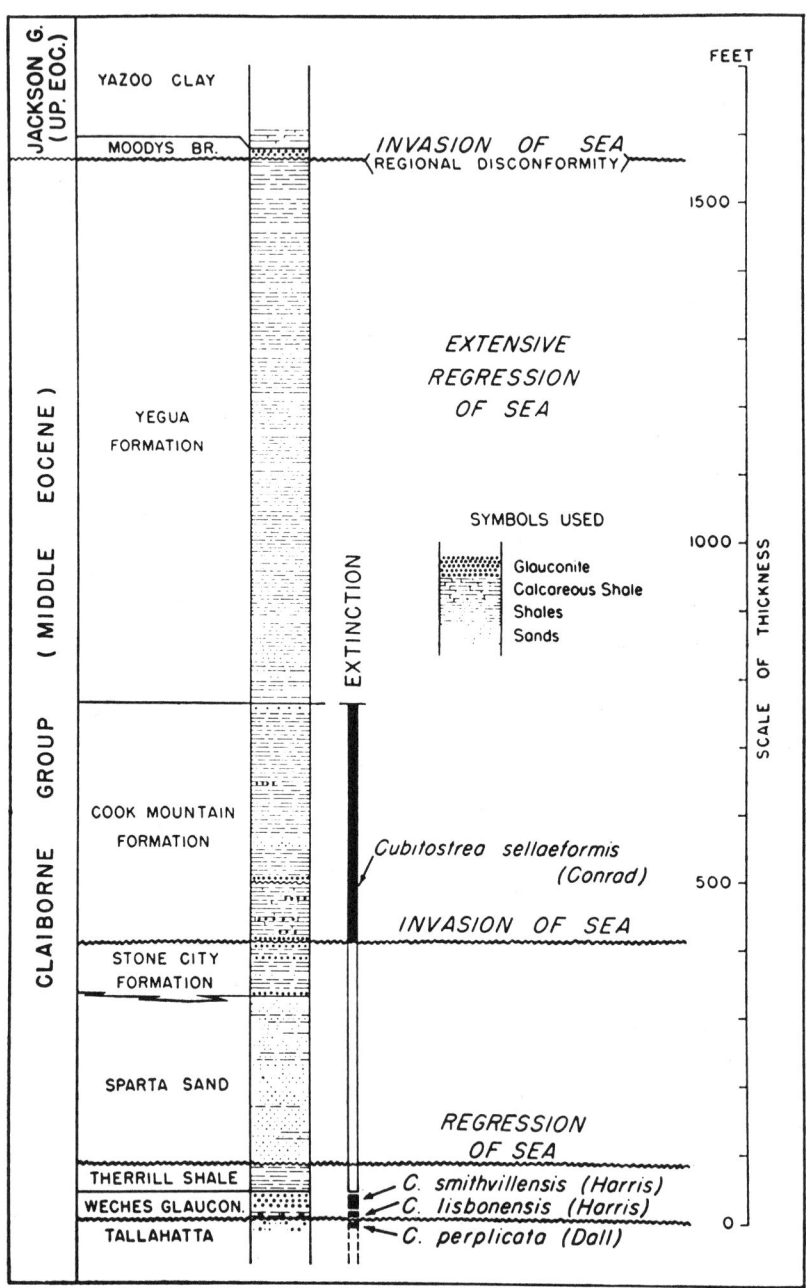

Fig. 14. Stratigraphic ranges of the chain of species comprising the *Cubitostrea sellaeformis* lineage in the Gulf Coastal Plain (Stenzel, 1949).

(Fig. 13). *C. sellaeformis*, the youngest known species, is characterized by large size, absence of ribs, thick shell, strongly developed auricles, and conspicuous arching. Gaps between these extremes are partly bridged by *C. lisbonensis* and *C. smithvillensis*.

This assemblage of species is viewed as the fossil record of a continuously evolving stock leading to *C. sellaeformis*. But the point to be emphasized here is that each of these species represents only a stage in this evolution and is a transient, not a successional, species. The stratigraphic range of each species is indicated in Fig. 14. Although *Cubitostrea* is represented almost continuously in beds ranging from the upper Tallahatta through the Weches formations, the morphological and stratigraphic ranges of the three species do not overlap. Similarly, transitional forms linking *C. smithvillensis* and *C. sellaeformis* are unknown. From independent stratigraphic evidence it is clear that the discontinuities just discussed are to be explained in terms of nondeposition, erosion, and migration associated with advances and retreats of the Gulf of Mexico.

*Successional Species.* All the difficulties discussed above in connection with transient species, with the single exception of the incompleteness of the fossil record, apply with equal force to successional species. In addition there is the theoretical taxonomic problem of selecting criteria for the subdivision of continuous lineages. From what has been said it will be clear that this problem is encountered far more often in the literature than in the laboratory. More extensive field work will undoubtedly bring forth more examples in the near future, but it is unlikely that the problem of successional species will ever be a burden to the student of fossils.*

One of the earliest and still the best example of successional speciation is the fossil spatangoid sea urchin *Micraster* originally described by Rowe (1899). A summary of *Micraster* evolution

---

* Successional species may, however, prove to be common in cores taken from deep ocean basins where deposition and preservation may be essentially continuous. Here lie fascinating and nearly untouched areas for taxonomic, evolutionary, and paleoecologic research.

148   THE SPECIES PROBLEM WITH FOSSIL ANIMALS

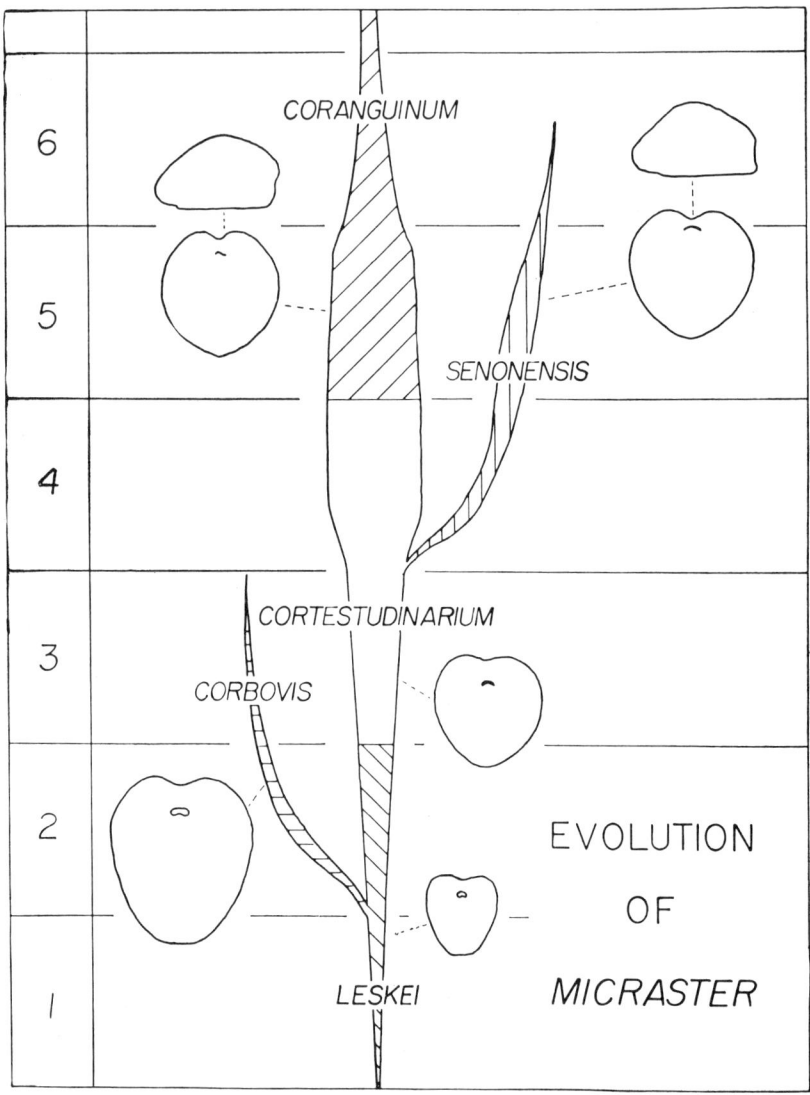

Fig. 15. Evolution of the spatangoid echinoid genus *Micraster* as recorded in six zones of the Cretaceous of southern England. Some of the characteristic features of the five designated successional species are indicated by ventral and lateral profiles. 1, Zone of *Cyclothyris cuvieri;* 2, Zone of *Terebratulina lata;* 3, Zone of *Holaster planus;* 4, Zone of *Micraster cortestudinarium;* 5, Zone of *Micraster coranguinum;* 6, Zone of *Marsupites testudinarius.* Constructed chiefly from data given by Kermack (1954).

will serve as a convenient focus for a theoretical discussion of the taxonomy of successional species. This review is based largely on Kermack's (1954) critical restudy of Rowe's materials.

From many horizons in the chalky limestones of the English Cretaceous large numbers of well-preserved *Micraster* have been collected. Most of the limestones are soft enough so that simple preparation techniques make it possible to free the tests from the enclosing matrix. Owing to the complexity of the echinoid test, a large number of unit taxonomic characters can be distinguished and interpreted functionally by analogy with living spatangoids. From many points of view, then, these fossils make ideal materials for evolutionary and taxonomic studies.

Some of the outstanding features of the fossil record of *Micraster*, as interpreted by Kermack, are shown in Fig. 15. The oldest known members of the genus (*M. leskei*) possess small tests with the widest portion of the test well ahead of midlength. The mouth, lacking a distinct labrum, is plainly visible in ventral aspect. The anterior ambital notch is shallow and the subanal fasciole small. Closely spaced collections of the main line of evolution from this form record a progressive shift in the average values of the skeletal characters just mentioned, and in other characters as well, giving rise to *M. coranguinum*. In this large species the widest portion of the test is approximately at midlength, the mouth is farther forward, and a strong labrum obscures the mouth opening in ventral aspect. The anterior ambital notch is deep and the subanal fasciole large. It should be emphasized that distinctions between successive population samples can be made only by comparison of the average values of overlapping frequency distributions.

Two side branches of the main *M. leskei-coranguinum* line can be distinguished. The first, *M. corbovis*, is a large form with a number of characteristic features not discussed here. The second branch, *M. senonensis*, differs from its contemporary *M. coranguinum* principally in its complete or nearly complete lack of a subanal fasciole. Frequency distributions of this character in a single population studied by Kermack show two nonoverlapping clusters. In addition, *M. senonensis* tends to be relatively taller than *M. coranguinum*.

Studies of living spatangoid echinoids make possible a reasonable interpretation of the data presented above in terms of environmental adaptations. According to Kermack's view, the evolution of the main *M. leskei-coranguinum* stock is to be interpreted as progressively better adaptation to a burrowing habit. This view is based primarily on known functions of the labrum and subanal fasciole. In burrowing urchins, the labrum increases the efficiency of the mouth as an organ of ingestion; and dense cilia associated with the subanal fasciole produce strong posteriorly directed water currents which aid in discharging water from the burrow.

The *M. senonensis* stock, on the other hand, seems to have evolved adaptations for living *on* rather than *in* the bottom. For this mode of life a subanal fasciole is unnecessary. The domelike form of the test, moreover, might facilitate cleaning of the surface; at least it would not hinder movement on the bottom.

We now come to the problem of expressing the facts and inferences about *Micraster* taxonomically. At least three possible criteria can be used for subdivision of such a continuum. First, boundaries between species (or subspecies) might be placed at those points in the phylogenetic tree where branching takes place. Assuming that the phylogeny has been adequately documented, this criterion has at least the merit of objectivity. But units so delineated may differ very greatly in scope. In the case of *Micraster*—assuming a relatively constant morphological evolutionary rate—the oldest lineage segment would be much smaller in morphological scope than the youngest.

Stratigraphic boundaries offer another possible basis for the delimitation of taxonomic units. *Micraster*, for example, might be split into eleven species with taxonomic boundaries placed at the limits of the six stratigraphic zones. Although to a biologist such a course might appear rather arbitrary, to a stratigrapher units so defined might be quite useful.

From a strictly biological point of view probably the most satisfactory basis for subdividing an unbranched lineage is morphology. For example, taxonomic boundaries can be placed in such a way that the total morphological difference between corresponding parts of successive species is of the same order as that between contemporary species. Application of this criterion of course involves a subjective evaluation of total morphological (and inferred genetical) differences.

The *Micraster* classification now in use and illustrated in Fig. 15 is based on a combination of the three criteria discussed above, with emphasis on the morphology. Note that the interpretation of the *M. leskei-coranguinum* lineage as the main line of *Micraster* evolution is reflected in the failure to subdivide *M. leskei* and *M. cortestudinarium*.

## Summary

A review of recent literature in paleontology indicates that the nature of fossil species remains a controversial matter, although the biological species concept based on variable, interbreeding populations has largely replaced the typological concept based on morphotypes. Data on morphology, association, biogeography, paleoecology, and biostratigraphy are used by the paleontologist to delineate fossil species. Most species actually defined by neontologists and paleontologists are transient species, i.e., essentially contemporaneous aggregates of interbreeding populations. In rare instances paleontologists describe segments of phyletic lineages. Groupings of this sort are called successional species.

The principal source of difficulty in applying the biological species concept to fossils is the incompleteness of the available fossil record. This lack is attributed to a combination of factors, including nondeposition, erosion, migration, nonpreservation, and inaccessibility. Taxonomic problems also arise from the limitations inherent in morphological data and the prevalence of biased frequency distributions. In dealing with successional species the problem arises as to the selection of criteria for subdividing continuous lineages. Taxonomic boundaries may be designated on the basis of phyletic branching, stratigraphy, or morphology.

## REFERENCES

Allen, G. M. 1938. The mammals of China and Mongolia. *Natural History of Central Asia*, Vol. 11, Pt. 1, American Museum of Natural History, New York, pp. 1-620.

Barrell, J. 1917. Rhythms and the measurement of geologic time. *Bull. Geol. Soc. Amer.*, **28**, 745-904.

Boucot, A. J. 1953. Life and death assemblages among fossils. *Am. J. Sci.*, **251**, 25-40.

Burma, B. H. 1948. Studies in quantitative paleontology: I. Some aspects of the theory and practice of quantitative invertebrate paleontology. *J. Paleontol.*, **12**, 725-61.

Burma, B. H. 1954. Reality, existence, and classification: a discussion of the species problem. *Madroño*, **12**, 193-224.

Colbert, E. H., and D. A. Hooijer. 1953. Pleistocene mammals from the limestone fissures of Szechwan, China. *Bull. Am. Museum Nat. Hist.*, **102**, 1-134.

Eagar, R. M. C. 1952a. Some problems in invertebrate paleontology. *Proc. Leeds Phil. Lit. Soc.*, **6**, Pt. 1, 50-53.

Eagar, R. M. C. 1952b. Variation with respect to petrological differences in a thin band of Upper Carboniferous non-marine lamellibranchs. *Liverpool and Manchester Geol. J.*, **1**, Pt. 2, 161-90.

Eagar, R. M. C. 1956. Naming carboniferous non-marine lamellibranchs. In P. C. Sylvester-Bradley (1956).

Imbrie, J. 1955. Biofacies analysis. *Geol. Soc. Amer., Spec. Papers*, No. **62**, 449-64.

Imbrie, J. 1956. Biometrical methods in the study of invertebrate fossils. *Bull. Am. Museum Nat. Hist.*, **108**, 211-52.

Kermack, K. A. 1954. A biometrical study of *Micraster coranguinum* and *M. (Isomicraster) senonensis*. *Trans. Roy. Soc. (London)*, **B237**, 375-428.

Kotaka, T. 1953. Variation of Japanese *Anadara granosa*. *Trans. Proc. Palaeontol. Soc. Japan*, n.s., No. 10, 31-36.

Kurtén, B. 1953. On the variation and population dynamics of fossil and recent mammal populations. *Acta Zool. Fennica*, **76**, 1-122.

Leitch, D. 1940. A statistical investigation of the *Anthracomyas* of the basal Similis-Pulchra Zone in Scotland. *Quart. J. Geol. Soc. London*, **96**, 13-37.

Mayr, E., E. G. Linsley, and R. L. Usinger. 1953. *Methods and Principles of Systematic Zoology*. McGraw-Hill Book Co., New York, pp. 1-328.

McKerrow, W. S. 1953. Variation in the Terebratulacea of the Fuller's Earth Rock. *Quart. J. Geol. Soc. London*, **91**, 97-122.

Newell, N. D. 1956. Fossil populations. In P. C. Sylvester-Bradley (1956).

Parkinson, D. 1954. Quantitative studies of brachiopods from the Lower Carboniferous reef limestones of England. I. *Schizophoria resupinata* (Martin). *J. Paleontol.*, **28**, 367-81.

Regan, C. T. 1926. Organic evolution. *Rept. Brit. Assoc. Southampton*, 1925.

Rowe, A. W. 1899. An analysis of the genus *Micraster*, as determined by rigid zonal collecting from the Zone of *Rhynchonella Cuvieri* to that of *Micraster coranguinum*. *Quart. J. Geol. Soc. London*, **55**, 494-546.

Simpson, G. G. 1941. Large Pleistocene felines of North America. *Am. Museum Nat. Hist.*, **No. 1136**, 1-27.

Simpson, G. G. 1943. Criteria for genera, species, and subspecies in zoology and paleozoology. *Ann. N.Y. Acad. Sci,.* **44**, 145-78.

Simpson, G. G. 1951. The species concept. *Evolution*, **5**, 285-98.

Spjeldnaes, N. 1951. Ontogeny of *Beyrichia jonesi* Boll. *J. Paleontol.*, **25**, 745-55.

Stenzel, H. B. 1949. Successional speciation in paleontology: the case of the oysters of the *sellaeformis* stock. *Evolution*, **3**, 34-50.

Sylvester-Bradley, P. C. 1952. In R. M. C. Eagar (1952a, pp. 52-53).

Sylvester-Bradley, P. C., ed. 1956. The species concept in paleontology. *Systematics Assoc. Publ. No. 2*.

Tasch, P. 1953. Causes and paleoecological significance of dwarfed fossil marine invertebrates. *J. Paleontol.*, **27**, 356-444.

Weir, J. 1950. Recent studies of shells of the coal measures, *Sci. Progress*, **38**, 445-58.

Westoll, T. S. 1950. Some aspects of growth studies in fossils. *Proc. Roy. Soc. (London)*, **B137**, 490-509.

# BREEDING SYSTEMS, REPRODUCTIVE METHODS, AND SPECIES PROBLEMS IN PROTOZOA*

T. M. SONNEBORN:† INDIANA UNIVERSITY,
BLOOMINGTON, INDIANA

The so-called modern biological species concept defines a species as a group of actually or potentially interbreeding populations. As the other papers in this symposium attest, this concept is widely accepted as the only valid species concept. Yet it is of admittedly limited applicability. In obligatorily self-fertilizing organisms, this concept of the species contracts to include but a single individual. In exclusively asexual organisms, the concept admits of no species at all. On this concept, perhaps half or more of the species recognized by students of the Protozoa are invalid and, in principle, cannot be modified so as to become valid. In like manner, practically every phylum of Invertebrates and all major plant groups contain numerous organisms in which species do not exist at all or in which each individual is a separate species. This greatly limits the applicability of the modern biological

Note: Superior numbers in text refer to Addenda, page 314.
* Contribution #618 from the Department of Zoology, Indiana University.
† The work of the author and his collaborators, much of which has not previously been published, was supported by grants from Indiana University, the Rockefeller Foundation, the American Cancer Society, and the Atomic Energy Commission (Contract No. AT(11-1)-235). The following people have critically read part or all of this manuscript: Mary L. Austin, G. H. Beale, T. T. Chen, Ruth V. Dippell, A. C. Giese, L. Gilman, E. Hanson, Mrs. Margaret Johnson, R. F. Kimball, E. Mayr, D. L. Nanney, Ursula Philip, J. R. Preer, and R. Siegel. I am immensely indebted to these experts for checking the accuracy of my statements, for generously providing me with unpublished information and permission to cite it, and for numerous criticisms and suggestions.

species concept. It has been maintained, not without some reason, that there are more species of Protozoa than of all other animals together, for each species of the higher animals is believed to be sole host to at least one species of parasitic Protozoa. If half or more of the Protozoa, and many other organisms as well, are in principle outside the domain of the modern biological species concept, it is applicable to only a minority of all organisms.

Although this is perhaps the greatest objection to adopting this species concept as the only valid one, there are also others. In some organisms, what is now recognized as a single species on morphological grounds would be broken up into fantastically large numbers of "biological" species. Moreover, very commonly these "biological" species would be to an appalling degree unrecognizable and unidentifiable by routine taxonomic procedures or by any other procedures that could reasonably be expected of working biologists.

The Protozoa illustrate on a lavish scale these difficulties and limitations of the modern biological species concept. There is need for a broader species concept which will embrace the whole of biology. But it should not sacrifice unnecessarily the very real values of the so-called modern biological species concept. The present paper will review the situation in the Protozoa and will inquire as to how there might be developed for obligatory inbreeders and asexual organisms a concept which embraces for them a level of biological organization comparable to the one which is embraced for outbreeders by the modern biological species concept. Both concepts would, it seems, have to be special cases of a more general concept applicable to all organisms. However, as will appear, such conceptual ordering does not in itself settle the question of whether the unit groups of individuals and populations designated by these concepts should always be called species and assigned specific names. This is a separate matter and is separately considered.

Until relatively recently, much of what now appears to be significant information about the Protozoa in relation to these problems was unknown. This is particularly true of the Ciliates which, because of the extensive literature now available on them, receive greatest attention. The pertinent modern literature is in great need of survey from a broad unitary point of view. That is attempted here in considerable detail. The basic facts about the six genera of Ciliates that have been investigated along modern lines are given species by species. This sort of work is being pursued more extensively now than ever before. The present review therefore will doubtless soon be outdated by new discoveries.[1] Nevertheless, it may serve a useful purpose not only in summarizing the situation as it now stands, but also in pointing up the gaps in knowledge and in suggesting a central idea to guide further investigation.

This central idea is that the system of breeding—inbreeding or outbreeding to various degrees—which an organism normally follows is closely correlated with a considerable number of its major biological features; that these features have significance primarily with respect to their consequences for the breeding system; and that herein lies the key to the species and evolutionary problems in the Protozoa. Attempts are therefore made to recognize, for each organism dealt with, those biological features that bear upon the breeding system and species problems and to assess their consequences. As will abundantly appear, although enough is known to validate this point of view in a general way, the attempt to apply it in detail to the organisms that have been studied reveals a great dearth of basic facts. One of the chief purposes of this paper is to call attention to the kinds of facts that are needed, in the hope that investigators will be stimulated to supply them. As a stimulus to such efforts, the paper attempts to portray the depth of understanding of fundamental problems and the unification of a field to which such efforts surely would lead.

Before proceeding to the modern work on particular species of Ciliates, which began in 1937, a few comments on some earlier observations concerning the breeding systems of Ciliates will serve to introduce it. Breeding in Ciliates has long been known to take the form of conjugation. The animals unite in pairs, undergo meiosis, fertilize each other, and separate; then both exconjugants of each pair multiply by repeated fissions. Maupas

(1888, 1889) discovered that a declining food supply was usually a necessary, but not sufficient, condition to bring about conjugation. He further maintained that outbreeding was the rule, partly because he obtained fruitful conjugation only when he had put into his aquaria collections from different natural sources. When conjugation took the form of inbreeding, it led only to death of the exconjugants.

These observations and conclusions of Maupas were controverted by the studies of Jennings (1910) on *Paramecium caudatum* and *P. aurelia*. He cultivated in the same container strains of the same species that could be distinguished by the size of the animals. When conjugation occurred in these cultures, each strain appeared to conjugate only among its own members, even when both strains were conjugating at the same time. Moreover, working with isolated single individuals of a single strain, he found that conjugation could occur among individuals produced in the course of a few fissions from a single ancestor and that such exconjugants produced normal viable clones. A number of successive closely inbred generations were followed with the same result. On the basis of these observations, he denied the validity of Maupas' generalization and held that conjugation in *Paramecium* could occur only between individuals of the same strain and that within a strain any individual could probably mate and mate fruitfully with any other.

Early workers on breeding systems thus came to diametrically opposed observations and conclusions. According to Maupas, the Ciliates are obligatory outbreeders; according to Jennings, they are obligatory inbreeders. As will appear, the observations of both investigators were entirely correct. The differences are in the Ciliates themselves. Some are outbreeders; some are inbreeders. On the question of who can and does mate with whom, and with what results, we shall have much to say below.

A few later workers (De Garis, 1935a,b; Müller, 1932) claimed to have succeeded in obtaining crosses between different strains and even between different species of *Paramecium*. Although the methods employed by De Garis were not unobjectionable, the fact that diverse strains of these species can be crossed has been amply confirmed. Objections can also be raised against the evidence for species crosses, but this needs to be reinvestigated with the better methods now available. In any case, no viable hybrids between species were claimed by these workers and that, from our point of view, is perhaps the main thing.

The modern work on Ciliates began with the discovery of mating types in *P. aurelia* (Sonneborn, 1937). As originally used, the term mating types refers to physiological differences between individuals that mark them off into mating classes. True conjugation does not ordinarily occur between individuals of the same physiological class or mating type; it occurs between individuals of complementary physiological classes or mating types. This is the sense in which the term has come to be used by students of Protozoa and Fungi. But it has a somewhat different meaning in Euplotes, as will be explained later, and the possibility that conjugation can occur between individuals of the same mating type in certain clones of some species or in certain stages of the life cycle has not yet been completely excluded. These special cases will be dealt with as they arise.

Commonly, but not invariably, the mating type of an individual is inherited by its asexually produced descendants. This makes it possible to use clones rather than individual animals in the study of breeding systems. Such studies quickly exposed fundamental relations in *P. aurelia* (Sonneborn, 1938) and in *P. bursaria* (Jennings, 1938) that have subsequently been widely confirmed in sufficiently studied Ciliates in which mating types have been found. If many collections are made from different natural sources (ponds, lakes, streams, rivers, and the like) and a strain from each source is established in the laboratory, the general features of their breeding relations are ordinarily found to be those described in the following paragraphs.

Clones of the same strain are either of the same or of complementary mating types. When complementary, they conjugate promptly and abundantly with each other under appropriate conditions. When of the same mating type, they cannot be made to conjugate with each other. Every clone of every strain has a definitely assignable mating type. When mating types of any one strain are matched against the mating types of any other strain, two very different results may be obtained. On the one hand, the two strains may contain clones manifesting identical sets of mating types. When this is so, the same rules of mating apply to combinations of clones from different strains that apply to combinations of clones from the same strain: conjugation occurs if the clones combined are of complementary mating types, not if they are of the same mating type.

On the other hand, two strains may manifest quite different sets of mating types. Then no conjugation will occur when any clone of the one strain is combined with any clone of the other strain. Thus, conjugation does not result from mere difference of mating type, but from complementariness of mating type. There are multiple sets of complementary mating types within a single species. Some strains have one set and these interbreed freely. Other strains have another set, and these also interbreed freely. But strains with the first set of types are sexually isolated from strains with the second set of types. When many strains of a species are examined, a number of such groups of strains is discovered. As a rule, each is sexually isolated from all the others. The mating types of one set usually will have nothing to do with mating types of the other sets.

From these results, it at once became evident that *P. aurelia* and *P. bursaria* each consisted of a number of "biological species." Both Jennings and Sonneborn were well aware of this and considered the problem in their first papers of 1938, which reported the basic facts. However, they were then unable to describe these biological species so that they could be readily distinguished and identified. They therefore agreed, for the time being, to refrain from assigning specific names. Instead, they designated the sexually isolated groups of strains within a species as varieties, assigning to each variety a number. This usage has persisted and spread, to the annoyance of some geneticists and doubtless to the relief of protozoologists. In this paper, I shall, among other things, inquire exhaustively—in view of subsequent additions to knowledge—into the questions of whether the time has come to recognize the varieties as species and of whether they should be given species names. Meanwhile, I shall continue to refer to them as varieties and will reserve the term species for the species now recognized by taxonomists.

In the next eight sections, I shall deal with six genera of Ciliates that have been examined with respect to mating types and varieties: *Paramecium, Tetrahymena, Colpidium, Euplotes, Stylonychia,* and *Oxytricha.* We begin with *Paramecium,* the first genus to be studied in this way and the one about which most is known. Pertinent information is available on six species of *Paramecium*—a good deal on *P. aurelia, P. multimicronucleatum, P. bursaria* and *P. caudatum;* less on *P. calkinsi* and *P. trichium.* The first four will be dealt with in extenso and in order in the immediately following sections; the other two will be only briefly mentioned in the last section on Ciliates, along with the other lesser studied Ciliates (Stylonychia, Oxytricha, and Colpidium), after sections that take up work on the more fully known Ciliates, Tetrahymena, and Euplotes. The sections on Ciliates will be followed by a section on obligatory inbreeding and asexually reproducing Protozoa, chiefly Flagellates and Rhizopods. The last two sections set forth the major conclusions and give a summary of the whole paper.

Before proceeding to the four intensively studied species of *Paramecium,* it should be emphasized that the six species of *Paramecium* to be discussed are, with one possible exception presently to be mentioned, readily recognizable and sharply set off from one another by combinations of a few morphological features: mainly, body shape and the number and structure of the micronuclei. Three species have the familiar cigar-shaped body: *P. caudatum, P. aurelia,* and *P. multimicronucleatum. P. caudatum* has a single, compact, relatively large micronucleus. *P. aurelia* has two smaller micronuclei which are vesicular in structure. *P. multimicronucleatum* has the same kind[2] of micronuclei as *P. aurelia,* but their number varies from 2 to 7. As will appear, I doubt whether a clear distinction[2] can be made between *P. aurelia* and *P. multimicronucleatum.* Pending clarification of this difficulty, I shall treat the organisms of both "species" as if they were all *P. aurelia.* The other three species have bodies shaped like a Dutch wooden shoe viewed from the side: *P. bursaria*, *P. calkinsi*, and *P. trichium*. *P. calkinsi* has two vesicular micronuclei. *P. bursaria* has a single, relatively large, compact micronucleus and is unique in being the only species that is normally colored; it carries a symbiotic, unicellular, green Alga. *P. trichium* also has a single, compact micronucleus; the possibility that it is related to *P. bursaria* in the same way as *P. aurelia* is related to *P. multimicronucleatum* needs to be investigated.

The sizes of the animals are also said to be characteristic for each of these six species. The figures given by Wichterman (1953) are 70 to 90 μ for *P. trichium*, 85 to 150 μ for *P. bursaria*, 120 to 170 μ for *P. aurelia*, 120 μ for *P. calkinsi*, 170 to 290 μ for *P. caudatum*, and 180 to 310 μ for *P. multimicronucleatum*.[2] Although these figures are the best available, they are not entirely satisfactory. Size varies greatly with nutritive conditions, stages of the fission cycle, and stages of the life cycle. No comparative study of the six species which properly takes such variables into consideration is available. On the whole, however, it seems reasonably certain that on the average *P. trichium* is the smallest species, that *P. caudatum* and *P. multimicronucleatum* are the largest, and that the others are intermediate.

## *Paramecium aurelia:* Specificity of Mating Types; Isolation of Varieties

In this section will be discussed not only the organisms that ordinarily go by the name of *P. aurelia*, but also those which are designated as *P. multimicronucleatum* and intermediate forms[2] which I have recently found. It is difficult if not impossible to draw a sharp line where one species ends and the other begins (Sonneborn and Dippell, 1956).[2] The older name, *P. aurelia*, will therefore be tentatively used for all these organisms.

This section begins with an enumeration of the varieties of *P. aurelia* and an account of the primary basis of their recognition, mating type specificity. It proceeds to a consideration of sexual and genetic isolation of the varieties, and then to their geographical distribution. Lastly is discussed the question of whether each variety may be considered to have a potentially common gene pool.

The next section enumerates the differences that distinguish the varieties and inquires whether these differences suffice, for practical purposes, in the recognition and identification of the varieties. At this point, the question of whether the varieties should be given specific names is considered. Then follows a section that deals with evolution in *P. aurelia* at various levels and attempts to relate the biological differences among groups of varieties, among varieties of a group, and among strains of a variety to the systems of breeding, the various degrees of inbreeding and outbreeding. This key section ends with a consideration of the relation of systems of breeding to the species problem.

*Mating Types and Varieties.* The primary basis for recognizing the existence of distinct varieties in Ciliates is the specificity of the mating types, i.e., the specific selectivity of the breeding relations. On this basis, 16 varieties of *P. aurelia* can now be recognized. Varieties 1 to 7 were discovered by Sonneborn (1938, 1942b), variety 8 by Sonneborn and Dippell (1946), variety 9 by Beale and Schneller (1954), varieties 10 to 15 by Sonneborn (1956a; unpubl.). What is here referred to as variety 16 was described under the name of *P. multimicronucleatum* by Giese (1941).[3]

The number of mating types per variety is remarkably uniform. As a rule, there are two and only two interbreeding mating types in a variety. Until recently, only one type was known in variety 7; but only one strain of that variety had been found. Sonneborn (1956a) found the missing mating type, along with the one previously known, in each of several new strains of variety 7. Only variety 16 remains as the exception to the rule. According to Giese (1941), it has four mating types, each of which can mate with the other three. Were it not for this unique feature, there would be reason to suppose that Giese's variety is identical with Sonneborn's variety 15.[3]

Mating relations between varieties follow a usual rule, to which

TABLE I. The System of Sexual Reactions among Mating Types of Those Varieties of *P. aurelia* in Which Intervarietal Reactions Occur[a]

| Variety | Mating type | 1 | | 3 | | 4 | | 5 | | 7 | | 8 | | 10 | | 14 | |
|---|---|---|---|---|---|---|---|---|---|---|---|---|---|---|---|---|---|
| | | I | II | V | VI | VII | VIII | IX | X | XIII | XIV | XV | XVI | XIX | XX | XXVII | XXVIII |
| 1 | I | − | +++ | − | − | − | − | − | ++ | ++ | − | ± | ± | − | − | − | − |
| 1 | II | C | − | + | − | − | − | ++ | − | ++ | − | ± | ++ | − | − | − | − |
| 3 | V | 0 | C | − | +++ | − | − | − | − | ± | − | − | − | − | − | − | − |
| 3 | VI | 0 | 0 | C | − | − | − | − | − | − | − | − | − | − | − | − | − |
| 4 | VII | 0 | 0 | 0 | 0 | − | +++ | − | − | − | − | − | +++ | + | − | + | − |
| 4 | VIII | 0 | 0 | 0 | 0 | C | − | − | − | − | − | ++ | − | − | − | − | − |
| 5 | IX | 0 | C | 0 | 0 | 0 | 0 | − | +++ | − | − | − | − | − | − | − | − |
| 5 | X | C | 0 | 0 | 0 | 0 | 0 | C | − | ± | − | − | − | − | − | − | − |
| 7 | XIII | 0 | 0 | 0 | 0 | 0 | 0 | 0 | 0 | − | +++ | − | ± | − | − | − | − |
| 7 | XIV | 0 | 0 | 0 | 0 | 0 | 0 | 0 | 0 | C | − | − | − | − | − | − | − |
| 8 | XV | 0 | 0 | 0 | 0 | 0 | C | 0 | 0 | 0 | 0 | − | +++ | − | − | − | − |
| 8 | XVI | 0 | 0 | 0 | 0 | C | 0 | 0 | 0 | 0 | 0 | C | − | + | − | + | ? |
| 10 | XIX | 0 | 0 | 0 | 0 | 0 | 0 | 0 | 0 | 0 | 0 | 0 | 0 | − | +++ | − | − |
| 10 | XX | 0 | 0 | 0 | 0 | 0 | 0 | 0 | 0 | 0 | 0 | 0 | 0 | C | − | − | − |
| 14 | XXVII | 0 | 0 | 0 | 0 | 0 | 0 | 0 | 0 | 0 | 0 | 0 | 0 | 0 | 0 | − | +++ |
| 14 | XXVIII | 0 | 0 | 0 | 0 | 0 | 0 | 0 | 0 | 0 | 0 | 0 | 0 | 0 | 0 | C | − |

[a] Symbols above the diagonal refer only to occurrence of mating reactions (adhesion, agglutination); symbols below the diagonal refer to the occurrence of complete conjugation. Varieties 10 and 14 are newly discovered and still incompletely studied: their reactions may be more extensive than now known. +++ maximal reaction. ++ reduced mating reaction. + weak mating reaction. ± barely detectible reaction. − no mating reaction. C conjugants formed. 0 no conjugants formed.

there are, however, a number of exceptions. The rule is that two mating types belonging to different varieties do not conjugate with each other. Indeed, they do not even react to each other in any observed way. Exceptions to this rule are shown only by certain combinations of mating types belonging to the eight varieties 1, 3, 4, 5, 7, 8, 10, and 14, as shown in Table I. These exceptions are of two kinds. First, certain combinations of mating types of different varieties react sexually to each other and form conjugating pairs. Second, other combinations give a weaker sexual reaction, adhering to each other briefly, but they never proceed to form conjugating pairs. These intervarietal reactions are often difficult to obtain except when the cultures are in optimal condition for conjugation. It is therefore possible that some combinations which have never been observed to react might do so and that others which have been seen to react only weakly might yield some conjugant pairs.

The system of intervarietal mating reactions leads clearly to the conclusion that, at least for the eight varieties listed in Table I, the two mating types of one variety are related to the two in other varieties. In the two pairs of varieties 1 and 5, and 4 and 8, the odd-numbered mating type of each variety mates with the even-numbered mating type of the other: both I and IX mate with II and with X; and both VII and XV mate with VIII and with XVI. Hence, similarities exist between mating types I and IX, between II and X, between VII and XV, and between VIII and XVI. Types V and VI can be brought into the system of similarities because V mates with both II and VI; and II mates with both I and V. Hence I is similar to V and II is similar to VI. The groups of similarities are enlarged by the facts that V mates with II, VI, and XVI; and XVI mates with V, VII, and XV. Hence V, VII, and XV form a similar set; and II, VI, and XVI form another similar set. By adding the other reactions in Table I, it can easily be shown that the odd-numbered mating types in these eight varieties form one similar group which may be designated the minus group; while the even-numbered mating types form another similar group, which may be called the plus group (Sonneborn and Dippell, 1946). Thus, each of these varieties has a plus and a minus mating type. The plus types of the different varieties are comparable, and the minus types of the different varieties are also comparable. Extension of this system to the other eight varieties which are not listed in Table I cannot be done on the basis of the same sort of evidence because these other varieties have never been observed to yield intervarietal mating reactions. General considerations and other less direct evidences, however, make it seem likely that the mating types of those varieties which have only two mating types (i.e., all except variety 16)[3] also are in each case plus and minus in the same sense.

In spite of the system of cross reactions and of intervarietal mating, the uniqueness of each mating type and therefore of each of the eight varieties in the table cannot be doubted. This is shown in two ways by the mating reactions themselves. First, the mating reaction is less intense when the plus type of one variety is mixed with the minus type of another. Even in mixtures involving the varieties 4 and 8, which show the strongest intervarietal mating reactions, one combination of odd and even types never yields as high a percentage of pairing as a comparable combination within a variety. The other combination of odd and even can yield as high a percentage of pairing as do the two types of one variety; but the conditions for such an optimal reaction seem to be stricter than those required within a variety. Hence, as a rule this intervarietal reaction is also somewhat reduced. All other intervarietal combinations yield considerably weaker reactions, usually very much weaker. Second, corresponding types of different varieties commonly differ in the intensities of their reactions with any type to which both react. Thus type XVI of variety 8 fully conjugates with type V of variety 3, but type VIII—the corresponding type of the closely related variety 4—does not even react with type V. Indeed, Table I shows that each mating type gives a set of reactions that is distinctly different from every other one. For example, type I of variety 1 reacts strongly with II, moderately with X, weakly with XVI, and not at all with VI, while the corresponding type V of variety 3 reacts weakly with II, strongly with VI, not at all with X, and moderately with XVI. The uniquely defined system of reactions of each mating type, including of course the total absence of reactions between mating types of certain different varieties, was the original, and remains the most reliable, basis for distinguishing the varieties.

*Isolation of the Varieties.* The preceding account has passed lightly over the important fact that conjugation *can* occur between certain different varieties. Since varieties 1, 3, 5, and 7 *can* interbreed, do they not have a common gene pool? Should they not all be grouped together as one biological species? Moreover, since variety 4 conjugates with variety 8 and the latter conjugates with variety 3, should not varieties 4 and 8 also be members of this species? These questions place squarely before us the problem of what the common gene pools in *P. aurelia* really are. To this problem, which includes the questions just asked, but is broader than they, we now turn.

First of all, there seems clearly to be no gene flow from certain varieties to certain other varieties. The seven varieties not listed in Table I have never been observed to give the slightest sexual reaction to animals of any variety other than their own. This may be due to inadequate study so far as the varieties which are newly discovered are concerned. However, others, like varieties 2 and 6, have been long known, and it is practically certain that neither of these reacts with the other or with any of the other long known first eight varieties; gene flow between each of these and the other six must be considered as nonexistent.

Secondly, even when the bar to interbreeding is let down, as among most of the varieties listed in Table I, other mechanisms interfere with gene flow among them. The only open question is whether the interference is completely effective or whether it is not quite completely effective. Crosses between varieties 4 and 8 almost always end in death after a few fissions (Sonneborn and Dippell, 1946; Melvin, 1949). Levine (1953) thoroughly studied the small proportion of F1 survivors. Although he could not prove their hybrid nature, about a dozen among over 300 F1 clones examined could have been true hybrids. Later generations, even by backcrosses, from these rare survivors were, however, as nonviable as the F1 generation. Thus, although a complete bar to gene flow was neither demonstrated nor controverted, introgression of genes from one variety into the other must be either very rare or nonexistent.

The other four possible crosses between varieties, 1 by 3, 1 by 5, 1 by 7, and 3 by 8, regularly yield a viable F1. The F2 generations and backcrosses prove, however, to be completely or nearly completely nonviable. Butzel (1953) found abnormalities even in the F1 of the cross between varieties 1 and 7. Like Levine, he was unable to prove that recombinants occur in the F2 or backcrosses. I have obtained a little evidence, not fully satisfactory, that the rare survivors in the F2 or later generations may sometimes be recombinants. Some clones behave as if they were the plus (or minus) mating type of *both* parental varieties, but a descendant of such a clone after a later fertilization may behave as if it had a mating type of only one of the parental varieties. Whether this is due to change of dominance relations in a heterozygote or to recombination is not clear. Further, descendants of the hybrids have been maintained for a year or more. They reproduced slowly, but mated avidly when given a chance. It seems unlikely that life could have been maintained so long without recurrent fertilizations, although it is possible, in view of some of our recent work (Sonneborn, unpubl.), that these may have survived by repeated regenerations of functional nuclei from the original hybrid macronucleus instead of developing macronuclei from the fertilization nucleus after each fertilization.

Thus, knowledge of the consequences of intervarietal crosses, in the five combinations in which they can occur, is still not sufficient to give a definite answer to the question of whether gene flow can occur among those varieties. On the other hand, the difficulties in analyzing the hybrids are due to the low viability of the later generations and to the abnormality of the survivors. This shows that, if genes flow between varieties, the flow can be at most but a trickle and in great danger of drying up. In short, the genes of every variety are virtually isolated from the genes of every other variety. Before the significance of this genetic isolation of the varieties can be properly evaluated, it is necessary to inquire how extended each variety is in space and whether potential gene flow exists within a variety.

*Distribution of the Varieties.* About 260 collections of *P. aurelia* from various parts of the world have been identified as to variety in my laboratory and in the laboratories of Giese (variety 16) and Beale. Most of these collections, about 200, were made in the United States. Of the rest, most were collected in Europe (chiefly by Beale), South America, and Japan. A few were collected in India (5), Australia (4), Madagascar (1), Puerto Rico (1), and Hawaii (1). Although the sampling is spotty, unbalanced, and grossly inadequate, a few important relations are clear (Beale, 1954; Sonneborn, 1956b).

At least three of these varieties, 1, 2, and 6, are probably distributed around the world. Varieties 1 and 2 have been found in Europe, North and South America, and Japan; variety 1 has also been found in Australia and Hawaii. Variety 6 has been found in the United States, in Puerto Rico, and in southern India. It is rare in the United States, except in Florida; but all five collections from the other two localities belong to this variety. In addition to these three, variety 4 is also of wide distribution, having been found in North and South America, Australia, and Japan, but not yet in Europe.

The other twelve varieties seem to be much less widely distributed. Variety 9 has been found only in Europe, variety 12 only in Madagascar, the other ten varieties only in the United States. These statements need some qualification. Varieties 12, 13, 15 and 16 are large paramecia; until recently we have paid little attention to paramecia of that size on the assumption that they were not *P. aurelia*.[2,3] So we have only one collection of large paramecia from outside the United States, and at present that is our sole strain of variety 12. It would be premature to suggest that 12 does not occur in the United States or that 13, 15, and 16 do not occur on other continents. However, even when these three varieties of large paramecia are excluded, this leaves the seven smaller varieties 3, 5, 7, 8, 10, 11, and 14 that have been found only in the United States. If other parts of the world were studied on a similar scale, it would not be surprising to find comparable numbers of varieties restricted to each continent. On the other hand, such a search might well reveal that some of the varieties found so far only in North America occur also elsewhere. Yet it is not to be assumed that we have already found all the varieties on this continent. Within the past year, three new small animal varieties (10, 11, and 14) have been found here; we have only begun to search for the large animal varieties; the far north (Canada and Alaska) has not been sampled at all; and we still have some large animal strains from the United States that we have not yet been able to identify or even to get to mate at all.

Without going into the quantitative details, which are given in Table II for the small animal varieties, it is clear that temperature is a major factor influencing the range of a variety on the continents where it occurs. Variety 6 has the highest temperature preference, being the only variety found in the hottest collecting

TABLE II. Distribution by Rough Temperature Zones of the Smaller Animal Varieties of *Paramecium aurelia*[a]

| Zone | Variety | | | | | | | | | | | Total number of collections |
|---|---|---|---|---|---|---|---|---|---|---|---|---|
| | 2 | 3 | 9 | 5 | 1 | 4 | 8 | 11 | 7 | 14 | 10 | 6 |
| Cold | 33 | 31 | 19 | 8 | 6 | 0 | 0 | 3 | 0 | 0 | 0 | 0 | 36 |
| Cool | 30 | 23 | 4 | 15 | 20 | 6 | 1 | 0 | 0 | 0 | 0 | 1 | 110 |
| Moderate | 26 | 3 | 0 | 10 | 45 | 16 | 0 | 0 | 0 | 0 | 0 | 0 | 31 |
| Warm | 12 | 0 | 0 | 0 | 18 | 18 | 29 | 6 | 6 | 4 | 2 | 4 | 49 |
| Hot | 0 | 0 | 0 | 0 | 0 | 0 | 0 | 0 | 0 | 0 | 0 | 100 | 5 |
| Total no. of collections | 59 | 37 | 11 | 23 | 47 | 21 | 15 | 4 | 3 | 2 | 1 | 8 | 231 |

[a] Varieties 12, 13, and 15 are the larger animal varieties; these are excluded because they have been little searched for. The numbers in the body of the table (except the bottom row and last column) are the percentages of all collections from a zone which belong to each variety. Cold zone: northernmost U.S.A., Norway, and Scotland; cool zone: England, northern France, Germany, and a strip across the U.S.A. just south of the most northern states; moderate zone: Japan (Honshu), southeast Australia, Switzerland, northern Italy, and the middle region of the U.S.A. (north-central California to Virginia and the Carolinas); warm zone: southernmost U.S.A. (New Mexico to Florida), Peru, north and central Chile; hot zone: Puerto Rico and southern India.

zone. Varieties 7, 8, 10, and 14 come next. With three exceptions, my strains of these varieties come from Florida; the exceptions are from Mississippi (variety 14), New Mexico (variety 8), and Maryland (variety 8). Variety 4 ranges from cool to warm climates, but it is absent from the coldest and hottest collecting sources; it is equally prevalent in moderate and warm regions, rarer in cool regions. Variety 1 has been found in all zones except the hottest, but its peak frequency is in the moderate zone. Variety 2 has the same range as variety 1, but its frequency decreases with rising temperature. Varieties 3 and 5 have never been found either in warm or hot regions; variety 3 has its peak in the cold zone, variety 5 in the cool zone. Variety 9 has a distribution in Europe somewhat like that of variety 3 in North America.

Much less is known about the distribution of the larger animal varieties. That is why they were not included in Table II. Giese's (1941, 1957) survey of variety 16 is the most extensive. He obtained 16 collections of this variety from various regions of the United States extending from Oregon and New Hampshire in the north to Texas in the south. From this it would seem that variety 16 has a wide temperature range. I have now four collections of variety 15 from the state of Washington in the northwest to Florida in the southeast. It too appears to have a wide temperature range similar to that of variety 16.[3] Of variety 13 I have four collections, all from southern United States (New Mexico and Florida), suggesting a restricted warm climate localization. Only the one strain of variety 12 from Madagascar is known, but no other large paramecia outside North America have been examined.

So far as present knowledge goes, it would appear that some varieties, both small and large animal varieties of *P. aurelia*, have wide temperature ranges while others—again both small and large animal varieties—have narrow temperature ranges. Independently of this, of the small animal varieties, some are restricted to one continent and others are distributed around the world in the appropriate temperature zones. Whether this is also true of the large animal varieties remains unknown.

*Do the Strains of a Variety Form a Potentially Common Gene Pool?* The distribution of each variety over a greater or lesser area raises the question of whether it is composed of subdivisions between which genes cannot flow or whether each variety, no matter how widely distributed, constitutes a potentially common gene pool. Laboratory analysis shows that normal vigorous recombinants can always be obtained in the F2 generation after crossing any two normal strains, i.e., samples of different local populations, of the same variety. This happens regardless of whether the two strains come from different continents or from neighboring ponds or streams. It has indeed become standard laboratory practice to make use of the genic diversity of different strains by introducing certain genes of one strain into the genetic background of another strain through a series of selected backcrosses. There thus can be no doubt that the genes of one variety form a potentially common pool.

That is the major fact, but there are also two other related facts of considerable significance in this connection. First, when strains of the same variety are crossed, at least in some varieties, the F2 and the first backcrosses invariably yield some nonviable clones. The proportion of nonviable clones depends upon which strains have been crossed. In variety 4, the frequency of nonviable clones obtained in the autogamous F2 generation after crossing certain strains may be as low as 15%; it may be as high as 98% after crossing certain other strains; and various combinations of strains yield percentages that cover the range between these extremes (Sonneborn, unpublished; Dippell and Hanson, unpublished). The same strain crosses regularly give approximately the same results. The magnitude of the mortality seems to vary independently of the geographical proximity or remoteness of the source of strains; but this needs to be investigated more fully.

Certain strain crosses within a variety thus yield about the same percentage of F2 mortality as intervarietal crosses. One wonders therefore whether gene flow is possible between such extreme pairs of strains. Margolin (1956a,b) has obtained normal, viable recombinations between strains 32 and 172 of variety

4, which yield 85% mortality in the F2, and also between strains 172 and 51, which yield almost the highest F2 mortality observed, about 96%.

Thus, with respect to F2 mortality alone, there appears to be a virtually uninterrupted series of gradations from crosses between varieties down to crosses between different clones of the same strain. The former approach 100% F2 mortality; the latter usually show 0% F2 mortality, at least in some varieties. But the series is not continuous when other features are taken into account. The survivors in the F2 or backcrosses show no improvement in viability at subsequent autogamies or backcrosses following intervarietal crosses; but the mortality decreases to zero at the second autogamy or after several backcrosses following interstrain crosses within a variety. The survivors in the later generations are not by any means of normal vigor after intervarietal crosses; they are quite normal in vigor after interstrain crosses within a variety. All intervarietal crosses that can be made give essentially the same low survival in the F2 and backcrosses; it is impossible for the genes of two varieties to be recombined in normal vigorous combinations. Although different strains of the same variety sometimes yield as low F2 survival as do diverse varieties, their genes can always be recombined in such combinations. The sharp break between varieties is thus never matched by the break between strains of the same variety. Among the latter, gene flow is readily possible; among the former, it is nonexistent or almost so.

## *Paramecium aurelia:* Differences among Varieties and Problem of Identification

The preceding section showed that each of the 16 varieties of *P. aurelia* has a potentially common gene pool which is effectively cut off from the gene pool of every other variety. Each variety therefore qualifies as a species according to the modern biological species concept. Two questions now arise. (1) Should the term "variety" be replaced by "species" in referring to them henceforth? (2) Should each former variety be baptized with a different specific name, while the old "species" *P. aurelia* becomes defunct as a species name and is raised to subgeneric status? The first question could be answered in the affirmative without necessarily giving an affirmative answer to the second question. There might be good grounds for objecting to this mass baptism, in spite of concluding that the varieties are nevertheless species. To this extent the two questions are independent. On the other hand, if each variety should be given a specific name, then this answer to the second question automatically answers the first question. The second question is therefore considered first and is alone dealt with in this section.

The decision as to whether the varieties should be given specific names depends, in my opinion—and I believe there would be general agreement on this—upon the feasibility of identification of the varieties by means available to working biologists, means that do not present unreasonable demands. Hence, before a decision can be reached, the known differences among the varieties must be summarized in sufficient detail to make clear whether they could reasonably be utilized in the practical job of identification. At the same time other probable differences, not yet known or not sufficiently known, will be considered in order to answer the related question of whether other useful differences might be available for the same purpose.

At this point the reader must be prepared to abandon some common, but grossly erroneous, ideas about the Protozoa. Many of them, including *Paramecium* are far from simple, either in structure or in life processes. Nor are all those which go by a single name pretty much alike. On the contrary, when one attends closely to those going under a single name, such as *P. aurelia*, studying in various ways samples of populations collected from far and wide around the world, one marvels at the complexity and variety of their lives. This section, which portrays this situation, will doubtless overwhelm the reader in much the same way as the investigator who devotes his life to studying these creatures is overwhelmed by their seemingly endless diversities which he slowly and with great labor discovers. So, reader, you can scarcely be more overwhelmed than I am. Indeed, not so much so. For I know, as you cannot, how pitifully small a fraction we have learned about the complexity, variety and marvels of this "simple, primitive, stereotyped animal." Yet there is no escaping the task of trying to comprehend the situation if we are to attempt to reduce it to significant order in relation to species problems. I shall therefore take up the various differences among the varieties in this section and shall try to reduce them to significant order in the next section.

*Mating Type Specificity.* The definite and sharp distinctions among the mating types uniquely mark off each variety, as described in the preceding section. No further comment is needed here. Toward the end of this section, the usefulness of these distinctions in identification will be considered.

*Distribution.* The distributions of the varieties in the major land masses and according to temperature were stated in the preceding section. The usefulness of this knowledge in relation to identification is limited of course by the inadequacy of the sampling which underlies our present knowledge. With due allowance for this, the source of a culture still has suggestive value in identification. For example, a sample from northern Europe in all probability would be found to belong to variety 1, 2, or 9, and there is no use considering other possibilities until these have been tested. Such considerations save much time in identification, but seldom if ever could a strain be reliably identified solely on the basis of its natural source.

*Temperature in Relation to Survival and Fission Rate.* The characteristic distribution of each variety with respect to temperature suggests the existence of varietal differences in temperature tolerance. Very little is known about this. Sonneborn and Dippell (unpublished data of 1942) found that strain P of variety 1 from Maryland grew well at 36.4°, but the strains of the other four varieties (2, 3, 4, and 7) tested did not survive at this temperature very long. Of the latter, variety 3 (strain 58 from Indiana) and variety 7 (strain 40 from Florida) grew well at 34°; but variety 2 (strain 50 from Oregon) and variety 4 (strains 29 from Maryland and 51 from Indiana) did not survive at 34°. These results indicate that predictions based on geographical distribution cannot be relied upon. For varieties 3 and 7 showed a similar temperature tolerance, although the latter has been found only in Florida whereas the former never has been found in warm climates and is most abundant in cold climates. However, this particular strain of variety 3 came from near the southern limit of its range, and there may be strain differences in temperature tolerance within a variety. Indeed, some strains of variety 4 grow well at 35° (Margolin, 1956b), while others die at that temperature. Preer (unpublished) has even found temperature mutants within a strain of variety 2. The mutant cannot survive at 31° while the wild type can. Of the tolerance for low temperatures even less is known. We have some indications that variety 4 strains become abnormal if long grown at 10°. There is much need for a systematic study of temperature tolerance. It may prove to be diagnostic for varieties with limited natural range of climates, but more variable in varieties with wide natural temperature distributions.

The relation of temperature to fission rate may be studied in several ways: comparing different varieties at a standard temperature, comparing the maximal fission rate at whatever temperature yields it, comparing the temperatures which yield the maximal fission rate, and comparing the entire curve relating temperature to fission rate over the whole viable range of temperatures. The latter method is the most revealing. Sonneborn and Dippell (unpublished data of 1942) did this for the representatives of varieties 1, 2, 3, 4, and 7 mentioned above. Up to about 23°, they found only minor differences in fission rate among these strains; but at higher temperatures the differences became strongly accentuated and each variety gave a distinctive curve. The question of the existence of strain differences within a variety was not satisfactorily analyzed in their work. For the strains examined, the maximal fission rates attained were in part related to the maximum temperature tolerated, for fission rate increases with temperature almost up to that point. But even at lower temperatures differences appeared. Thus both at 28.5° and 30.5°, the order of the varieties in fission rates, from high to low, was 4, 1, 7, 3, 2, covering a range (at 28.5°) from 5.8 down to 4.2 fissions per day. Varietal differences in fission rate thus unquestionably exist, but much more needs to be known before this could become of much value in identification.

*Morphological, Anatomical, and Cytological Differences.* The most obvious morphological features are of course size and shape. Unfortunately, however, these vary with cultural conditions, stage of the division cycle, stage of the life cycle from one fertilization to the next, and so on. Nevertheless, under comparable conditions there are some striking differences, particularly in length and breadth, among the varieties. Some unpublished observations by Sonneborn and Dippell, made in 1942, may be compared with unpublished observations by Elizabeth Powelson made 14 years later on different strains of the same varieties. Only the mean lengths of overfed vegetative animals are given here:

|  | *Variety 1* | *Variety 2* | *Variety 4* |
|---|---|---|---|
| Sonneborn and Dippell (1942) | 139 $\mu$ | 145 $\mu$ | 112 $\mu$ |
| Powelson (1956) | 131 $\mu$ | 149 $\mu$ | 112 $\mu$ |

This remarkable agreement inspires much confidence in the existence of very real size differences among the varieties, so that others which were made at only one time may be added: Variety 3, 123 $\mu$; Variety 7, 140 $\mu$; Variety 12, 204 $\mu$. The measurements on variety 12 are by Powelson, those on varieties 3 and 7 by Sonneborn and Dippell. Measurements on other varieties are not available, but long familiarity with the material permits me to estimate that the characteristic size in varieties 8, 10, and 14 is essentially the same as in variety 4, i.e., about 112 $\mu$. Likewise, the animals of varieties 13 and 15 and, judging by Giese's report, 16, are very large, surely 170 $\mu$ or more. The other varieties are intermediate and roughly in the range 125 to 150 $\mu$. This leaves open the question of strain differences within a variety. Such differences doubtless exist, at least in some varieties, but on the whole they probably never blur the distinctness of the three major varietal size classes listed above. The latter provide an easy means of at once placing an unknown strain in one of three groups of varieties. No measurements are necessary to do that, for the differences are considerable to a practiced eye.

A cytological feature of taxonomic importance is the number of micronuclei. As mentioned earlier, the number is supposed to be two in P. aurelia and two to seven, but usually four, in P. multimicronucleatum. Because we include both here as P. aurelia, varietal differences in this respect might be expected.[2] A serious difficulty, however, is that much variation in micronuclear number can occur in a single strain, as has been repeatedly observed by a number of investigators. Some of this variation seems to be correlated with the age of the clone under observation. Observations of number of micronuclei in clones of unknown age are insecure grounds for identification. Thus, the clones of large paramecia which we have thus far obtained had mostly two or no micronuclei; but further study, as set forth below, showed clearly that young clones of the same variety normally have four micronuclei. Experience of this sort leads to the conclusion that micronuclear number in this material becomes a reliable diagnostic trait only when used for young clones and only then when a considerable sample of clones is available to indicate the usual condition.

The young stage, which is in my experience most critical, is the stage shortly before the first fission after fertilization. At this time there are present the differentiated micronuclei and macronuclear anlagen that develop from the products of the first few divisions of the syncaryon. In varieties 13 and 15 the normal condition is to have four micronuclei and four macronuclear anlagen. Giese does not give the number for variety 16, but from other information he provides I conclude that this variety will also be found normally to develop four nuclei of each sort.[3] All other varieties normally form two nuclei of each sort. The persistent qualification with the word "normally" is necessary, for variations occur in all varieties. But the normal condition is found in the overwhelming majority of animals at this stage.

Other cytological and anatomical features are not as yet sufficiently known to be discussed. Among these, the greatest need is for study of the silver line system and chromosome number and morphology. The former is under investigation by Powelson

(1956). At present, only variety 4 has been carefully studied in respect to chromosome variation (Dippell, 1954). I have reason to suspect that very different results might be obtained with varieties 15 and 16, and perhaps certain others.

In sum, three groups of varieties may be readily distinguished on the basis of size, and the largest varieties may be further subdivided on the basis of the number of nuclei just after fertilization. Variety 12 regularly has two micronuclei and two macronuclear anlagen at this stage whereas the other large animal varieties regularly have four of each. All the smaller varieties regularly have two of each.[2,3]

*Killers and Sensitives.* Some strains of *P. aurelia* liberate a poison to which they are resistant, but which kills other strains of paramecia. Only certain varieties of *P. aurelia* include strains of killers. They have been reported in variety 2 (Sonneborn, 1939) and variety 4 (Sonneborn, 1943). A different kind of killer, which liberates no poison but kills by prolonged contact during mating, was discovered in variety 8 (Siegel, 1954). I have recently found a killer in variety 8, which kills without mating, and also killers in variety 6. Thus killers of one kind or another are now known in varieties 2, 4, 6, and 8, but not in any other variety.[4]

As a rule, all nonkiller strains of all varieties are sensitive to all killers. Only strains that can conjugate with mate-killers can be tested for sensitivity, but this restriction does not apply to other killers. Sensitive strains of variety 2 alone show heightened resistance to killers; the degree of resistance varies from strain to strain, but it is sometimes very great (Preer, 1950). The significance of this variation in resistance is somewhat complicated by Austin's (1951) discovery that variations in resistance occur within a single sensitive strain of varieties 4 and 8 in correlation with the serotype they happen to be expressing at the time (see under "Serotypes"). Nevertheless, the commonly greater resistance of variety 2 may well be independent of this.

For purposes of identifying strains, the preceding information is of some value. If one has a killer strain, it probably belongs to variety 2, 4, 6, or 8; if it is a mate-killer, probably it belongs to variety 8. If a nonkiller strain is extraordinarily resistant to a known killer, it almost surely belongs to variety 2.

*Serotypes.* As is well known, serologic techniques have been used in the study of evolutionary relationships, and they are currently used in routine identification of certain microorganisms. The possibility of using such techniques in identification of the varieties of *P. aurelia* therefore needs to be considered, particularly since serological traits have been much studied in several of the varieties. These studies have exposed a serologic situation which is far from simple and not yet completely known; but enough is known to indicate that serologic identification of the varieties of *P. aurelia* may become possible.

Present knowledge is confined almost exclusively to antigens which are detected by immobilization of the paramecia in the presence of appropriate antisera obtained from injected rabbits. Any one individual seems to manifest, as an almost invariable rule, one and only one immobilization antigen at a time. This antigen defines its serotype in the sense in which that term is here used.

The serotype of an individual commonly persists among its progeny; but, as in phase variation of bacteria, changes of serotype may occur within a single line of descent. However, there are more phases or serotypes expressible in a line of *P. aurelia* than in a line of bacteria. As many as 14 different serotypes have been brought to expression in different individuals of a single line of descent (Margolin, 1956b), and nearly as many serotypes have been found in other lines by others. Change of serotype occurs without genetic recombination or gene mutation. The whole array of serotypes characteristic of a line of descent appears in individuals of identical—and even entirely homozygous—genotype. The changes of serotype are reversible and can be experimentally controlled and directed by appropriate manipulation of environmental conditions. In order to discover the array of serotypes producible by any line or strain, prolonged study with a wide variety of cultural conditions is required.

Different strains of the same variety show similarities in their systems of serotypes. If a serotype, A, is found in one strain, the homologous anti-A serum is found often to be capable of immobilizing one of the serotypes in each of many other strains of the same variety. These are therefore all designated as serotype A. In like manner, one finds many strains have serotype B, C, and so on. Genetic analysis reveals another kind of similarity between strains of the same variety. The serologic specificity of each serotype is determined by a single gene. Different strains may carry different alleles. The whole array of specificities controlled by alleles at one locus is given the same serotype designation regardless of whether cross reactions are detectable serologically. There are thus genetic as well as serologic similarities between serotypes of different strains.

There are also differences between the serotypes of different strains of the same variety. These are of three main sorts. First, corresponding serotypes of the different strains, although designated by the same letter, may differ more or less in serologic specificity. For example, the immobilization titer of an anti-D serum against the D serotype of the homologous strain may be greater than against the D serotype of another strain. This is not always true; but when it is, the differences in titer vary in extent from very small to very great. In the extreme case, there may be no cross reaction at all, and the inference as to correspondence then depends upon genetic considerations. Second, the cultural conditions required to bring a particular serotype to expression may be alike or different for different strains. For example, serotype B may require for its expression cultivation at high temperature in some strains, cultivation at low temperature in others. Third, some strains may be capable of expressing a serotype which other strains cannot be made to express at all under any of the many conditions of culture employed. This means then that the discovered array of serotypes may differ somewhat from strain to strain.

As might be expected, different varieties also show similarities and differences in their systems of serotypes, and the degree of resemblance and differences runs parallel to the closeness or remoteness of evolutionary relationship as inferred from other lines of evidence. These independent evidences of evolutionary relationship will be discussed later. Figure 1 anticipates that discussion by diagramming the inferred relationship of the six varieties that have been studied serologically: varieties 1, 2, 3, 4, 8, and 9. The serologic results are summarized in Beale (1954), except for the subsequent publications by Finger (1956; 1957a, b) on variety 2; by Melechen (1955) on variety 3; by Austin et

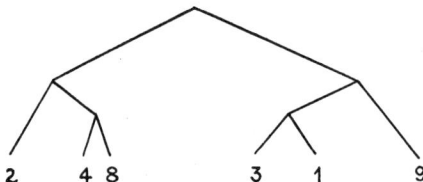

Fig. 1. Inferred evolutionary relationships of varieties 1, 2, 3, 4, 8, and 9 of *Paramecium aurelia*, the only varieties that have been studied serologically.

al. (1956), Margolin (1956a,b), and Skaar (1956) on variety 4; and by Pringle (1956) on variety 9; and unpublished work of Preer on variety 2 and Sonneborn and associates on varieties 4 and 8.

The most closely related varieties are 4 and 8 and their serotype systems show the greatest resemblance. For the most part, each serotype found in one variety is matched by a serologically recognizable corresponding serotype in the other variety; but a few serotypes have thus far been found only in one or only in the other variety. Further, the corresponding serotypes are seldom if ever serologically identical although they commonly cross-react. The cross reactions are usually weaker than those involving corresponding serotypes of different strains of the same variety.

The differences between these two varieties and variety 2 are much more marked. Several serotypes (A, C, E, F, G, and H) have been recognized in variety 2 which correspond to serotypes of varieties 4 and 8, but they show marked differences in specificity. There are, moreover, some general resemblances in the system of serotypes, which, as will appear, distinguish these three varieties from the other three more remotely related ones (i.e., varieties 1, 3, and 9). First, in varieties 2, 4, and 8 each strain usually possesses several alternative serotypes that can be stably maintained under appropriate conditions. Second, so far as now known, corresponding serotypes of different strains of the same variety seldom if ever totally fail to cross-react to each other's homologous antiserum. Third, in any one strain, usually more than one stable serotype may be expressed in different sublines under the same conditions; that is, there is a strong cytoplasmic element controlling the persistence of a given serotype. Fourth, corresponding serotypes of different strains may require very different conditions for their expression.

In all four of these respects the varieties 1, 3, and 9 of the other group seem to be like each other but different from varieties 2, 4, and 8. The phenomena have been most fully worked out in variety 1 (Beale, 1954). Although each strain of variety 1 has a considerable array of potentially producible serotypes, only three serotypes, D, G, and S, are very stable, and these are regularly found in all natural strains. The specificities of any one of these serotypes in different strains may be much alike, but often the differences in specificity are so great that no cross reactions occur and correspondence is detectable only by genetic and physiological means. Further, the expression of these three serotypes is in a uniform way in all strains closely dependent upon temperature. Thus, in any strain serotype S requires for its expression lower temperature, and serotype D higher temperature, than serotype G. This results regularly in the expression of only one stable serotype in any one strain under given conditions and evidence of a cytoplasmic element in the persistence of a serotype is as a rule inconspicuous or lacking. In so far as evidence is available, these features of the serotype system appear also to be operating in varieties 3 and 9.

The differences among varieties 1, 3, and 9 concern mainly the array of serotypes in each and the specificities of corresponding serotypes when such are found. These differences are comparable to those found among varieties 2, 4, and 8. The two varieties 1 and 3 (which are more closely related than either is to variety 9, but which are not as closely related as are varieties 4 and 8) show greater resemblances than either does to variety 9. Further, their serologic resemblances do not seem to be as close as those which exist between varieties 4 and 8.

In spite of the profound differences between the two groups of varieties, some serologic similarities are still detectable. Thus, serotypes D and G are found in both groups. Moreover the D serotype of a strain of variety 1 may be more like the D of a strain of variety 4 than it is like the D of another strain of variety 1. These serotypes may be ancient and widely distributed. Preer (unpublished) has found serotype G even in the distinctly different species P. caudatum.

The body of knowledge now available on the serotypes of P. aurelia, although limited to only 6 of the 16 varieties, is already of some practical value in identification. If one has a battery of antisera against the various major known D, G, and S serotypes of variety 1 and grows an unidentified strain at low, intermediate, and high temperatures, the strain could be identified with reasonable confidence as belonging to variety 1 if the animals are immobilized by one of the D antisera at high temperature, one of the G antisera at intermediate temperature, and one of the S sera at low temperature. Even a less complete series of responses would be highly indicative. In like manner identifications of strains belonging to varieties 3 and 9 could probably be made with appropriate antisera; but at present work on these varieties is less extensive and doubtless a fuller battery of antisera than now available would be required. The procedure with varieties 2, 4, and 8 would have to be somewhat different because of the absence of the regular temperature relations and, consequently, the great strain differences in serotypes which are stable at any particular temperature. One would have to use all available antisera against the whole array of serotypes known in the variety. In variety 4, there are 17 such serotypes and a number of variants of some of them. Moreover, new serotypes are continually being found as more strains are studied. Because of this great variability and the incompleteness of present knowledge, the probability of success in attempts at serologic identification is perhaps less than in variety 1. However, I have already repeatedly proved the usefulness of this serological method of distinguishing closely related varieties, such as 4 and 8, which are the most difficult to distinguish in the usual way by their mating specificities, and the serologic identification has never failed to be confirmed by later mating tests. Beale (unpublished) has had similar success in serologically distinguishing varieties 1 and 9.

There can be little doubt that the practical usefulness of the serological method of identification of varieties will increase as our knowledge of the serotypes is extended to more strains and more varieties, thereby providing more complete sets of antisera for use in this task. In principle, it should become possible eventually to identify the varieties of P. aurelia in this way with as much assurance as the so-called species or serotypes of *Salmonella* are now identified. But no one should underestimate the magnitude of the task, even if all of the necessary antisera were available. Their number would be very great, perhaps about 20 for each of the 16 varieties, or a total of over 300.

*Mating Type Inheritance.* Two groups of varieties, A and B, are distinguished by their systems of mating type determination and inheritance. Group A includes varieties 1, 3, 5, 9, and 11 and also probably the varieties 7 and 15 which have been little studied. Group B includes varieties 2, 4, 6, and 8 and also probably the little studied varieties 10, 12, and 14. Nothing is known about mating type inheritance in variety 16.[3] Variety 13, as will appear later, cannot yet be assigned to either group A or B.

The outstanding features of the group B system are (1) the conspicuous role played by the cytoplasm in the determination of mating type and (2) the consequent rarity of change of mating type after fertilization. A fertilized animal usually gives rise to a clone of the same mating type as the parent. The group A system shows no conspicuous effect of the cytoplasm on the determination of mating type. Consequently there is no significant correlation between the mating type of parent and offspring arising after sexual processes. Instead of being a clonal trait, as in group B, mating type is a caryonidal trait in group A; that is, each new macronucleus that arises in a fertilized animal is independently determined for mating type. Hence, the unit of mating type inheritance is the caryonide, the descendants that carry products of division of one original macronucleus. Since usually two new macronuclei arise in a fertilized animal and one passes to each of the two products of the first fission, a clone includes two caryonides that arise at the first fission after fertilization. Because of the independent determination of the macronuclei, the two caryonides from a fertilized animal often are different in mating type. In group B this also happens sometimes, but so rarely as to be inconspicuous. For a detailed account of the similarities and differences between the group A and group B systems, as well as of a number of exceptional phenomena that are omitted here, the reader is referred to a review by Nanney (1954). Further details and a more general interpretation will also be given below in other connections.

As a rule, the differences between the group A and group B methods of determination and inheritance of mating type are so striking that an unidentified strain could be assigned to one or the other group on the basis of a study of half a dozen or so pairs of conjugants and their vegetative descendants. It would be required only to isolate the pairs, then to isolate the exconjugants, to cultivate separately the products of the first (better, the first two) fissions, and finally to ascertain the mating types of the various cultures. If the cultures from the two exconjugants of a pair are usually of different mating types, the strain belongs to a variety of group B. If the two caryonides from a single exconjugant often differ in mating type, or if the same mating type frequently occurs among the descendants of both exconjugants of a pair, the strain probably belongs to a variety of group A. There is, however, one serious source of possible misinterpretation. In group B, if the two conjugants of a pair exchange much cytoplasm, they may produce clones that are alike in mating type. This difficulty can be avoided by watching the conjugating pairs as they conclude conjugation and eliminating those that remain long united posteriorly after separating at the anterior ends. Only those that remain long united posteriorly exchange enough cytoplasm to affect the inheritance of mating type.

*Life Cycle Differences.* Evidence for the existence of a life cycle in Ciliates was first obtained in abundance by Maupas (1888, 1889) and was subsequently confirmed by many investigators. But it was immediately challenged on theoretical grounds by Weismann and eventually on experimental grounds by many investigators. The chief objection grew out of the experimental prolongation of the cycle as cultural conditions were improved and the maintenance of some Ciliates in culture for very long periods without indications of deterioration or ultimate extinction. From my own observations on *P. aurelia* (Sonneborn, 1954a) and my analysis of the reports of others on other Ciliates, I have concluded that there is indeed a life cycle in these organisms but that it may vary greatly in extent and in its manifestations in different species and varieties. These conclusions are well illustrated by the diversities in life cycle among the varieties of *P. aurelia*. The same fundamental pattern may be seen in the life cycle of all of them, but there is great variation in the duration of the various stages and, in some respects, in the processes that mark them.

The life cycle, as Maupas discovered, begins with fertilization and proceeds successively with immaturity, maturity, senility, and death. Each of these stages will be taken up in order, and the differences among the varieties will be pointed out. At the start, it is necessary to emphasize that the course of the life cycle depends upon the conditions of culture, and especially upon the abundance or scarcity of available food. What happens at a particular stage may be decisively determined by this and by other conditions which will be mentioned later. In order to follow the entire potential cycle, it is necessary to maintain an excess of food at all times, and this is best accomplished by the laborious method of re-isolating single animals daily. On the other hand, since depletion of food is so decisive and since this is doubtless commonly recurrent in nature, to get a more complete picture of the cycle it is also necessary to starve subcultures taken frequently from the overfed daily re-isolation lines and to note their responses to starvation. Complete studies of this sort have been made on but few varieties, yet partial information is available on all of the varieties. To the extent that is now possible, all this information will be utilized in the following comparison of the varieties of *P. aurelia*.

1. Fertilization. Varietal differences with respect to fertilization are of several sorts: (*a*) the kinds of fertilization processes that occur; (*b*) the tempo of the processes associated with fertilization; (*c*) the environmental conditions required for the occurrence of conjugation; (*d*) the number of interbreeding mating types; and (*e*) the number of micronuclei and macronuclear anlagen formed from products of the fertilization nucleus. In previous sections, (*d*) and (*e*) have been set forth and will not be further mentioned.

(*a*) The kinds of fertilization processes. Basically there are only two kinds of fertilization processes, conjugation and autogamy. Conjugation involves union of the animals in pairs. Each mate undergoes meiosis and one of its reduced nuclei divides to form a male and female gamete nucleus. The male nucleus of each conjugant migrates into the mate with whose female nucleus it unites to form a diploid fertilization nucleus. Autogamy involves exactly the same nuclear processes as conjugation except that cross fertilization is impossible because the animals are not united in pairs. Instead, the fertilization nucleus is formed by union of the male and female gamete nuclei of one and the same individual. Both conjugation and autogamy can initiate a new life cycle, but an appreciable period of immaturity never occurs after autogamy. Conjugation occurs in all varieties. Autogamy has been found in all varieties except varieties 13, 15, and 16. It may yet be discovered in variety 13;[6] but it probably will not be found to occur as a regular normal event in varieties 15 and 16.

(*b*) The tempo of the processes associated with fertilization. The time that conjugants remain united, i.e., approximately until the time of the first division of the fertilization nucleus, varies inversely with temperature. At any one temperature, the time varies with the variety. Thus, at 22°, conjugants remain united for about 7 to 8 hours in variety 4, about 12 hours in varieties 13 and 15. Correspondingly, the time from separation of the conjugants until the first fission is about 15 hours in variety 4, 24 to 25 hours in varieties 13 and 15. In both cases the interval is much longer than an ordinary interfission interval. Also prolonged, but not so much so, is the time from the first to the second fission. This is about 10 hours in variety 4, about 16 to 17 hours in varieties 13 and 15. These are probably the extreme differences among the varieties. Similar but smaller differences may well exist among certain other varieties. Information is at present lacking on that point. In general the timing of autogamy seems to be about the same as for the comparable nuclear stages of conjugation in varieties in which both processes occur.

(c) Environmental conditions for conjugation. The temperature optima and ranges for the occurrence of conjugation differ for certain varieties (Sonneborn, 1938, 1939, 1941, 1956a, unpublished; Sonneborn and Dippell, 1943, 1946). Varieties 2, 3, 11, and possibly a few other less studied varieties, have relatively low optimal temperatures for mating and conjugate with reduced frequency or not at all at high temperatures within the tolerated range. These varieties mate avidly at 19° or less, poorly or not at all at 27° or more. Other varieties, such as 1, 4, 8, and 9 conjugate avidly at 27°, or even higher temperatures, as well as at 19°. Sonneborn and Dippell (1946) compared the sexual reactivity of representatives of several varieties over a wide range of temperatures and found for each one a unique pattern of responses to temperature. Possibly extension of such a study to all varieties would reveal unique features for each. Nothing is yet known as to the possibility of strain differences in these respects in a variety—like 1, 2, 15, or 16—which occurs in nature over a wide range of temperatures.

Visible light inhibits sexual reactivity in variety 3 (Sonneborn, 1938, 1939). Under the usual alternation of day and night, this variety is sexually reactive only between about 1 A.M. and 1 P.M. Diurnal periodicities also exist in some other varieties; the period of reactivity is during the night in variety 2; from about 4 P.M. to 10 A.M. in variety 11; and from about 8 P.M. to 10 A.M. in variety 15, peak reactivity after 11 P.M. (Sonneborn, 1956a; unpublished).[3] There are also indications of periodicities in varieties 12 and 13, but the details are not yet known. Some of these varieties can become sexually reactive at any hour after they have been grown for some days in darkness, and they totally lose sexual reactivity if grown several days in continuous illumination. Other varieties show no diurnal periodicity.

Variety 16 conjugates better when grown in test tubes than when grown in culture slides (Giese, 1939, 1957). Variety 15 shows similar relations.[3] I have observed that animals from a highly reactive tube culture lose their capacity to mate in an hour or so after transfer to culture slides, while the animals left in the tube remained highly reactive. These observations suggest a decisive role of dissolved gases. Other varieties do not require cultivation in tubes to become sexually reactive; they become strongly reactive when grown in depression slides as well as when grown in tubes. Perhaps, however, detailed study would reveal quantitative differences among the varieties in this respect.

None of the conditions which have just been discussed is known to affect the occurrence of autogamy, nor are there known any varietal differences in external conditions required to induce autogamy. In the proper stage of life only starvation seems to be needed in order to induce autogamy. More or less depletion of the food supply is also necessary to bring about conjugation. The same stimulus leads to these two different results at different stages of the life cycle, as will appear.

2. Immaturity. Following the initiation of a new life cycle by conjugation there may be a longer or shorter period during which fissions occur, but during which the animals so produced are unable to conjugate even when provided with mature potential mates under conditions most favorable for mating. This defines the period of immaturity. It is doubtful whether the first fission or two after conjugation should be included in this period, for during them the nuclear reorganization consequent upon fertilization is still in progress. If they be excluded, it can be said that immaturity has never been found after autogamy, but only after conjugation. Even without this exclusion, as has long been known, animals of some strains can conjugate again while still in the process of nuclear reorganization following the last conjugation or autogamy.

An immature period appears to be totally lacking in varieties 4, 8, 10, and 14, and perhaps in some others. On the other hand, some varieties are not uniform in this respect. Some strains may totally lack an immature period and others may have a short one, its duration varying from strain to strain. Thus, Siegel (unpublished) finds that the period of immaturity in variety 1 lasts about four to five days in most strains, although in some strains it may last eight or nine days. I was unable to find any immature period in strain R of variety 1. Comparable variations among strains seem also to occur in varieties 2 and 3. Translated into fissions, these longest periods of immaturity represent about 35 or more successive fissions. Much longer immature periods occur in varieties 15 and 16. Giese (1957) reports that some strains of variety 16 did not begin to conjugate until they had been in the laboratory for 6 months (fission rate during this period not stated). In variety 15, I have thus far studied only a relatively small number of crosses, but among them considerable variation in the extent of the immature period was noted even among different caryonides of the same exconjugant. In some, it extended for only 8 to 12 days, in others for a little more than two months. However, these cultures were being grown slowly, and this represents a range of about 20 to 100 fissions or more. Curiously, there were relatively few immature periods of intermediate duration. This variability needs further careful study.[5] There thus appear to be at least three different conditions as to immaturity in different varieties of *P. aurelia:* some have no period of immaturity, some have variable but short periods of immaturity, and some have much longer periods of immaturity, at least in most lines.

3. Maturity. The onset of maturity is marked by the acquisition of the capacity to mate. When there is a preceding immature period, the capacity to mate develops gradually, as evidenced by mating reactions of increasing intensity, i.e., involving larger proportions of the animals. When a period of immaturity is lacking, maximal mating reactions occur as early as adequate tests can be made, certainly by the 8th fission after fertilization. Of course, in order to bring about mating, the food supply must be

more or less depleted and the other conditions necessary for certain varieties, which were mentioned above, must be provided. The period of maturity terminates when the animals fail to respond to these conditions by mating, but instead give one of the responses characteristic of senility, which will be set forth in (4) on senility. The transition from maturity to senility is not sharp, but gradual, like the transition from immaturity to maturity. It is marked by decreasing intensity of the mating reaction under conditions most favorable for mating. Maturity will be considered to have ended when conjugation with the complementary mating type of another culture becomes difficult or impossible to obtain. The following account of varietal differences in the period of maturity takes up first the variations in duration of this period and then variation in the system of mating during maturity.

The period of maturity is exceedingly brief in some varieties. In varieties 10 and 14 it lasts only until about the 15th fission and in varieties 4 and 8 until about the 25th fission. Under the standard conditions of culture, the time from the end of postzygotic nuclear reorganization to the end of maturity (no immature period existing in these varieties) is about 3 days in varieties 10 and 14, about 6 days in varieties 4 and 8. Maturity lasts longer in most varieties, much longer in some. It lasts until about the 10th to 13th day (30 to 40 fissions) in variety 2 and until about the 17th to 25th day (50 to 75 fissions) in variety 1. According to observations of Powelson (unpublished), it may last more than 30 days (more than 80 fissions) in variety 12. Current and still incomplete observations on variety 15 indicate that it has a still longer period of maturity. The first cultures to be followed are still mature after three months since the end of immaturity. Giese's (1957) observations on variety 16 indicate that its mature period may last more than a year. There are reasons to believe that the duration in time of the various periods of life depends largely upon the fission rate, the characteristic feature being a certain average number of fissions. Eventually, therefore, comparisons will have to be made in terms of number of fissions.

Varietal differences in the system of mating during maturity concern the incidence of "selfing" caryonides. A selfing caryonide, or selfer, is one in which conjugation can take place between members of the same caryonide in the absence of mixture with another caryonide. Selfers occur in all varieties, but with very different frequencies. Kimball (1939) found that about 4% of the caryonides of strain S of variety 1 are selfers, while Sonneborn found a higher frequency in strain R of the same variety. In the group B varieties, or at least many of them, selfers are relatively rare in exautogamous caryonides that have not recently been derived from conjugants, but are much more common in exconjugant caryonides. Variety 5 (group A) presents what is thus far a unique situation. Caryonides pure for either mating type (IX or X) are relatively rare, the great majority being selfers. Of the latter, most are predominantly type X. In these selfers, usually the great majority of the animals are of one type, so that relatively few selfing pairs are formed, and the predominant type remains characteristic for any caryonide. The most extreme condition is found in variety 13, every caryonide, so far as now known, being a selfer.[6] Moreover, although I have demonstrated that two mating types occur, the predominant type does not remain the same from day to day in a caryonide. It seems as if mating type is continually liable to change in both directions during multiplication by fissions. Whether this inconstancy of mating type appears at the very beginning of maturity or only after a period of constancy as one type is not yet known. However, if there is a constant period, it must be brief. The inconstancy of mating type in variety 13 is the reason for present inability to assign it to group A or B.

4. Senility. In varieties that undergo autogamy, senility begins when the animals respond to starvation by undergoing autogamy instead of becoming ripe for conjugation. Senility extends until death of the culture. At the beginning of senility, fission rate is high and virtually all animals that undergo autogamy survive well. As senility progresses, the fission rate declines and the frequency of survival after autogamy decreases. Eventually no animals can survive autogamy, and still later fissions cease, the culture dying out. In variety 4, the period of senility

may extend through nearly 300 fissions or three to four months. It is by far the longest stage of the life cycle. By the time it has lasted for 170 to 200 fissions, no animals can survive autogamy. During advanced senility the autogamy response to starvation also declines and eventually disappears; concomitantly, the response of conjugation reappears but in weak form. However, survival after conjugation is low and approaches zero. The clone thus becomes genetically dead about 100 fissions before it ceases to exist. The length of senility is not greatly different in varieties 1 and 2 from what it is in variety 4, when measured in number of fissions, but the lower fission rates of these varieties make it longer in time.

The period of senility in varieties 15 and 16 is very different from that in the other varieties. The response to depletion of food is selfing, not autogamy. The phenomena are most marked in the usual kind of caryonide which is pure for one mating type during the long period of maturity and then becomes a selfer during senility. Caryonides that are selfers during maturity have not yet been studied in this respect; but it is possible that there are characteristically different features of selfing in the two periods of life. During senility, it is impossible to isolate from a caryonide individuals that yield cultures entirely or even predominantly of different mating types. Moreover, as senility progresses, several changes take place which do not occur during maturity. First, the proportion of animals that self when the food supply declines increases slowly but steadily. Second, the amount of depletion in food necessary to bring about selfing decreases. Third, the percentage of viable clones produced from selfer pairs decreases: at first survival is high, eventually it decreases to zero. Giese (1957) describes these changes which he had observed in variety 16 as early as 1941. My recent observations on variety 15, though far from complete, agree as far as they go with his observations on variety 16.[3]

Of particular interest are two strains of variety 15 which are very old but which came into my hands only a few months ago. One strain came from the University of Washington about 4 years ago through Dr. Vance Tartar. The other came from Stanford

University where it was isolated about 1941. This 15-year-old culture is the well-known strain long maintained in axenic culture under the name of *P. multimicronucleatum* by Dr. Willis Johnson (1952), who gave it to me. Both of these strains I found at once to be predominantly mating type XXIX but selfers, and in both all exconjugants died. Dr. Ruth Dippell stained these for me and found, as I suspected, that both are now amicronucleate. These strains are thus genetically dead. How long they have been so, I do not know. But it is clear that the total life cycle in variety 15 is at least 15 years and that life persists, perhaps very long, after a culture has become genetically dead. As mentioned above, the same sort of relation exists on a smaller scale in varieties with shorter life cycles, e.g., clones of variety 4 may live for about 100 fissions after they are unable to survive sexual processes. In both cases, deterioration and loss of the micronuclei, the germ plasm, mark genetic death which precedes somatic death by a considerable period. Whether variety 15 and variety 16 clones ever reach the stage of somatic extinction is as yet unknown. If they do, it must be after ages measured in decades and therefore rarely if ever discovered by an investigator.

*Should the Varieties of P. aurelia Be Assigned Species Names?* The preceding detailed account of the known and, so far as can now be foreseen, the knowable differences among the varieties of *P. aurelia* had as its first objective marshalling all evidence which might be pertinent to the question: Can the varieties of *P. aurelia* be readily identified? A positive answer to this question would justify changing the status of the varieties to species and assigning to each a species name. A negative answer would seem to require that we abstain from assigning species names even if we regard the varieties as species.

So far as now known or now foreseeable, all the 16 varieties could be uniquely identified either by mating type tests alone or by serologic means alone, but not by any other single characteristic. To carry out the mating type tests, it is necessary to employ successfully standard cultures of each of the 34 mating types. This involves having available enough strains to yield all these types, being able to obtain and isolate the various mating types, and having the knowledge and skill required to get all the types in sexually reactive condition. This is a colossal task and beset with many difficulties. Beyond question, it is too much to expect of a zoologist who merely wishes to conform to the mores by attaching the correct species name to the organism about which he wishes to publish his observations. The only alternative is for him to send his culture to an expert for identification. Although this could be done now—it is no easy job even for the expert and could take weeks or months of work in the case of varieties with long immature periods—the solution is one of limited dependability. There is no guarantee that there will always be a laboratory maintaining a full battery of live standard cultures. If identification depends upon such a hazardous procedure, this basis for setting up species is much too insecure. Type specimens would have to be kept perpetually alive. I therefore reject mating type specificity as a basis for defining named species.

At present, serologic specificity cannot serve the purpose either. However, this may be largely due to insufficient knowledge and to unavailability of enough diagnostic antisera. If these lacks were ever supplied and the diagnostic antisera were routinely maintained by a number of established institutions well distributed around the world, perhaps identification of the varieties might become practicable and within the limits of routine taxonomic procedure. The situation would then be comparable to the one that now exists for certain microorganisms such as *Salmonella*. But *P. aurelia* is not *Salmonella*. There is no reason to believe that the demand for these means of identification would ever justify the great labor and expense of maintaining a supply of the 300 or more necessary antisera. I therefore also reject serologic characteristics as a basis for defining named species in this material.

To aid in judging whether there are any combinations of characteristics which could routinely be used to achieve successful identification, Table III presents the differential traits of the varieties in summary fashion. Most of the differential characters involve for their recognition prolonged study and highly elabo-

TABLE III. Some Characteristics of the 16 Varieties of *Paramecium aurelia*[a]

| | \multicolumn{16}{c}{Variety of *P. aurelia*} |
|---|---|---|---|---|---|---|---|---|---|---|---|---|---|---|---|---|
| | 1 | 2 | 3 | 4 | 5 | 6 | 7 | 8 | 9 | 10 | 11 | 12 | 13 | 14 | 15 | 16 |
| 1. Where found | Cosm | Cosm | N.Am. | Asia N. S.Am. | N.Am. | Cosm | N.Am. | N.Am. | Eur. | N.Am. | N.Am. | Mad. | N.Am. | N.Am. | N.Am. | N.Am. |
| 2. Temp. range | Cold Warm | Cold Warm | Cold Mod. | Cool Warm | Cool Mod. | Warm Hot | Warm Warm | Cool Warm | Cold Cool | Warm Warm | Cold Warm | Hot Hot | Warm Warm | Warm Warm | Cold Warm | Cold Warm |
| 3. Temp. mode | Mod. | Cold | Cold | Warm | Cool | Hot | Warm | Warm | Cold | Warm | Warm | Hot | Warm | Warm | Warm | Warm |
| 4. Optimal temp. for mating | 20°-38° | <21° | <24° | 20° 30° | | | | | <25° | | <24° | | | | | |
| 5. Daily period of mating reactivity | N | 6P-6A | 1A-1P | N | N | N | N | N | N | | 4P-10A | ? | | | 8P-10A | |
| 6. Mating reactions to varieties | 3, 5, 7, 8 | N | 1, 7, 8 | 8, 10, 14 | 1, 7 | N | 1, 3, 5, 8 | 1, 3, 4, 7, 10, 14 | N | 4, 8, 14 | N | N | N | 4, 8, 10 | N | |
| 7. Nuclear reorganization number | 2 | 2 | 2 | 2 | 2 | 2 | 2 | 2 | 2 | 2 | 2 | 2 | 4 | 2 | 4 | 4 ? |
| 8. Mean length (μ) | 135 | 147 | 123 | 112 | (M) | (M) | 140 | (v4) | (M) | (v4) | (v2?) | 204 | (v12) | (v4) | (v12) | (v12) |
| 9. Mating type system | A | B | A | B | A | B | A | B | A | B | A | B? | S | B | A | |
| 10. Serotype system | A | A | A | B | | | | B | A | | | | | | | |
| 11. Killers (K) Resistants (R) | N | K R | N | K | N | K | N | K | N | | | | | | N | |
| 12. Immature period D | 0-9 | 0-7 | 0-9 | 0 | | | 0 | 0 | 0 | 0 | | | 0 | 0 | 60+ | 180+ |
| F | 0-35 | 0-25 | 0-35 | 0 | | | 0 | 0 | 0 | 0 | | | 0 | 0 | | |
| 13. Mature period D | 10-20 | 3-13 | 10-20 | 6 | | | 7-11 | 6 | 12 | 3 | 30+ | | | 3 | 90+ | 365+ |
| F | 40-75 | 10-40 | 40-75 | 25 | | | 25-40 | 25 | 48? | 15 | 80+ | | | 15 | | |
| 14. Fertilization in senility | A | A | A | A | A | A | A | A | A | A | A | | S | A | S | S |

[a] Key to symbols and abbreviations, by number of row. Row 1: Cosm cosmopolitan; N. North; S. South; Am. America; Eur. Europe; Mad. Madagascar. Rows 2 and 3: Mod. Moderate. Row 5: P P.M.; A A.M.; N none; ? periodicity exists, but limits not yet ascertained. Row 6: N none. Row 7: the number given is both the number of micronuclei and the number of macronuclear anlagen formed from products of the fertilization nucleus before the first fission of the animal. Row 8: symbols in parentheses represent estimates not based upon measurement; v2, v4, and v12 roughly the same as varieties 2, 4, or 12, respectively; M intermediate size, estimated between 125 and 150 μ. Rows 9 and 10: see text for characterization of the A and B mating type and serotype systems; B? B indicated by preliminary study only; S all caryonides self. Row 11: signifies whether killer or resistant strains have been found in the variety; N none found. Rows 12 and 13: D days; F fissions. Row 14: A autogamy; S selfing. In all rows, blank spaces indicate information is lacking. See text for additional characteristics known for only a few varieties. These include: temperature tolerance, fission rate, certain conditions for sexual reactivity, duration of conjugation and time until first and second fissions, frequency of selfing caryonides, and duration of senility. See Addenda 2-6 for further information.

rate and technical procedures: obtaining conjugants, following the life cycle, genetic studies, serological procedures, and other physiological procedures. These are totally unsuitable for use in routine identification. Mean body length alone, among all the traits listed, could be used routinely. And this does not identify all the varieties; it could only limit an unknown to one of three or four groups of varieties. I therefore also reject combinations of the traits in Table III as a basis for assigning species names to the varieties.

Nevertheless, the information summarized in Tables I, II, and III make it *possible* to define the varieties and to prescribe a procedure for identifying them without recourse to living standard cultures. Sonneborn (1950) explained this in detail for the 8 varieties known at that time. The same general approach, supplemented by much new information, would make possible identification of all 16 varieties now known. The first step would be to obtain many strains from regions that have provided the known varieties. From each strain mating types would have to be isolated and, in the way originally done, the various strains would have to be sorted into varieties. When 16 have been obtained, they could be identified with the 16 now known in the following ways. The only variety with 4 mating types would be variety 16.[3] Varieties 2, 3, 11, and 15 could be identified by their distinctive diurnal periods of sexual reactivity and their sizes. The remaining large animal varieties would be 12 and 13. Variety 12[2] could be identified as the only large variety that regularly formed 2 macronuclear anlagen and 2 micronuclei after fertilization. Variety 13 could be identified by its regular selfing.[6] Of the 4 smallest animal varieties (4, 8, 10, and 14), 8 could be singled out because of the reaction of one of its mating types with one of those of the already identified variety 3. Variety 4 would then be identifiable by its two-way sexual reactions with variety 8. The remaining two varieties (10 and 14) have been but recently discovered and I do not yet know enough about them to prescribe how to differentiate them. The system of reactions in Table I shows how variety 3 could be used to identify varieties

1, 5, and 7. This leaves only varieties 6 and 9. These could be identified both by their nonoverlapping distributions and by the fact that 6 shows group B inheritance of mating type, while 9 shows the group A pattern. Thus, without either living standard cultures or diagnostic antisera, the complete identification of all varieties (except 10 and 14) could be accomplished.

It is a great satisfaction to know that now or centuries from now, anyone who is willing to give the time and effort needed for the job would be able to identify strains in conformity with our descriptions. This, however, would involve a major research effort over a period of years. I think every reader will agree that this is too much to require of anyone who merely wishes to know to what variety the one stock on which he has been working belongs. To make matters worse, it seems all too likely that we do not yet have in hand by any means all the varieties of *P. aurelia* that exist in nature. Until we do, there is always the chance that anyone who attempts to follow the procedure outlined above will come out with 16 (or more) varieties some of which are different from those we have tried to define. Under such unhappy conditions, I contend that assigning species names to varieties of *P. aurelia* is indefensible because it is totally impracticable.

*Should the Varieties of P. aurelia Be Referred to as Unnamed Species?* Even the most devoted adherents of the modern biological species concept would probably agree with the conclusion that the varieties of *P. aurelia* should not—or at least not yet—be assigned species names; but they doubtless would also tenaciously hold that the varieties *are* species nevertheless and should be referred to as numbered species, not numbered varieties. This is a much more fundamental and important question than the one considered in the previous section. The answer to be given to it must be based on reason and judgment. Let us review the major considerations pro and con.

The chief argument for the modern biological species concept is that it refers to a real level of biological organization, a group of populations among which gene flow is potentially possible.

This contrasts with the typological or morphological species concept which is to a degree arbitrary, depending upon what traits the taxonomist can recognize and what ones he believes to be of taxonomic importance. This arbitrariness, the existence of intergrades, and other difficulties have often led to the conclusion that species so defined are but figments of the imagination and have no objective validity. The definition of species in terms of a potentially common gene pool provides an objective criterion which has the great merit of delimiting a natural entity with evolutionary significance.

On the other hand, there are strong objections to this otherwise attractive point of view. First of all, the task of discovering what constitutes a potentially common gene pool and the relation of such groups to morphological species is very great. For example, after 20 years of research on *P. aurelia*, we know it consists of at least 16 distinct potentially common gene pools and suspect that very many more remain to be discovered. Obviously only a very small proportion of the enormous number of morphological species can in the foreseeable future be sorted out into common gene pool species. This means inevitably that there will always be a double standard of species, relatively few ever being defined on the modern biological species concept. Second, as pointed out at the beginning, a very considerable proportion of organisms, those which are obligatory self-fertilizers and those which totally lack sexual reproduction, are totally outside the possible domain of the modern biological species concept. Potentially common gene pools simply have no meaning in connection with them. This leads logically to the view that such organisms cannot be sorted into species. Proponents of the modern biological species concept, when faced with this situation, retort that difficulties and exceptions do not destroy the value of a concept, but this is not the point at issue. I for one certainly do not challenge the value of the biological species concept. It has great value. But its value is very narrowly limited, in the first place to outbreeding organisms in principle and in the second place to the very small proportion of them that will in the foreseeable future be studied sufficiently. What this boils down to is a division of opinion on the most desirable usage of the technical term species. Until recently it has been used as a universally applicable designation for pigeon-holing all organisms. Proponents of the gene pool species concept have used the term in a second sense which is inapplicable in principle to a large proportion of all organisms and which can strictly be extended to but an infinitesimal fraction of the eligible organisms because of the great labor involved in applying it. The last difficulty is minimized on the assumption that taxonomists have probably already defined most species of outbreeders in such a way that, if breeding studies were made, the taxonomists' species and the biological species would be found to be identical. However, in the absence of breeding studies, this remains only a more or less reasonable conclusion. In any case, the gene pool species is more restricted in applicability than the species of routine taxonomy. Many taxonomists are unwilling to abandon the universal use of the term for a limited use, and I agree with them.

The alternatives seem clear. (1) Accept the double standard of applying the biological species concept whenever information justifies it and using routine taxonomic procedure in all other cases, i.e., in the great majority. (2) Restrict the term species to the limited group and find a new term for the majority. (3) Continue to use the term species in its universal pigeon-hole sense and find a new term for the limited number of cases in which common gene pools are known or will be discovered. I reject the first alternative on the ground that a technical term should have a single usage, and the second alternative on the grounds of priority and generality. Accepting, therefore, the third alternative, I propose the term "syngen" for the potentially common gene pool, for organisms capable of "generating together." I recommend that it or a better term, if one can be found, should be adopted. I am therefore prepared to abandon the use of the term variety in *P. aurelia* in favor of *syngen* or whatever other new term meets with the approval of the experts. But I am not willing to abandon the term variety in favor of species, for the reasons just given.

## *Paramecium aurelia:* Breeding Systems and Evolution

Disagreements concerning terminology must not be permitted to obscure the biological problems that underlie them. The question discussed at the end of the preceding section—whether the varieties of *P. aurelia* should be called species or should be given some other designation such as syngens—is of relatively minor importance. More important are the evolutionary questions raised by the recognition of these biological units of organization and by the nature of the diversities among and within them. These are the questions to be dealt with in the present section. In doing so the biological units will have to be referred to by some term. I shall continue to call them varieties so that the reader will not have to change horses in the middle of the stream. However, I urge that a term such as syngen eventually be agreed upon both for the varieties of Ciliates and for the so-called biological species of other organisms.

*Basic Biological Frame of Reference in P. aurelia.* Probably the most fundamental evolutionary requirements of any organism are to take full advantage of opportunities for reproduction and to possess and maintain mechanisms for providing enough genetic variability to meet the demands imposed by changes in the environment. The common mechanism for the latter is of course mutation. But organisms differ in the way mutations are handled. Among lower organisms, two major ways are observed. Some combine haploidy with large and rapidly produced populations. This permits both an adequate supply of mutations and an effective way of bringing them to immediate or rapid expression. Other lower organisms, like most higher organisms, combine diploidy with genetic recombination by sexual means. This permits storing an accumulation of mutations and selecting among their various combinations. The latter is the alternative generally adopted by Ciliates, including *Paramecium*. It is further reinforced by the occurrence of asexual reproduction between sexual processes and by the existence of nuclear dimorphism. The new mutations are stored in the micronucleus, which has little or no effect on the phenotype, and they are multiplied during asexual reproduction so that they become widely distributed prior to recombination.

In *Paramecium,* and in Ciliates generally, the occurrence of sexual processes and genetic recombination is linked to decline in the food supply. This must be an inevitable and frequently recurring event in the lives of such organisms. The logarithmic increase in numbers resulting from repeated binary fission quickly yields populations that no conceivable food supply could support. Not even an amount of food equal to the bulk of the earth could support reproduction from a single individual for as much as 100 successive fissions. With due allowance for predation and other causes of accidental death, it is difficult to see how asexual reproduction could long continue without the onset of famine. The prevailing pattern of life must be to increase, multiply, and spread while the harvest lasts, forgetting about sex while enjoying food to the full, and to use hard times when reproduction comes to a virtual standstill as occasions for grasping opportunities to carry out sexual processes.

Little is yet known about the alternation of asexual reproduction and sexual processes in nature. We are at present largely limited to inferences based on laboratory study. What is worse, we systematically carry out the laboratory studies under conditions that differ in important respects from those which probably exist in nature. For example, we routinely and at once transfer fertilized animals from the starvation conditions that induced fertilization to conditions of excess in available food. In nature, the conditions that induce fertilization would hardly be expected to change so radically in the course of the hours occupied by the processes of meiosis and fertilization. Yet we know very little about the effects of this difference between natural and laboratory conditions. That it does have some effects on the nuclear processes following fertilization and even on the subsequent breeding relations is already indicated. The whole matter is in great need of investigation.

Nevertheless the laboratory studies seem clearly to show that there are two distinct and independent functions of sexual processes. First, they have the usual function of genetic recombination. Second, they serve to initiate new life cycles. In the absence of sexual processes, paramecia grow old and either die or become genetically extinct by losing their germ plasm (micronuclei). Survival depends upon the occurrence of fertilization before senility has gone too far. The fertilized animals are reinvigorated and start new, young clones.

Different sexual processes may occur in different stages of the life cycle when food is depleted. This sequence of responses to famine is a clue to the relative values and functions of the different sexual processes. Invariably, the response of ripeness for conjugation precedes the response of autogamy when both occur in the life cycle. In varieties that undergo selfing instead of autogamy, ripeness for cross conjugation precedes ripeness for selfing. Cross conjugation thus has priority over selfing, and conjugation has priority over autogamy. These invariable priorities indicate the judgment of natural selection as to which sexual processes best fulfill their functions. Since all are equally operative in initiating new life cycles, the advantage of conjugation must lie in its greater effectiveness in bringing about genetic recombination.

A priori it might have seemed possible to question this verdict of natural selection. Autogamy results in conditions approaching those associated with haploidy: there are no hidden recessives masked by dominant alleles; all loci are homozygous in the one case and hemizygous in the other. Selfing also approaches the same condition, but less efficiently. Why then would not autogamy or even selfing work as well as haploidy? The answer lies in the decisive differences from the conditions in haploids. In a diploid with autogamy, the clone is the unit of expression of a genotype, not the individual animal. This results in a considerable lapse between the occurrence and expression of a mutation. More important, the number of units or *clones* in a population must be exceedingly small in comparison with the number of units or *individuals* in the usual successful haploid organism. Hence, neither selfing nor autogamy in *Paramecium* can be as efficient as haploidy in large and rapidly produced populations. To achieve the necessary genetic variability, natural selection has in *P. aurelia* given first priority to cross conjugation, which however cannot be guaranteed to occur; and second choice to genetically less satisfactory, but physiologically equally satisfactory sexual processes, which can be completely depended upon to occur. The second choice is essential to short-term survival; the first choice may be essential to long-term survival. The universality of the occurrence of a first and a second response in time indicates that this is not a mere laboratory phenomenon, but a double assurance mechanism which operates in nature and has been developed and preserved by natural selection. Incidentally, if this is so, the second choice response must occur often enough in nature for selection to be able to preserve it.

However, the existence of these two sorts of responses in all varieties of *P. aurelia* does not mean that they are equally well developed in all. As will appear, most of the varietal differences we have set forth in an earlier section bear importantly on just this point. They have the effect of giving greater or lesser emphasis to one or the other of two successive sexual responses to famine. The life processes of some varieties favor and prolong the possibility of cross conjugation, making them outbreeders. The life processes of other varieties contract these possibilities almost to the vanishing point, making these varieties inbreeders. And other varieties show intermediate conditions. Since outbreeding may be considered advantageous for long-time survival of diploid free-living organisms, it may be assumed that this was the breeding system of the ancestors of the *P. aurelia* now living. Further, those living varieties that now maintain the outbreeding habit are most likely to provide the solid core of persistence on the evolutionary time scale. Relatively few varieties of *P. aurelia* conform to this exacting requirement; but they are the ones with the widest distributions in nature. On the other hand, inbreeding suffices well enough for short-time survival while conditions remain relatively constant. This mode of life is less exacting. The life cycles succeed one another rapidly so that the system approaches the one in haploids. Evolutionary divergence can and does take place at a rapid rate; but at the sacrifice of some of the genetic plasticity needed for rapid adaptation. These are the views I shall now attempt to develop in detail by considering the bearing of varietal differences on breeding systems and evolution.

*Varietal Differences in Breeding Systems.* As just mentioned, most of the varietal differences are related to their breeding systems. Aside from this relation, they would appear, as they have until now, as mere chaotic brute facts of nature. Recognition of their relation to breeding systems makes them fit together, like the pieces of· a jigsaw puzzle, into a meaningful picture. The nature of the relations will be best illustrated by contrasting first the most extreme inbreeders with the most extreme outbreeders, even though less is known about them than about some of the less extreme varieties. I shall then review the situation in the intermediate varieties progressing from those that appear to be least, to those that appear to be most, committed to inbreeding.

1. The extreme inbreeders, varieties 10 and 14. These two varieties have been found only once each in nature, in the southernmost part of the United States. They are both apparently rare and highly localized. This alone would limit narrowly their possibilities for cross breeding. Everything else that is known about them points in the same direction. They have no detectable immature period and the shortest known mature period which extends to about the 15th fission after conjugation or autogamy. This means two things: they are able to cross-conjugate immediately after they complete the reorganization following fertilization; but only for a short time thereafter. After that the response to depletion of food is autogamy, the closest possible form of inbreeding.

In estimating the chances for cross conjugation during the short period in which it is possible, several considerations are pertinent. The processes of meiosis and fertilization and the subsequent process of nuclear reorganization are relatively rapid in these varieties, and they afford little time for dispersal while they are in progress. The organisms occur in warm regions, and this also hastens the processes. The fission rates are among the highest, and this would lead to the most rapid exhaustion of available food and cut down on time available for migration before starvation again leads to fertilization. These varieties belong to group

B. Usually, therefore, each pair of exconjugants produces two clones of complementary mating types. Since the factors previously mentioned operate to make the animals quickly ripe for mating, this will happen when complementary mating types of common descent have had little time to migrate. They will therefore be apt to mate with each other, if they find mates at all. Failing this, they would have little opportunity to find mates other than those available from the same population. Conjugation with individuals that are not closely related is thus almost impossible except as a rare event; conjugation with close relatives or inbreeding by autogamy would be expected as the almost universal forms of fertilization. Possibly in a sparse local population provided with much food, maturity could be passed before starvation occurred and induced autogamy. The same might occur in a sparse population with little food, as a result of the difficulty of finding mates. Only in a dense population would conjugation be likely on a large scale. The various characteristics of these varieties would then conspire to bring about mating between close relatives or within the population.

2. The extreme outbreeders, varieties 15 and 16.[3,5] These varieties are widely dispersed from cold to warm regions of North America; they have not yet been searched for elsewhere. There are thus many local populations, the first prerequisite for outbreeding. Most important is the fact that these are the varieties with the longest known periods of immaturity, lasting for months, perhaps for a year, at the low reproductive rates that may usually prevail in nature. This gives time for the animals to scatter and migrate. The period during which they are most likely to be near close relatives is the period in which they cannot mate. As a corollary, the method of mating type inheritance is less important when there is so long a period of immaturity. Even if complementary mating types arise among the descendants of a single pair of conjugants, they are not so likely to meet each other when they are mature if there is an intervening long period of wandering about. Actually the method of mating type inheritance is unknown in variety 16; but the group A method is indicated in variety 15.

The period of maturity is probably even longer than the immature period in these varieties. An outbreeder, especially one with no known self-controlled method of getting from one body of water to another, doubtless needs much time to find a suitable mate. The chance of finding a suitable mate is increased in variety 16 because there are three other mating types. In all other varieties, there would be only one. Obviously if the chance of finding any stranger belonging to the same variety is low, there is a selective advantage in a system of multiple mating types which greatly increases the probability that the first stranger encountered will, if mature, be a suitable mate. No such advantage attaches to multiple mating types in inbreeders. Lacking immaturity and having either the group A or B method of mating type inheritance, they would still have prospective mates at hand even with a two-type system.

Two considerations apply to both the periods of immaturity and maturity. First, varieties 15 and 16 have lower fission rates than varieties 10 and 14. This makes both immaturity and maturity last longer in time than is indicated by merely comparing the numbers of fissions in these periods with those in the inbreeders, for at low fission rates it obviously takes longer to use up the number of fissions allotted to each stage of life. The times available for migration and for finding a suitable mate are thus to this extent increased. Second, various collectors (e.g., Pringle, 1956) have reported that some varieties of *P. aurelia* can be found in nature only during summer and fall, in certain regions, and that the numbers found in samples of equal volume have an annual rise and decline. In varieties 15 and 16, with immaturity and maturity lasting through one, two, or more such cycles, this might drastically reduce the number of caryonides inhabiting a particular body of water and so would perhaps reduce the probability that caryonides of complementary mating type and common descent would persist in the same body of water until they reached maturity and attained a sufficiently high population density to be likely to encounter one another.

Finally, varieties 15 and 16 are the only ones in which autogamy appears to be replaced by old-age selfing. Thus, even this second choice method of fertilization, which is required to forestall clonal death, takes a form in which inbreeding is less intense. There is no published evidence to decide with assurance whether immaturity follows old-age selfing, but Giese's (1957) comment about obtaining a series of successive inbreedings in variety 16 following selfing suggests that immaturity after this sort of conjugation may be lacking as it is after autogamy. If so, this would permit rapid restoration of the heterozygosity lost by selfing. The episode of selfing would thus serve to rejuvenate and give more time for crossing without sacrificing heterozygosity.

The contrast between these two inbreeding and two outbreeding varieties illustrates how differences in some of the major life processes bear on the breeding system. The same features, with mainly quantitative differences, and some additional ones, are encountered in the other varieties, as will now appear.

3. *Variety 12.* Very little is yet known about the only available strain of this variety. It has an immature period of undetermined length and a long mature period that probably extends for more than 80 fissions. Yet its fertilization process during senility is autogamy. Thus it seems to be a somewhat less extreme outbreeder than varieties 15 and 16.

4. *Varieties 1, 2, 3, 7, 9, and 11.* These varieties are all less extreme outbreeders than varieties 12, 15, and 16, and less extreme inbreeders than varieties 10, 14, and those to be discussed later. However, not enough is yet known about them to rank them in a definite order. Little is known about variety 11, of which only two strains have been found. It belongs to group A, has a moderately short period of maturity, and undergoes autogamy. Immaturity has not been studied in it; but the relatively short total interval until senility begins precludes any but a very short immaturity. Variety 11 could not be a strong outbreeder. Variety 7 has long been known only as a single strain pure for mating type XIII. Recently a few more strains have been found. These are not pure for mating type, but show the group A method of mating type determination. All come from Florida. Preliminary observations indicate no immaturity and

moderately short maturity. The little information available may appear to be somewhat conflicting. Purity of a strain for one mating type limits its conjugation to outbreeding. Yet it can live on indefinitely without conjugation. I have had this strain for 18 years during which it has undergone successive autogamies at intervals of some 30 to 40 fissions. It is still as vigorous as ever. Thus, purity for one mating type in a strain that undergoes autogamy has double significance: it conjugates only by outbreeding, but without conjugation it undergoes the most intense form of inbreeding. Further, the purity for type XIII depends upon a recessive gene. Therefore, once an outbreeding has taken place, this purity disappears except in segregants in later generations, and these at once have available for inbreeding plenty of related segregants of complementary mating type. On the whole, therefore, purity for one mating type in this context cannot be considered as an important outbreeding feature. All other available facts (group A, no immaturity, short maturity, narrow range) indicate that variety 7 is prevailingly an inbreeder.

Variety 9 has been found only in western Europe. It belongs to group A, has a moderately short mature period, with senility beginning by about the 50th fission, and it undergoes autogamy. With so short a period until the onset of senility, immaturity, if it occurs, could only be brief. These features point to an inbreeder. In this case, the inference can be supported by analysis of Pringle's (1956) field studies. He made collections from a small part of a large pond at approximately 20-day intervals over a period of 15 months. At each collection, 24 to 55 samples of 30 ml. each were brought into the laboratory and searched for variety 9. Not more than one animal of this variety per sample was studied. It and its descendants were tested for the alleles present at two loci. At each locus three alleles were found, but one was very rare. The results on the two loci were similar and may be combined. For the two principal alleles at each of these loci, 151 homozygotes and 7 heterozygotes were found. At one of the two loci, no heterozygotes were found. In a randomly mating population, altogether 116 homozygotes and 42 heterozygotes would be expected. Pringle cautiously remarks that his detection of heterozygotes might have been imperfect, so that his figure is to be considered minimal. Knowing the methods employed, I believe the error cannot be very great. Even if it is assumed, as is by no means justified, that only one-fourth of the heterozygotes were identified, their deviation from the expectation on random mating is still significant at the 1% level of confidence. Accepting the data at their face value, it can be calculated that at least 78.5% of the homozygotes arose by inbreeding. Altogether I consider these field results to constitute strong evidence for the occurrence of considerable inbreeding in nature in variety 9.

The preponderance of homozygotes could be due to conjugation between relatives or to autogamy or both. Some calculations which seem to bear on this can be made from data given by Pringle (1956) in connection with the observations mentioned above. Applying the Poisson formula to the proportion of 30-ml. samples which contained no animals of variety 9, the peak densities, found in late summer, amounted to only 25 animals per liter or 1 animal per 40 ml. This calculation is surely subject to some error for it assumes unjustifiably that the animals were randomly distributed in the area sampled; but it is a good first approximation, in spite of the occasional finding of local aggregations by other collectors. During a considerable part of the year, winter and spring, few or no variety 9 animals could be found. With such low population densities, even at their peak, the chances of animals of complementary mating type meeting and mating would seem to be low except in so far as local aggregations occur. Yet the occurrence of some heterozygotes proves, as Pringle points out, that conjugation must occur sometimes between animals of different clones. The low population densities and the excess of homozygotes on the other hand suggest that autogamy probably also takes place. As mentioned earlier, the invariable existence of autogamy or selfing during senility in the laboratory indicates that there has been strong natural selection to maintain them, and this could be true only if the processes do occur in nature.

The preceding analysis applies only to what goes on in a

single pond. Pringle's data also throw some light on what goes on in the range occupied by variety 9. The pertinent data appear in his table that gives information on segregation of serotype alleles in the F2 by autogamy after crosses between homozygotes for different alleles. Here the proportion of nonviable F2 clones is the significant information. A cross between a Scottish and French strain gave 16.9% F2 mortality. Crosses between Scottish strains from different sources gave 7 to 20% F2 mortality, mean 13.4%. Crosses between different isolates from the one intensively studied pond gave 2 to 10% F2 mortality, mean 6.3%. As far as they go, these data indicate that F2 mortality is less following conjugation within a local population than following conjugation between different local populations. This implies to a certain degree that inbreeding has a selective advantage over outbreeding. But the impediment to outbreeding imposed by F2 mortality is much less in this material than in some varieties to be discussed later. Taking all the facts into consideration, variety 9 appears to inbreed to a marked degree, even within a single local population; but it is capable of and apparently undergoes some small measure of outbreeding, at least within the confines of a local population.

Varieties 1, 2, and 3 show several features in common. All of them show variations among strains with respect to the occurrence and duration of the immature period, from none at all up to about a week or more. There is also some variation in the duration of the moderate mature period, up to about 20 days or about 75 fissions. These periods appear to be somewhat shorter in variety 2. All undergo autogamy when starved during senility. These common features indicate that inbreeding and outbreeding occur to different extents in different strains of each of these varieties. The main variable involved is the immature period. In its absence, inbreeding is favored; in its presence, outbreeding is favored in proportion to its duration. Each of these varieties thus appears to be in process of evolutionary divergence with respect to its breeding system and it may become necessary to subdivide each variety into groups of strains differing with respect to the immature period in order to obtain a satisfactory picture of the breeding relations.

The diversity of conditions within variety 1 may here be noted also for certain other features. In one stream in Woodstock, Maryland, I found many years ago a strain pure for mating type I. No such strain has been found among the many collections of variety 1 from the rest of the world, including Asia, Australia, South America, and Europe. Its significance for the breeding system is the same as set forth above for the pure strain of type XIII in variety 7. Of special interest, however, are two further pieces of information. First, I collected from this same source several times over a period of years and always found there this same strain and no other variety 1 (although variety 2 was also always found with it). This shows that the strain can and does persist for years in nature by repeated autogamous inbreeding, a conclusion supported by 18 years of maintenance in the laboratory. The same facts also indicate how constant a local population can remain for a long period. Second, Beale has examined the serotype loci of representatives of this population and finds them to be entirely homozygous at all three major serotype loci. This is in marked contrast to his observations of heterozygosity at these loci in representatives of other populations that contain both mating types. It is therefore clear that within variety 1 there are populations that inbreed closely and almost exclusively by autogamy and other populations that at least occasionally undergo conjugation. The constancy of the Woodstock population further suggests that the conjugations in other local populations are probably as a rule within the population. Migrations and interbreeding between populations seem to be relatively rare.

There are differences, as well as similarities, in the life features of varieties 1, 2, and 3. Variety 2 belongs to group B, whereas varieties 1 and 3 belong to group A. Variety 2, like variety 9, seems to show relatively little F2 mortality in the strain crosses that have been examined. On the other hand, variety 1 shows exceedingly high F2 autogamous mortality after crossing certain strains. For example, Beale (1952) found more than 60% F2 mortality after crossing strains 60 and 90. Field studies by Beale (1954, and unpublished) indicate that heterozygotes occur in nature in variety 1 less commonly than expected from random mating. Both close inbreeding and some measure of outbreeding doubtless occur in these three varieties. Evaluation of the net effect of the complex of their features on the breeding system must await fuller study; but it is already evident that these three varieties have breeding systems intermediate between those of the most extreme inbreeders and those of the most extreme outbreeders.

5. Varieties 5, 6, and 13. Nothing is yet known with precision about some of the most important features of the life cycle in these three varieties, e.g., their periods of immaturity and maturity. Variety 5 belongs to group A, variety 6 to group B, and variety 13 cannot yet be assigned to either group. Every caryonide of variety 13 becomes a selfer;[6] most caryonides of variety 5 also do; and so do many caryonides of variety 6. However, the frequency of selfing pairs in the selfing caryonides of varieties 5 and 6 is usually very low, while it is very high in variety 13. Further, the predominant mating type in selfing caryonides of varieties 5 and 6 remains constant for a caryonide, but it does not in variety 13. Finally, autogamy occurs in varieties 5 and 6, but it has not yet been found in variety 13.[6] On the basis of the widespread occurrence of selfing in these varieties, they may all be considered as strongly predisposed to inbreeding; but, since selfing occurs universally in all caryonides of variety 13 and with high frequency in each of them, variety 13 is by far the most extreme inbreeder of this group. This picture may have to be changed somewhat when variety 13 becomes more fully studied.[6] The possibility has not yet been excluded that a short immature period exists and that the frequency of selfing increases with age, though rather rapidly. If these things do occur, they would operate to some extent toward favoring early outbreeding.

6. Varieties 4 and 8. These last two varieties to be taken up are clearly very strong inbreeders, almost as extreme in this re-

spect as varieties 10 and 14, which were used at the start as the examples of the most extreme inbreeders. They both belong to group B and lack a period of immaturity. Hence they are capable of mating when they are most likely to be near close relatives of complementary mating type. Their period of maturity, however, is short, about 6 days or 25 fissions, but this is twice as long as in varieties 10 and 14. Such a short mature period might be considered to provide relatively little opportunity for conjugation to occur. Yet, if all the individuals lived and food remained abundant, the two clones of complementary mating type from a pair of conjugants would yield many millions, possibly even billions, of progeny before famine would induce 100% autogamy rather than ripeness for mating. Applying the maximum density of population calculated from Pringle's data on variety 9, it would take a lake with a surface area of a square mile and an average depth of 25 feet to house so large a population. It would seem, therefore, that even with so short a period of maturity, the progeny of a single pair of conjugants would be likely to deplete the food and be ripe to conjugate before maturity had passed. This would certainly happen in the second generation for it would start, not with a single pair of conjugants, but with a large population. This and all such calculations fail to take into account predation and other factors that eliminate part of the population and permit the residue to multiply further before exhausting the food supply. On the other hand, it also ignores competitors for food. Such errors will have to be corrected before conclusions as to the relative frequency of conjugation and autogamy in nature can be placed on a firm basis. Meanwhile, considerations based on the limited available knowledge have value in exposing needed types of further investigation. For example, it is known that although mating type changes relatively rarely at autogamy and conjugation, changes are more frequent in one direction at autogamy and in the opposite direction at conjugation in variety 4 at 27°. There is need to try to discover the various factors involved in determining the relative frequencies of the two mating types in a population, their quantitative effects, and the equilibria attained under various conditions. Such information should be of great importance in estimating the relative frequency of occurrence of conjugation and autogamy.

Unlike the extreme inbreeders (varieties 10 and 14) which have been found once only, variety 8 has been found in numerous localities, all but one in the extreme southern part of the United States; and variety 4 occurs abundantly in cool to warm parts of the United States, South America, Japan, and Australia. The materials for outbreeding are thus much greater in variety 8 than in varieties 10 and 14, and they are enormously greater in variety 4.

Nevertheless, in both varieties crosses between strains from different sources invariably lead to considerable mortality in the F2 obtained by autogamy. In variety 4 this varies from about 15% to about 98%, depending upon which strains have been crossed. This is in striking contrast to the results of repeated autogamies within a strain. Sonneborn and Schneller (1955) passed a strain of variety 4 through 50 successive autogamies without an intervening conjugation and without increasing mortality which remained throughout virtually negligible (0.7%). Variety 4 thus survives the closest inbreeding very well, but invariably it suffers from outbreeding. This imposes a strong selective disadvantage on crosses between different local populations.

The probable basis of F2 mortality following crosses of representatives of different populations of variety 4 was revealed by Dippell's (1954) karyological studies. Every strain of variety 4 that she examined had a different karyotype. The karyotypes differed in number of chromosomes in an aneuploid series from about 33 to 51 chromosomes in the haploid set. Further, there were regularly differences in the shapes and sizes of the chromosomes of different strains. In the laboratory, a strain remains as a rule constant for all features of its karyotype. Whether variations occur in detectable frequency within a natural population remains unknown. Finally, Dippell roughly correlated the degree of F2 mortality with the degree of karyotypic diversity.

If, as now appears, F2 mortality has this basis, then the high F2 mortality found in other varieties, such as varieties 1 and 8, may be taken as prima facie evidence of the existence in those varieties of karyotypic diversities among local populations. Clearly this situation could develop only if crosses between local populations were rare events. Otherwise, interbreeding would swamp out existing karyotypic diversities and prevent new ones from arising, as we shall later find to be true in certain outbreeding Ciliates. Thus Dippell's studies lead to the view that F2 mortality following strain crosses may be one of the most potent evidences of the existence of regular close inbreeding in a variety.

The general picture of variety 4 that now emerges is of a variety which is a loose assemblage of innumerable highly isolated and highly inbred local populations. Grouping these together as a variety is to some degree artificial. Although it is true that they are *potentially* able to interbreed and do so in the laboratory, everything we know about them points strongly to the conclusion that such interbreeding occurs very rarely in nature, almost negligibly except perhaps on a very long-range view. In effect, each local population of variety 4 just falls short of having the same status as a variety among outbreeders. The same conclusions apply even more extremely to varieties 8, 10, and 14, perhaps less so to variety 1 and some others.

The importance of these conclusions in relation to problems of species, breeding systems, and evolution in *P. aurelia* is self-evident. Some varieties are in effect virtually obligatory inbreeders, and each isolate in nature is an entity of some biological stature. For such varieties, the local population is the natural biological unit of organization, although still a part of a larger unit (the variety) the members of which are in actuality all but completely cut off from one another.

This situation raises many problems about which we can only speculate at present; most of the information needed for more solid analysis is lacking, though at least much of it is obtainable. Some speculation may stimulate efforts to supply the needed information. The problem of how paramecia get from one body of water to another, as their distribution proves they must, is an old but still unanswered one. Cysts capable of withstanding desiccation have never been satisfactorily proved to occur in *Paramecium*. Human transportation of water must be a minor and recent factor. Transportation by animal carriers, such as insects, amphibians, and terrestrial vertebrates, needs to be looked into far more than in the past. Perhaps even occasional hurricanes and floods and subterranean water connections between surface waters play a part. The pressing problem with respect to migration is not just how or how far, but how often.

The next problem is what happens when a migrant succeeds in getting away from home base into another body of water. This is of course a complex of problems. What happens will depend upon whether or not the new locale contains a population of the same variety. If it does and the migrant is an outbreeder, the events that follow will differ from those that would take place with an inbreeder. Special problems arise if the intruder finds a local population, not of the same variety, but of a variety with which it nevertheless can mate.

It is clear that such interbreeding varieties can coexist for some time. For example, I have found varieties 4 and 8 in a small sample of water from a single source. Since in the laboratory the hybrids are readily produced but either die or are very weak, one would expect equilibrium to be reached only when one variety, the less numerous one, dies out completely as a result of such crosses. Persistent coexistence would seem possible only under one of two conditions. (1) If corresponding mating types of the two varieties were in excess, they could not interbreed, and the minority types would be eliminated or kept down by crossing. This would also enforce inbreeding by autogamy in each variety because only one mating type would be present. (2) If there is nonrandom mating between the two varieties and each preferentially mates with its own variety, this would reduce the elimination which results from crossing.

No great problems are posed by the migration of an outbreeder into waters occupied by another strain of the same variety. To the extent that crossing occurs, a new equilibrium of the now common gene pools would occur. The low or nonexistent mortality from strain crosses of outbreeders would impose little or no bar to amalgamation. The situation for an inbreeding intruder is of course very different. Karyotypic diversity and F2 mortality would tend to wipe out the intruder by reason of its small proportion in the population, if extensive mating with the indigenous strain took place. The degree to which such interbreeding would occur depends upon the mating type frequencies in the two strains. If some does occur, the progeny will include some viable combinations, and these will give rise in further fertilizations to a supply of genotypes and karyotypes for selection to work upon. Eventually this could lead to a new relatively homogeneous population of a single selected best adapted karyotype or to a balanced mixture of diverse karyotypes and genotypes. Until field studies of local populations of strongly inbreeding varieties and laboratory studies of mixtures of diverse strains of such varieties are available, it is not worth pursuing the possibilities further.

If a migrant enters a new body of water not already occupied by its variety or one with which it can mate, there is no difficulty in seeing how both an inbreeder and an outbreeder could soon give rise to a population containing both mating types. Both the group A and group B methods of mating type inheritance provide ready-made mechanisms for this except in the exceedingly rare cases in group A (varieties 1 and 7) of strains pure for one mating type. The only real problem raised by migrations of this sort is how to account for the development of a distinctive karyotype by the inbreeders. This may happen very slowly by selection of rare spontaneously arising chromosomal aberrations. In the latter case, two factors present themselves as possibilities. First, one thinks of the invariable production of chromosomal aberrations with advancing senility (Sonneborn, 1955; Dippell, 1955). Ordinarily these would not be expected to occur in nature. In established populations, famine would be expected to set in long before the age at which the aberrations occur. This would lead to fertilization and thereby to setting the age clock to zero (Sonneborn, 1954a). However, if the migrant is already no longer young and the new waters provide food for much reproduction by a single intruder (or very few), the requisite age might be reached. A second possibility arises from some indications that fertilization at extreme temperatures (10° and 35°) also results in chromosomal aberrations in certain variety 4 stocks. If paramecia migrate via animal carriers such temperatures might prevail, e.g., on the moist surfaces of insects during flight or in the bodies of warm-blooded animals. The small amount of available nutrient under such conditions might at the same time induce autogamy. Both of these possibilities may be wide of the mark, but they and others need to be looked into. The existence of regular karyotypic diversities among local populations of inbreeders is a phenomenon that needs to be accounted for.

7. Conclusions and general comments. This survey of the bearing of varietal characteristics on the breeding system is no doubt incomplete. Increase of knowledge and further reflection will probably bring to light the relevance of other characteristics. Among those we have discussed, the period of immaturity stands out as of major significance. The longer it lasts, the more opportunity close relatives have for dispersal before they can mate; hence, the less likely that they will mate with each other. A very long period of immaturity is thus an indicator of an outbreeder. The absence of a period of immaturity marks an inbreeder. Being ready at once to mate again, they are apt to mate with those nearest at hand, i.e., their close relatives.

The duration of the mating period is also of much importance. This limits the period for finding a suitable mate. Probably no great period of time is required to find a mate in the same local population. In accordance with this, a short mature period marks an inbreeder. If it fails to find a local mate in time, it undergoes autogamy. Outbreeders need more time to find a suitable stranger to mate with and they have long periods of maturity. Their chance of success is enhanced, as in variety 16, by multiple mating types.

Certain considerations apply to these two periods taken together. First, the available evidence indicates that migration from one body of water to another is a relatively rare event; hence, short combined periods of immaturity and maturity will as a rule prevent breeding between different local populations and conjugation would occur between inhabitants of the same body of water. The longer the duration of the two periods, the more time for migration into or out of the local population and consequently the more opportunity for outbreeding. Under the most favorable conditions of the longest immature and mature periods, the chance of outbreeding may still be small. But the occurrence of selfing during early senility in varieties with such long periods starts a new cycle and extends the time available for finding suitable mates.

Second, if the total life prior to senility occupies a year or more, the annual rise and fall in population density may reduce the chance of inbreeding by rendering the population less heterogeneous in mating type.

Third, the immature and mature periods must be considered together sometimes to avoid making an erroneous judgment based on one of them alone. According to old observations of mine on stock S of variety 1, it has an immature period that lasts almost until the beginning of senility and that leaves but a very brief mature period. The immature period alone suggests an outbreeder, but the mature period is so brief as to militate strongly against mating with members of another population and even against mating at all. Unless famine comes at just the right moment, its consequence would be autogamy or no sexual process at all. However, since no immature period follows autogamy and the mature period is then correspondingly longer, there is more opportunity for the occurrence of conjugation after an autogamy in such a strain. This example shows how the interplay of various aspects of life must be taken into account before arriving at a judgment of the breeding system.

In addition to the duration of the periods of immaturity and maturity, there are two other important signs of the breeding system. One is a direct examination of the frequencies of various genotypes in a natural population. An excess of homozygotes and a deficiency of heterozygotes, in comparison with expectations based on random mating, is of course a sign of inbreeding even within a local population. The other sign is the karyological picture or inferences about it based upon the nature of the F2 after crosses between representatives of different populations. Karyological differences between local populations and increased F2 mortality following crosses between them indicate that the populations are isolates of inbreeders. The indication is confirmed if each population alone survives inbreeding very well. Outbreeders would be expected to show inbreeding mortality as a result of lethal and detrimental genic recombinations, but this should happen both when the inbreeding follows crosses and when it occurs within a single population.

Certain other indices of the breeding system are perhaps relatively minor in comparison with those already discussed, but they are still significant. First, the range of distribution of a variety is both a rough measure of the number of populations available for interbreeding and an index of the genetic heterogeneity of the variety. Outbreeders generally have a wide range of distribution both in latitude and longitude. Inbreeders are generally restricted either in latitude or in both directions. Second, the rates at which the processes of meiosis, fertilization, postzygotic nuclear reorganization, and particularly asexual reproduction occur are relatively slower in the extreme outbreeders than in the extreme inbreeders. The slower the tempo, the more time for dispersal before maturity and before senility, and therefore the more favorable this is for outbreeding. Third, strong outbreeders become selfers during senility; strong inbreeders undergo autogamy during senility. Nevertheless, some degree of outbreeding is compatible with the occurrence of autogamy. This is true because autogamy is never followed by a period of immaturity. Hence, the homozygosity resulting from autogamy can at once be compensated by restoring heterozygosity through conjugation within the population. The episode of autogamy serves to prevent the progression of senility and to restore vigor, and it can do this without undue sacrifice of heterozygosity, thanks to the clever device of suppressing immaturity after autogamy.

This summary of the life features that affect the breeding system has thus far failed to refer to what might, a priori, have been supposed to be the most important single feature, namely, the mating types and their inheritance. In spite of the large amount of information available on this complex subject, I find it the most difficult one to relate to breeding systems. Before attempting to do so, the units with which we shall have to deal will be reviewed. The basic unit is the caryonide, the group of individuals that possesses macronuclei descended from the same ancestral macronucleus. The clone, embracing the vegetative descendants of a single autogamous individual or a single conjugant, usually includes two caryonides in most varieties but four in varieties 13, 15, and 16. A new term, synclone, is a convenient designation for the vegetative descendants of two animals that have conjugated with each other. Each of these three terms, caryonide, clone, and synclone, designates a group of individuals sharing a common genotype. (The two conjugants of a pair normally acquire identical genotypes.) Nevertheless, in spite of genic uniformity only the caryonide is (aside from selfers) always phenotypically uniform with respect to mating type. The two caryonides of the same clone are often unlike in mating type in group A varieties but usually alike in mating type in group B varieties. The two clones of a synclone are usually unlike in mating type in group B varieties. The only larger unit we need to consider is the local population which consists of unknown numbers of caryonides, clones, and synclones.

The difficulty in judging the bearing of diverse systems of mating type inheritance on the breeding system stems from the fact that in both the group A and the group B varieties, as well as in the unclassified variety 13, any one synclone as a rule has a reasonably high probability of including both mating types. Whether inbreeding will occur or not thus seems to be independent of the method of mating type inheritance at conjugation. The same difficulty is encountered if we start, not with synclones, but with a single caryonide. When it goes into autogamy (and also presumably if it selfs in senility), both mating types will arise in varying relative frequencies. The only known exceptions are the very rare strains pure either for mating type I or mating type XIII; as already pointed out, they are thereby forced, not to outbreed, but to inbreed by autogamy. Thus, so far as the methods of mating type inheritance alone are concerned, it would seem that all varieties of *P. aurelia* should be close inbreeders. Outbreeding between local populations would be favored if each whole population were pure for one mating type, but only if there were also a very long period of maturity and no autogamy. Outbreeding *within* a local population, that is, mating between different synclones, would require a mechanism for preventing mating until there has been time for dispersal, i.e., by a sufficient immature period, or for a mechanism to make the synclone uniform in mating type.

In fact, both groups A and B have mechanisms for bringing about synclonal uniformity in mating type. In group B, the mechanism is massive cytoplasmic exchange between mates. It has been studied to some extent in variety 2 (with intermediate breeding system), but most fully in variety 4, a strong inbreeder. In this variety, the mechanism usually gives a synclone of mating type VIII at high temperatures, while at low temperatures both types of synclones are produced in high frequency (Nanney, 1954). Remarkably, there are strain differences in the capacity to experience cytoplasmic exchange during mating. Wood (1953) found a single major gene difference between the two kinds of strains. He also found that the gene for cytoplasmic exchange has much higher penetrance if conjugation takes place at low temperatures and if the conjugating clones are very young. These facts could well be of importance in relation to the breeding system. Outbreeders would be expected to have the genotype for cytoplasmic exchange and to conjugate in nature under conditions favorable for the penetrance of the relevant gene. Inbreeders would be expected to lack this gene or to conjugate in nature under conditions in which it rarely comes to expression. It is not known whether these alternatives actually occur in association with inbreeding and outbreeding.

The mechanism in group A is quite different. Here cytoplasmic exchange plays no part whatever, for there is in group A no cytoplasmic difference between the two mating types. Each caryonide is independently determined for mating type and the probability for being determined as the even-numbered mating type increases linearly with the temperature prevailing during pre- and postzygotic reorganization (Sonneborn, 1939). At high temperatures, the even type has so high a probability that a considerable proportion of synclones are uniform for it. However, it seems likely that at the temperatures prevailing in nature, the probability of each type may not be far from 0.5; and with this probability only one-eighth of the synclones would be uniform for mating type. However, this is the minimal proportion of synclones uniform for mating type; all other probabilities give higher proportions. Since the minimum is virtually zero in group B varieties under some conditions, the group A method of mating type determination is, under these extreme conditions, more efficient in promoting outbreeding than the group B method. In agreement, the most extreme outbreeder in which the method of mating type inheritance is known (variety 15) occurs in group A and the most extreme inbreeders (varieties 10 and 14) in group B.

The potentiality of the group A method for promoting outbreeding is further indicated by some other facts. Different strains of the same variety, e.g., variety 1, have characteristically different proportions of the two mating types under the same temperature conditions (Sonneborn, unpublished; some data quoted by Butzel, 1955). Butzel (1955) presented evidence that these differences are due to genes other than the mating type genes. This probably explains why hybrids between certain strains show in the F1 a much higher frequency of type II than either parent strain (Sonneborn, 1942a), while hybrids between other strains do not (Butzel, 1955). If the group A method of mating type inheritance is ever a strong factor for promoting outbreeding, one might expect outbreeding strains or varieties to yield high frequencies of one mating type under the conditions in which they conjugate in nature. As this discussion has attempted to bring out, there are aspects of mating type inheritance which bear on the breeding system, but little is known as to whether they do in fact operate this way in nature.

With due allowance for the points on which we are still regrettably ignorant, the mass of information now in hand on the varieties of *P. aurelia* has acquired for the first time coherence and significance by the perception of its bearing on systems of breeding. Instead of being merely so many interesting but isolated facts of natural history or experimental research, they are now the meaningful elements of the grand design by which this organism achieves its primary function of self-perpetuation. In view of its surpassing importance for every species, no surprise or suspicion should be aroused concerning the apparent convergence of so many aspects of the biology of P. aurelia upon it.

*Evolution of Groups A and B and Their Varieties.* (1) Originally groups A and B were recognized only on the basis of their different systems of mating type inheritance (Sonneborn and Dippell, 1946). In the discussion of the serotypes, it was pointed out that all the group A varieties examined (varieties 1, 3, and 9) show one system of serotypes and that all the group B varieties examined (varieties 2, 4, and 8) show a different system. Thus, group A varieties may equally well be distinguished from group B varieties by either the mating type or the serotype system. There is also a third difference between the two groups (Sonneborn, 1956c). Killers are common in all group B varieties in which many strains have been observed, and have been found in every group B variety of which more than one strain is available, i.e., varieties 2, 4, 6, and 8. Not a single killer strain has been found among the large number examined in the varieties of group A.[4] Curiously, these three independent sets of traits show in their inheritance a conspicuous cytoplasmic involvement in group B, not in group A; but the mode of cytoplasmic involvement seems different for each of the three sets of traits. One can hardly refrain from suspecting some at present unrecognized common element in the situation which leads to this nonrandom concentration of cytoplasmically inherited traits in group B.

The differences between group A and group B varieties in three sets of traits suggest that the cleavage into these two groups of varieties represents an ancient and important evolutionary divergence. If so, then one might expect the differences between the group A and group B alternatives for each of the three sets of traits to be rather simply derivable from one another. The alternatives with respect to the killer trait are genetically simple. The presence of this trait depends upon the presence of one major gene (Sonneborn, 1943) and a few minor genes (Sonneborn, unpublished; Balbinder, unpublished). Without these genes, the cytoplasmic particles, kappa, mu, or pi, which determine the killer trait and its various mutant forms, cannot be maintained in the animals. The differences between groups A and B with respect to the serotypes and mating types have been commented upon by several authors (Beale, 1954; Nanney, 1954, 1956b; Sonneborn, 1947, 1954b). Here I shall discuss only the mating types, for they are of special importance in relation to breeding systems and evolution.

The group A and group B systems of mating type determination and inheritance have been characterized in detail from various points of view at several places in the present paper. On the basis of these facts an attempt will be made here only to formulate these systems in an abstract and general way in order to see how many and what kinds of changes would be required to evolve one from the other. In making this formulation, I shall pass rapidly over the points on which evidence is good and interpretation is fairly straightforward, but shall say more about certain other points.

On a few points, common to both systems, there can be little if any doubt. These are: (1) the mating types are controlled by the macronucleus; (2) except in certain mutants, all macronuclei carry a genotype for *both* mating types; (3) usually a macronucleus becomes differentiated early in its development from a product of the syncaryon so that only one of its mating type potentialities will thereafter come to expression; (4) this macronuclear differentiation is permanently or long irreversible and, as a result, a caryonide is commonly pure for one mating type. Nanney (1956b) interprets the macronuclear differentiation as an intranuclear steady state between products of the alternative genes for different mating types.

The differences between the group A and B systems involve the agents that bring about macronuclear differentiation and the genetic control of these agents. In group B, the cytoplasm clearly is the agent of macronuclear determination, and the cytoplasmic agent is clearly controlled by that macronuclear mating type genotype which is being expressed. This is what results in the usual correlation in mating type between the two caryonides of a clone and between both of them and the mating type of the sexual parent. Superficially, the group A system seems very different: there is no correlation between mating types of the two caryonides of a clone or between these and the sexual parent. This lack of correlation gives the impression that macronuclear determination is not brought about by a cytoplasmic agent, an impression which is strengthened by the marked effect of environmental conditions, such as temperature, on macronuclear determination.

However, the facts are not contradictory to the interpretation of macronuclear differentiation by the cytoplasm or to the control of the cytoplasmic agent by the macronuclear genotype. Further, some facts point to just this sort of mechanism. Such a hypothesis will therefore be developed in order to have a single unified interpretation for both the group A and group B systems. There will obviously have to be some differences between the two systems, but the hypothesis will attempt to reduce these to the minimum and preserve maximal similarity. If it be assumed that there is a different cytoplasmic agent corresponding to each possible nuclear differentiation, i.e., two in a variety with two mating types, four in a variety with four mating types, there need be only two differences between the group A and B systems. (1) In group B, the cytoplasmic agents or differentiators of the macronuclei are subject to a steady state relationship; but in group A they are not. This means that in group B one or the other of the cytoplasmic differentiators is always in relatively high effective concentration as compared with the other or others; but that in group A a continuous series of relative concentrations is possible. (2) In group B, the macronuclear genes for the expressed mating type also as a rule determine that the corresponding cytoplasmic differentiator will have a relatively high effective concentration; but in group A, although the macronuclear genotype also controls the relative concentrations of the cytoplasmic differentiators, the pertinent part of the genotype is not that, or not only that, part which controls the expressed mating type. The genes involved in group A control of the cytoplasmic differentiators are assumed to be the same in both mating types.

With this interpretation, group A would show the absence of parent-offspring correlation that is observed. Further, the observed linear relation between temperature and the frequency of differentiation of macronuclei determining the even-numbered mating type would be explicable as due to the rise in relative concentration of the corresponding cytoplasmic differentiator with increasing temperature. Exactly the same rise has been noted by Nanney (1954) in group B when conjugants exchange cytoplasm and thus set up an approximately equal concentration of the two cytoplasmic differentiators. However, because of the steady state mechanism in group B, a rise in the concentration of the even-numbered differentiator leads to a steady state in which it greatly predominates. Finally, Sonneborn (unpublished) and Butzel (1953) showed that different genotypes in group A result in determination of different proportions of the mating types at the same temperature. Hence, the genes seem to be determining something on which temperature also acts. On the present hypothesis, this something is the relative concentrations of the cytoplasmic differentiators.

Viewed in this way, the evolution of the differences between the two mating type systems needs involve but few changes: the establishment or loss of the steady state condition for the cytoplasmic differentiators, and the predominant control of these differentiators by the expressed mating type genes versus control by genes which are identical in both mating types. If Nanney's (1956b) conclusion as to the existence in both groups of varieties of an intranuclear steady state among the products of the mating type genes is correct, the extension of this steady state mechanism into the cytoplasm in group B and not in group A may represent a relatively simple difference.

The group B system seems to be the more ancient one, occurring in all investigated varieties of other species of *Paramecium* and in one variety of *Tetrahymena pyriformis*. It is associated in some of these organisms with outbreeding and, as already mentioned, outbreeders were probably the ancestors of current inbreeders. However, one variety of *T. pyriformis* shows the group A system. Thus, either we are on the wrong track in supposing groups A and B of *P. aurelia* diverged early, or such divergence has occurred repeatedly and independently in different genera.

Since the group A system can arise merely by loss of two features of the group B system, the repeated independent evolution of the A from the B system seems not unreasonable.

2. Evolutionary relations within groups A and B. If the three different features which characterize all group A varieties and the alternative three which characterize all group B varieties indicate a fundamental and ancient evolutionary divergence of the two groups, we should expect to find evidences of this even in those features which distinguish varieties of the same group. For them, we should expect closer relations among varieties of a group than, on the average, between varieties belonging to different groups. This is indeed found with respect to the two sorts of chemical specificity expressed as antigens and as mating types.

In the account given earlier of the serotypes, it was pointed out that the specificities of the immobilization antigens were different for each variety, but that the differences were least for varieties 4 and 8 of group B. Most of the antigens that occur in one variety are matched by somewhat different but related antigens in the other. Resemblances are detectable between several of the antigens in these varieties and several of those in variety 2, the other investigated group B variety, but the resemblances are fewer and weaker. Of the three principal antigens in variety 1, two bear some resemblance to the group B antigens, but these resemblances are recognizable serologically in relatively few strains. Melechen (1955) reports that correspondences to all three principal antigens of variety 1 are indicated by antigens in variety 3. One of the two antigens found in variety 9 corresponds to one of the three main antigens in variety 1. The serologic relations are certainly closer for some varieties of the same group than for some varieties of different groups. Whether this relation will prove to be general remains to be discovered.

The relations of mating type specificity are indicated by the capacity of mating types of different varieties to interact sexually. Table I shows these interactions. Within group A, they occur among varieties 1, 3, 5, and 7, most strongly between 1 and 5. Within group B, they occur among varieties 4, 8, 10, and 14, most strongly between 4 and 8. The close relationship of varieties 4 and

8 is thus attested by both their serologic and their mating type specificities. The only sexual reactions between groups A and B involve variety 8 of group B. It reacts and mates with variety 3 and gives barely perceptible reactions but no mating with varieties 1 and 7. Sexual reactions between complementary mating types of different varieties, without regard to their intensity, occur in 6 of 42 possible combinations within group A, 6 of 42 within group B, and 4 of 84 between groups A and B. They are thus three times as frequent within each group as between groups. If only those that lead to mating are counted, there are 6 among the 84 possibilities within groups, and only 1 of the 84 possibilities between groups. The mating type specificities are thus on the whole less different within a group than between groups.

What light, if any, does the geographical distribution of the varieties throw on their evolution? On this subject, knowledge is of course meager, but it is suggestive. In group A, only variety 1 is known to be cosmopolitan; variety 9 has been found only in Europe, varieties 3, 5, 7, and 11 only in North America. Variety 15 will be disregarded for little is known about it. These facts of distribution suggest that varieties 3, 5, 7, 9, and 11 may have arisen from variety 1. The antiquity of variety 1 is indicated by its occurrence in Australia and Hawaii as well as in both Americas, Asia, and Europe. The relative recency of origin of the others is suggested by the fact that each is, so far as now known, confined to a single continent. And the close relationship of varieties 3, 5, and 7 to variety 1 is indicated by sexual reactions that occur between them. Serological evidence relates varieties 3 and 9 to variety 1. Significantly, varieties 3 and 5 are localized near the northern, and variety 7 near the southern limit of the range of variety 1 in North America. Altogether it seems reasonable to conclude that the other group A varieties (3, 5, 7, and 9) arose from variety 1 relatively recently.

Several varieties of group B have a wide distribution, but only variety 2 is cosmopolitan. Variety 6 occurs in India, Puerto Rico, and southernmost North America and it may be assumed to be subtropical around the world. Variety 4 has been found in Australia, Japan, and both Americas, but not in Europe. These three widespread varieties of group B give no sexual reactions with each other. Of the varieties of group B with more limited distribution, variety 12 has been found only once (in Madagascar) and it seems to be very different from the other group B varieties. The three remaining varieties, 8, 10, and 14, have been found only in North America, almost exclusively in its southernmost parts. As has been pointed out in detail, these three varieties are much alike in a number of ways and are also much like variety 4. These resemblances and distributions suggest that varieties 8, 10, and 14 evolved relatively recently from variety 4 directly or indirectly.

3. Gaps in knowledge. Many of the gaps in present knowledge have already been mentioned. A most important one is the paucity of information on the large animal varieties, both those that form two micronuclei and two macronuclei after fertilization and those that form four of each. Whether the latter constitute a distinct and readily recognizable species (*P. multimicronucleatum*) or whether they intergrade into the others or cannot be readily identified is one problem that needs solving.[2] Another is the possibility of mating and of serologic relations with the small animal varieties, a subject barely touched as yet. A third problem concerns how many varieties of large animals occur and what their distributions are. Attention is just beginning to be directed to this. When these matters are cleared up, a good deal more about evolution in these organisms will doubtless become apparent and some of the inferences in the preceding section may well have to be changed. It may, for example, turn out that certain large animal varieties are the closest to the ancestors of both group A and group B. I have long suspected, on a priori grounds, that the ancestors of these groups were more extreme outbreeders than any small animal variety yet found; and some of the large animal varieties are extreme outbreeders.

*Isolating Mechanisms.* Geographic isolation appears to be the most potent isolating mechanism in *P. aurelia*. Although it is obvious that the organisms have become spread over the face of the earth, the means by which they accomplish this remains unknown. Moreover, observations such as those I made over a period of years on the population of the Woodstock stream, together with Beale's observations on the occurrence of only one allele at each serotype locus in the samples taken in various years, show how rare must be the effective migration of a foreign strain into a preexisting population. In this case, moreover, the observations were made on a population of variety 1 in the midst of a region well within the range of the variety and where many other populations of the same variety were known to occur. If migrations from one body of water to a neighboring one are so rare, then migrations between continents must be exceedingly rare or nonexistent except on a geologic time scale. The fact that strains of the same variety (such as 1, 2, 4, and 6) from different continents can still interbreed freely when brought together in the laboratory implies that the widely distributed varieties must be evolving very slowly in spite of long isolation. Nevertheless, with each local population virtually locked up in the tight compartment of a river system or body of still water from which leakage is slow and erratic, the stage would seem to be set for abundant evolutionary divergence.

Differences in mating type specificity and differing conditions for mating reactivity (temperature, light), once they arise, constitute such effective barriers to interbreeding that different varieties can and do intimately coexist in the same body of water without losing their integrity. Even when the mating type specificities are only slightly different and interbreeding is possible, as in the case of varieties 4 and 8, the varieties are found to coexist in the same body of water in nature. From these observations, it would seem that even relatively slight changes in mating type specificity could lead to isolation of a new variety. Probably the inbreeding propensities discussed earlier assist such isolation. Outbreeders might not be so easily isolated in this way. Another related possible isolating factor is assortative mating with respect to body size (Jennings, 1911). There is some evidence that body size is in a general way correlated inversely with the rate at which the processes of meiosis and fertilization occur and marked differences in these rates in two mates leads to abnormality (Sonneborn, 1954b). Such abnormality would put the hybrids between the two types at a selective disadvantage.

The role of the killer traits in bringing about isolation is perhaps not as great as might at first be supposed. Killers that liberate paramecin might be expected to eliminate or reduce other strains in the population, especially when the population density is high. To the degree that mating is at random, the excess of killers would foster inbreeding among the killers, but outbreeding among surviving sensitives of the same variety. The mate-killers, known only in variety 8, present a special situation, which has been discussed by Siegel (1954).[4] Gene flow occurs readily from sensitive to killer, but not directly in the reverse direction. However, since mate-killers readily outgrow the cytoplasmic basis, $\mu$, of the trait sensitives of the same strain are likely to be present and through them some gene flow from the killer strain to other sensitive strains would be possible. Even so, gene flow in the two directions would probably be very unequal unless the proportion of mate-killers in the strain were low.

Whatever degree of isolation is due to killers is cytoplasmic in basis; and cytoplasmic involvement in isolation has also been found in another case. Levine (1953) showed that the death of hybrids between varieties 4 and 8 is due to the lethal action of the cytoplasm of each variety on the nuclei of the other. This is also to be noted as an additional evidence of the importance of the cytoplasm in the varieties of group B. I doubt whether the same sort of mechanism plays any role in isolation of varieties of group A. Butzel (1953) was unable to find any trace of such cytoplasmic action in hybrids between varieties 1 and 7. Cytoplasmic differentiation may regularly be a stage in genetic isolation in organisms with well-developed cytoplasmic genetic systems like group B, not in other organisms like group A.

Probably the most important isolating factor in *P. aurelia*, aside from geographical isolation, is the widespread and strong inbreeding habit. This is so important in relation to the species problem that the two are considered together in the next section.

*Breeding Systems and the Species Problem.* Among the 16

varieties of *P. aurelia* only three varieties, 12, 15, and 16, seem to be adapted primarily for outbreeding. Unfortunately, these are among the varieties which have been least studied from the point of view of speciation. It might be expected that these outbreeders and others like them, if such are yet to be discovered, constitute the main line of descent in *P. aurelia* from the past and the main line that will long continue into the future. They above all have the mechanisms for maintaining the supply of genetic variability needed for long-time survival. We may also expect such varieties to show the minimal differentiation of local populations. Among them, we may also eventually expect to find the best evidence for the existence of races embracing the local populations in a considerable land mass.

At present, evidence for such races is meager and is limited to Beale's (1954) survey of serotype alleles in strains of variety 1. This variety is intermediate in breeding habit. Beale's Table 19 shows that one serotype allele, designated $G^{90}$, is confined thus far to North American strains of variety 1: 7 of the 16 North American strains examined possessed it, but it was found in none of 17 strains from Japan, Europe, or South America. This is the only shred of evidence I can find for the existence of racial differences in *P. aurelia;* but there is as yet little available information pertinent to the subject.

The evidence that most varieties of *P. aurelia* are prevailingly inbreeders has already been fully set forth. This, together with microgeographic isolation, serves to go far toward making each local population a species in the making. The setup is precisely one which Wright (1940) considers so favorable for speciation. Evolution is indeed in progress on a lavish scale in these organisms. There must be many thousands or millions of such differentiated local populations of *P. aurelia* in the world. Many or all of them may be karyotypically diverse, so that laboratory crosses among them would yield the characteristic F2 mortality. This is one of the major things that makes the species problem so complex and difficult in *P. aurelia*. And it is in this sense that the species problem is tied so closely to the breeding system.

The species problem becomes more and more acute as one

passes from one breeding system to the next, in this sequence: obligatory outbreeder, facultative out- and inbreeder, obligatory inbreeder, exclusively asexual reproduction. Nearly the whole sequence is found among the varieties of *P. aurelia;* they embrace the range from nearly obligatory outbreeder to nearly obligatory inbreeder. Exclusively asexual reproduction is not found, for survival depends upon fertilization or something associated with it. The near-obligatory outbreeders have a wide geographic range both in longitude and latitude and relatively few closely related varieties of this sort are found in the same territory. The more one approaches obligatory inbreeding, the larger the number of varieties and the more restricted their range, especially in latitude. Toward the end of the series, each local population becomes almost as distinctive, differentiated, and isolated as a variety of outbreeders. Correspondingly, the further one proceeds down the sequence of breeding systems, the more difficult the species problem becomes until, in terms of the biological species concept, it bursts into nothingness when the closest inbreeding becomes obligatory or when sexuality is altogether abandoned.

Toward the end of this paper, we shall take up these extreme situations. But first we turn to the other species of Ciliates in which mating types are known. The points that have been made in the detailed analysis of *P. aurelia* will serve as a touchstone for orientation and interpretation in reviewing the status of knowledge in the other Ciliates.

## Paramecium caudatum

*The Varieties of P. caudatum and Their Distribution.* In general, the situation in *P. caudatum* bears many resemblances to the one just set forth in detail for *P. aurelia;* but there is, as will appear, at least one feature of special interest. Like *P. aurelia*, *P. caudatum* consists of a number of varieties or syngens. Each of these varieties has two mating types. At least 16 varieties have been found: varieties 1 and 3, independently by Gilman (1939) and by Hiwatashi (1949); varieties 12 and 13 by Hiwatashi (1949); and at least twelve more by Gilman (1939, 1941, 1949, 1950, 1954, 1956a,b). Gilman (1950) reported the relation of the four varieties found by Hiwatashi to those he had independently found. Giese and Arkoosh (1939) found in western United States a pair of interbreeding mating types which doubtless represented a variety. Y. T. Chen (1944) found four varieties with two mating types each in China. Vivier (1955) found a variety with two mating types in France. The relation of the varieties of Giese and Arkoosh, of Chen, and of Vivier to those of Gilman has not been reported.

Of the 16 numbered varieties, 1 and 3 have been found in Japan and the United States, 12 and 13 in Japan, 2 in North and South America, 8 in North America and Scotland, 15 and 16 in Europe, and the rest only in the United States. Most varieties seem to be scarce or absent in warm regions, but varieties 1 and 3 and probably 12 and 13 appear to be concentrated in warm to moderate regions. Two varieties may have a very restricted range (Gilman, 1956a and unpublished): variety 7 has been found only on the island of Martha's Vineyard and neighboring Cape Cod; variety 11 was found only once, in Connecticut.

*Varietal Differences.* Varietal differences are being investigated by Gilman (1941, 1956b), but relatively little has been published on this subject. There are some differences in the temperatures that give optimal mating reactions. Varieties 4 and 5 are resistant, while varieties 1, 2, and 3 are sensitive, to the killer strain H of variety 2 of *P. aurelia*. The animals of variety 3 appear to be smaller than those of certain other varieties. A number of characteristic varietal differences in the dimensions of the animals and of the micronuclei appear to exist, but most of the details are yet to be published. Thus far, it would be quite impossible to identify a variety by any features yet made known, except their mating types. Hence, the situation is in general the same as in *P. aurelia* with respect to the question of whether the varieties should be given specific names. This would be quite useless.

*The Question of Gene Flow within and between Varieties.* Nine of the 16 varieties, which may be designated group 1, are completely isolated from each other and from the other varieties by inability to interbreed. Varieties 3 and 6, in one combination of their mating types (VI with XI), give a faint sexual reaction which does not lead to conjugation. These varieties, together with varieties 1, 4, 5, 7, 11, 12, and 13, show no other sexual reactions except between the two mating types of the same variety. Strains of the same variety do interbreed, although little or no information is in the record concerning later generations. Presumably, however, each of these nine varieties constitutes a group of populations among which gene flow is possible, but between which there is no gene flow.

The remaining seven varieties, 2, 8, 9, 10, 14, 15, and 16, constitute what may be called group 2. (Gilman has spoken of it as the "variety 2 complex.") Certain combinations of mating types of different varieties in group 2 give sexual reactions and all the varieties of this group are interlinked in this way directly or indirectly. The intensities of these intervarietal reactions vary from the weak sort, mentioned above as occurring between types VI and XI, up to intensities equal to the maximal reaction normally obtained between complementary types of the same variety. Nevertheless, according to Gilman, each mating type of each variety has a uniquely defined pattern of mating reactions.

Whether the varieties of group 2 are sharply marked off from one another genetically, as the varieties of group 1 are sexually, seems to be as yet unsettled. Gilman has been studying this, but has not yet published his results. Meanwhile, Pringle (1955) and Johnson (1955) have been obtaining remarkable results on a considerable number of European strains. All these strains belong to group 2, but many of them have not yet been assigned to definite varieties. Indeed Pringle and Johnson find it difficult or impossible to sort their strains into varieties. Whether Gilman can do this, as he is now attempting to do, remains to be seen. The two former workers have the impression that matings between members of the same local population usually give a fair proportion of viable F1 clones, whereas crosses between representatives of different local populations give highly variable results but usually a less viable F1. Pringle suggests that each local population may have a gene pool which is to a considerable degree isolated from the gene pools of other local populations. Certainly the results at present are difficult to fit into a system in which there is free flow of genes among populations of a variety, but none between varieties. The extent to which the same difficulties exist in the American "varieties" of group 2 is unknown.

The preceding relations constitute the feature of special interest in *P. caudatum* to which reference was made at the start of this section. It is the possibility to which Pringle referred, namely, that group 2 or a part of it may be a group of differentiated local populations more or less isolated by F1 mortality. This, as he emphasizes, is different from the F2 mortality which serves to aid in isolating local populations of some varieties of *P. aurelia*. It is obviously most important for our understanding of evolution and speciation in *P. caudatum* to have this situation clarified.

*Breeding Systems.* The observations on the European populations belonging to group 2, as already indicated, point to inbreeding as the prevailing system in that material. There are other indications that inbreeding is widespread among the varieties of *P. caudatum*. With few possible exceptions, no immature periods have been found thus far, though the methods Gilman has employed would not detect periods of immaturity of much less than one month. Selfing is common in some varieties (Gilman, 1941) and it may occur more often in cultures of one prevailing mating type than in cultures of the other. The method of mating type inheritance is apparently the kind characteristic of the group B varieties of *P. aurelia*, at least in six of the seven varieties examined (Gilman, personal communication). This would assure the presence of animals of both mating types in a synclone. All these features speak for inbreeding. Of the varieties thus far examined, all fall into the same pattern with the possible exception of variety 3. The latter may show a number of outbreeding features, but the picture is not yet clear.

With such widespread occurrence of inbreeding features, autogamy might be expected. It, or what can now be interpreted as autogamy, was reported by Erdmann and Woodruff (1916), by Jollos (1916), and by Giese and Arkoosh (1939). On the other hand, Gilman (1941 and unpublished), Pringle (1955), and Johnson (1955) have searched for autogamy and have been unable to find it with certainty. These conflicting reports suggest that autogamy may occur in some varieties and not in others; but further investigation is called for. The relation of selfing to autogamy seems also to differ in different materials. Giese and Arkoosh (1939) found that autogamy was followed by the occurrence of conjugation in a culture previously pure for one mating type; autogamy had resulted in the differentiation of stable complementary mating types. Gilman (1941) studied cultures descended from single individuals and could find no autogamy, although selfing occurred within single cultures. It would seem that, as with different varieties of *P. aurelia*, some varieties of *P. caudatum* differentiate mating types without prior autogamy, while others do not. However, a special peculiarity in *P. caudatum* seems indicated. Gilman (1939, 1941) observed that the cultures which selfed, presumably without prior autogamy, in some cases contained animals which, when isolated, gave cultures pure for different mating types. In other cases, the selfing was apparently due to transient change of mating type, for animals isolated from the culture, even those that were coming together to mate, all gave rise to the same mating type or to predominantly the same type from each isolate. These varied observations indicate that there may be several mechanisms in *P. caudatum* which bring about mating between progeny of the same ancestral individual.

A different type of selfing was observed by Hiwatashi (1951) in variety 3, the variety that may be an outbreeder. He marked cultures of the two mating types with different vital dyes and observed that about 5% of the conjugant pairs obtained by mixing the two cultures consisted of two animals of the same color. This is selfing induced by mixture of pure types. It is not mediated by detectable hormones in the medium. I doubt whether this sort of induced selfing plays much of a role in nature, for it depends upon the formation of an agglutinated group consisting of at least two animals of one mating type and one of the other. It is unlikely that the formation of agglutinated groups occurs in the diluted populations of natural waters to anything like the extent observed with concentrated cultures in the laboratory. Yet this sort of selfing too remains as a natural possibility.

In brief, nine of the 16 varieties, composing group 1, are sexually isolated from each other and all others, and apparently they conform to "biological species" or syngens. The other 7 varieties, composing group 2, are not so sharply isolated from each other. The status of these varieties is still unsettled. It is possible that each local population is to a marked degree isolated from all others. The inbreeding with which this extreme evolutionary diversification is correlated is supported by a number of other adaptations for inbreeding in both groups of varieties, though some varieties may be outbreeders. Because of the widespread close inbreeding, the species problem in P. caudatum is even more difficult than in P. aurelia.

## Paramecium bursaria

*Mating Types and Varieties.* Six varieties of this organism have been reported: varieties 1, 2, and 3 by Jennings (1939), varieties 4 and 5 by Jennings and Opitz (1944), and variety 6 by Chen (1946b). The putative variety 5 is known from only one collection, and its existence is suspected only from the fact that the clones attributed to it do not mate with any other mating type standards. However, they also did not mate with each other. It is therefore dubious whether they were immature or impotent members of a known variety or whether they were of a distinct variety but all of one mating type. Because of the uncertain status of variety 5, it will be largely ignored in the present account.

Of the five remaining varieties, only one, variety 4, has a system of two mating types. The rest have systems of multiple mating types. Each of the varieties 1, 3, and 6 has four mating types, and variety 2 has eight. The breeding relations are the same in all: each mating type can conjugate with all the others in the same variety, but cannot mate with animals of its own mating type.

One and only one combination of different varieties can react sexually and conjugate; this is the combination of varieties 2 and 4. One of the two mating types of variety 4, type R, reacts and conjugates with four of the eight mating types of variety 2, the types E, K, L, and M. Chen (1946a) and Sonneborn (1947) pointed out that this indicates similarity between R of variety 4 and F, G, H, and J of variety 2, because all these mate with E, K, L, and M; and between S of variety 4 and E, K, L, and M of variety 2, because all mate with R. Obviously one cannot express such similarities in terms of a basic binary system of plus and minus types.

Metz (1954) suggests that mating reactions nevertheless depend upon reactions between pairs of complementary mating substances, even in varieties with multiple mating types. According to this hypothesis, there are three pairs of complementary substances in a system of eight interbreeding mating types: alpha substances A and a, beta substances B and b, gamma substances G and g. Each mating type would have one alpha, one beta, and one gamma substance. The formulas for the eight mating types would therefore be: ABG, ABg, AbG, aBG, Abg, aBg, abG, and abg. A system of two mating types would have only one pair of substances, e.g., A in one type and a in the other. The mating type with the A substance would thus be complementary to the four mating types with the a substance in an eight-type set. This hypothesis is consistent with the observation that one mating type in the two-type system of variety 4 mates with four mating types of the eight-type system of variety 2.

*Isolation of the Varieties.* The isolation of the varieties of *P. bursaria* is partly sexual. All combinations of varieties, except 2 and 4, are completely unable to mate with each other. (Jennings, 1939, p. 223, remarks that very rarely a single pair of conjugants is found when an immature clone is mixed with a mature clone of a different variety. He suggests that such pairs are not crosses, but selfing in the mature clone.)

In the exceptional combination of variety 4 with variety 2 Jennings and Opitz (1944) found that all four of the possible reactive combinations of mating types of these two varieties (R with E, K, L, and M) led to the same result: most of the conjugants died without separating; those that eventually separated died within four days without undergoing a single fission. Chen (1946a) examined in detail the cytology of the invariably fatal crosses. Although the nuclear processes proceed normally for a considerable period, abnormality suddenly develops. The chromosomes clump, meiosis does not proceed beyond the first division, and the nuclei acquire abnormal shapes and behavior. These abnormalities are not consequences of an intervarietal hybrid genotype, for the nuclear processes do not reach the stage of fertilization. As Chen pointed out, death is due to an interaction between the conjugants, which is dependent upon contact and possibly upon migration of substances from one mate to the other. The situation has features in common with the mutual nucleocytoplasmic incompatibility reported by Melvin (1949) and Levine (1953) for the cross between varieties 4 and 8 of *P. aurelia*. Melvin reported that when cytoplasmic exchange was induced (by exposure to antiserum), the pairs died before separating. The isolation of varieties 2 and 4 of *P. bursaria* is also probably to be understood as due to mutual nucleocytoplasmic incompatibility.

A viable F1 is thus never achieved in crosses between the known varieties of *P. bursaria*. On the other hand, as in *P. aurelia*, different strains of the same variety can interbreed freely and yield (under proper conditions, see below) viable, normal progeny. Hence, the varieties of *P. bursaria*, like those of *P. aurelia*, conform to the modern biological concept of species. In our terminology, each variety is a syngen.

*Varietal Differences.* As with *P. aurelia* and *P. caudatum*, the question now arises as to whether the varieties of *P. bursaria* could be identified without recourse to standard living cultures of the various mating types and if so, whether such identification is sufficiently practicable to warrant assigning species names to the varieties. In order to answer this question, the varietal differences will be summarized.

1. Geographical distribution. Variety 1 occurs in China and the United States (Chen, 1956); varieties 2 and 3 in the United States; varieties 4, 5, and 6 in Europe. Barring new extensions of range and the discovery of additional varieties in the regions explored, the problem of identification would be much simpler in *P. bursaria* than in the other organisms thus far considered, for not more than three varieties are known from any continent. If

the natural source of a strain is known, its identification is seemingly limited to but few possibilities.

2. Karyotypes. Chen (1946b) reported that the chromosomes of varieties 2 and 4 are thin and short while those of variety 3 are larger and much longer. However, similar differences distinguish strains of variety 6 collected from different parts of Europe, England and Czechoslovakia. The usefulness of this criterion is hampered not only by the possibility of differences within a variety, but also by the fact that one must obtain conjugants to examine the chromosomes satisfactorily. Late prophase or prometaphase of the first meiotic division is the stage employed.

3. Number of mating types. As stated above, varieties differ as to whether they have 2, 4, or 8 mating types. With regard to the usefulness of this criterion, the pertinent question is: Can the number of mating types in a variety be readily ascertained? Success in answering this question depends upon finding at least two different mating types in the variety, for Jennings (1941) reports that all types of a variety can be obtained among the sexually produced descendants of any two. The most reliable way of finding two types is to make a number of collections from nature in the same region. If only one type is found in nature, it is still possible to obtain all types, but this depends upon the occurrence of a rare event: self-differentiation of an additional mating type in a culture previously pure for one mating type. According to Jennings (1941), this happens on the average once in about 2000 culture-days. Once a culture yields a second type, conjugation between the two types can yield all the others in the variety. Lee (1949) reported that exposure to high doses of x-rays induced selfing in a previously pure culture. He suggests that the irradiation induced self-differentiation of the sort that occurs rarely without irradiation. If this proves repeatable and general, a method of finding all types in a variety would be in hand. However, the process of doing this would still take much time and labor because immaturity is apparently a regular occurrence after conjugation, and it may last a long time. Incidentally the difficulty of getting all types from one (without Lee's technique) is attested by the fact that in years of observation the one collec-

tion called variety 5 never yielded more than one mating type.

4. Other differences. At 26.5°, conjugation is completed in 20 to 22 hours in variety 1, but it lasts for 32 hours or more in varieties 2, 4, and 6. Variety 3 shows no daily periodicity in sexual reactivity; but varieties 1 and 2 are most reactive around noon and are not reactive at night (Jennings, 1939; Wichterman, 1948; Ehret, 1953). Reliable varietal differences in size, shape, or fine anatomy have not been reported. Strains of the same variety appear to differ more commonly in size, shape, and certain physiological traits than do strains of the same variety in either *P. aurelia* or *P. caudatum*.

5. Status of the problem of varietal identification. All the differences among varieties occurring in the same region involve mating types and conjugation. This imposes great labor and long-continued laboratory investigation on the part of anyone who sets out to identify a strain and therefore is, in my opinion, not suitable as a basis for species naming. However, it is true that with labor and time the varieties could be identified without recourse to living standard cultures. The requirements are perhaps not so great as they would be for a comparable accomplishment in *P. aurelia* or *P. caudatum*, but they are still much too great for routine taxonomic purposes.

*Life Features and the Breeding System.* The life features of *P. bursaria* form a striking contrast to those which characterize most varieties of *P. aurelia* and *P. caudatum*. The differences are by no means fortuitous. They are all clearly understandable as adaptations to or consequences of a different breeding system. In brief, *P. bursaria* is an outbreeder, while most varieties of the other two species are inbreeders to a greater or lesser degree. In this frame of reference, the life features of *P. bursaria* and their contrast with those of most varieties of the other two species make excellent sense.

The life history of *P. bursaria* is documented in great detail in a series of papers by Jennings (1944a,b,c,d; 1945). Most of this work was done on variety 1. My comments are therefore to be understood as applying chiefly to that variety. There is no indication in these papers of any differences in the major life features among the varieties, but that possibility needs to be kept in mind. Particularly, one would like to know more about variety 3, which is rarer and less widely distributed; and about variety 4, which has only two mating types. A priori, I would expect that these, if any, varieties might be less committed to outbreeding. The main features of life and the breeding system in variety 1 follow.

1. Immaturity. Conjugation is invariably followed by an immature period. It never lasts less than twelve days and usually lasts three to five months when the animals are well nourished. If the organisms are placed for a time under poor nutritive conditions during immaturity, this may disproportionately delay the onset of maturity. For example, 18 days of such conditions, with good nutritive conditions during the rest of the time, resulted in a total immature period of 10 to 14 months, and some clones failed to mature during a period of years. Translating these laboratory findings into what might be expected in nature, immaturity probably lasts at least a year, for under natural conditions poor nutritive conditions are hardly avoidable for long and this, combined with the lower fission rate usually to be expected, would extend immaturity at least to the 10-to-14-month range. This feature of its life is alone sufficient to mark *P. bursaria* as an outbreeder for the features that could balance it are lacking, as will be shown.

2. Maturity and mating types. The capacity to mate develops gradually and reaches a peak at which it remains for very long periods. In the next paragraph I shall define what brings maturity to an end. Under laboratory conditions, maturity last about 3 years. In nature, it probably lasts much longer because of the lower reproductive rate. As was mentioned in connection with the outbreeding varieties 15 and 16 of *P. aurelia,* a very long period of maturity is an adaptation to increase the probability of finding a suitable mate. Further, like variety 16 of *P. aurelia,* the varieties of *P. bursaria* (except variety 4) have multiple mating types which serve to increase the probability of mating when a mature stranger is encountered. Finally, unlike any variety of *P. aurelia,* in about 97.4% of the cases all four caryonides of a synclone are alike in mating type (Jennings, 1942). More-

over, caryonides that mature as selfers are totally unknown in *P. bursaria*. Both the absence of selfing caryonides and the usual synclonal uniformity in mating type serve to reduce the probability of mating between close relatives.

3. *Senility, autogamy, and selfing.* Senility is marked by decreasing fertility, as in other species. However, unlike most varieties of *P. aurelia*, variety 1 of *P. bursaria* is able to mate readily throughout senility, almost until the clone dies. The probability of death after conjugation is low, usually less than 10%, during maturity; but with the onset of senility, as a rule at an age of 30 to 40 months, the probability of death after conjugation begins to rise and continues to rise, often eventually approaching 100%. There is much variation in the details from clone to clone, but representative results are about 80% mortality after conjugation at 4 years of age and 100% or nearly so at 5 to 5½ years of age and thereafter. In the laboratory the period of senility lasts 4 to 5 years, making the total life cycle 8 to 9 years. (Of course many clones are weak and die immediately or soon after conjugation; the present account refers only to clones of full vigor.)

Comparable decreasing fertility with age is found in other Ciliates. Most varieties of *P. aurelia*, those that undergo autogamy, exhibit an age-correlated mounting death rate after autogamy; and varieties 15 and 16 show the same after selfing during senility. The varieties that undergo autogamy during senility can also conjugate during that stage, but conjugation is difficult to obtain. In this case, the parallel to the situation in *P. bursaria* is exact, for the probability of death after these conjugations also increases with the advance of senility.

The long period during which conjugation, but only cross conjugation, can occur in the life cycle of *P. bursaria*, 7 to 8 years, suggests how rare opportunities for crossing may be in nature. Long ago Weismann suggested that the length of the reproductive period in organisms is adjusted by natural selection to permit the production of just about a sufficient number of offspring to perpetuate the species. Even allowing for sexual progeny that fail to survive either because of inherent defects or because of extinction due to external causes, Weismann's suggestion, if valid, would lead to the conclusion that *P. bursaria* "needs" a long reproductive period because sexual reproduction rarely occurs.

Senility in *P. bursaria* may be marked by another feature, which also has its analogy in the other species that have been discussed. This is the phenomenon, mentioned earlier, which Jennings (1941) discovered and called "self-differentiation": a clone previously pure for one mating type comes to contain two mating types which can be isolated and shown to reproduce true to type. Self-differentiation is a very rare event, occurring once in about 2000 culture-days. It was not assigned to a definite stage in the life cycle by Jennings; the available data are actually not sufficient to make such an assignment with assurance. The uncertainties are due to two facts. First, many of the observed cases occurred in clones begun with a wild individual of unknown age. Second, the clones of known age, in which self-differentiation occurred, are heterogeneous with respect to the kind of matings that produced them and there may be life cycle differences associated with this. For example, clones produced by conjugation between two mating types that arose by self-differentiation within a single clone may themselves reach earlier than usual the stage at which self-differentiation can occur. There is some indication of this in the data of Jennings, but the matter has not been systematically explored. Tentatively, and largely by analogy with other species, self-differentiation in *P. bursaria* may be considered as a feature of senility which assures rejuvenescence for clones that have failed to find suitable mates, by permitting conjugation to occur within the clone.

Jennings (1941) assumed that self-differentiation was probably a consequence of prior autogamy. There were two reasons for this assumption. First, the differentiation of stable mating types within a single line of descent is due to autogamy in *P. aurelia*. Second, Erdmann (1927) briefly noted the occurrence of nuclear reorganization within a clone of *P. bursaria*, and her observations have been later interpreted as representing autogamy. However, neither of these reasons now seems cogent to me. Self-differentiation of mating types within a caryonide in the absence of autogamy has been demonstrated in varieties 15 and 16 of *P. aurelia* and in *P. caudatum*. In the latter, stable types may arise in this way. In the former, the differentiation can occur in variety 15 in amicronucleate caryonides which are therefore obviously incapable of undergoing autogamy. Further, Erdmann's observations were fragmentary and not well documented, and all could have been stages of reorganizing exconjugants, the conjugation of which had been overlooked. Finally, Chen's long and careful observations on *P. bursaria* have never included a single instance of autogamy in unpaired, i.e., non-mating, animals. It might be possible now, if Lee's (1949) method of x-irradiation actually induces self-differentiation, as he suggests, to put to a direct test the question of whether autogamy precedes self-differentiation. Previously, this was impossible because of its great rarity and unpredictability. Meanwhile, until better evidence is forthcoming, autogamy in single animals must be considered as not only undemonstrated in *P. bursaria*, but as probably nonexistent.

Thus, as matters now stand, *P. bursaria* conforms to the pattern of outbreeders by lacking autogamy and by using in its place self-differentiation and conjugation between the mating types thus differentiated, to restore vigor to clones that have reached senility without finding opportunities for outcrossing.

4. *Survival after inbreeding and outbreeding.* The comments made above about the relation of parental age to the frequency of death after conjugation are correct, but the situation is greatly complicated by the fact that other factors also influence the death rate after conjugation. A most important factor is the relationship of the parent clones themselves. In general, with one exception to be noted later, outcrosses yield better survival than crosses between relatives. Jennings noted 5 to 14 or more times as much death after conjugation between sibs (i.e., crosses between different synclones derived from the cross of the same two parental clones) as after conjugation between clones derived from different natural populations. Such comparisons were of course made with clones of similar age. These relations obviously reinforce the conclusion that *P. bursaria* is an outbreeder. They suggest that the organisms are normally heterozygous for lethal and detrimental genes which come to expression through inbreeding and are

masked to a considerable degree by outbreeding. Jennings' efforts to carry on series of successive inbred generations by sib matings were frustrated by rapidly increasing death rates which approached 100%.

The exception referred to above concerns the closest possible inbreeding, matings between the two types that self-differentiate within a clone. This will be referred to as selfing. No greater mortality appeared after such selfing than when one or the other type was outbred. However, there is some indication that the clones produced by selfing age more rapidly than clones produced by outbreeding: the former may give at ages of 12 to 18 months as much death after conjugation as the latter do at ages of 3 to 4 years. This, like the earlier occurrence of self-differentiation, indicates shorter stages in the life cycle after selfing than after outcrossing. The higher death rate from sib matings than from selfing or outcrosses led Jennings to remark, "There is in clonal self-fertilization some biological relation that prevents it from resulting in high mortality" (Jennings, 1944c, p. 195).

What this biological relation may be is of great interest, and this will be considered later. Here it suffices to note that it operates adaptively. The organism, as an outbreeder, survives inbreeding but poorly; yet, to survive at all, it requires some form of inbreeding as a last resort to forestall death of those members of a clone that have failed to find a suitable mate. It has managed to accomplish this by facilitating selfing through self-differentiation and it has somehow endowed this saving sort of mating with protection against the usually disastrous effects of other sorts of inbreeding. Whether something like this also operates in the selfings that occur during senility in variety 15 of *P. aurelia* needs to be investigated. It is even possible that a similar basis underlies the perplexing observations of Pringle (1955) and Johnson (1955) on inbreeding and outbreeding deaths in *P. caudatum*.

*The Mechanism of Mating Type Determination and Its Bearing on the Breeding System.* Jennings (1942) maintained that the mating types are determined by the chromosomal constitution. He based this conclusion on two facts: (1) conjugation brings about identical genotypes in the two clones of a synclone;

(2) in nearly all cases, these two clones are alike in mating type. Yet, when he tried to formulate genotypes which would agree with the observed results of crosses, he was totally unable to do so. The chief difficulty was that synclones of all possible mating types were producible from the cross of any two mating types, indeed from any single type which underwent self-differentiation. For, like other matings, selfings also could produce all possible types of synclones.

Sonneborn (1947, p. 340) pointed out that the inheritance of mating type in *P. bursaria* conforms in principle to the group B pattern in *P. aurelia* when there is massive exchange of cytoplasm between mates. The only difference is that there are four or eight alternative types in most varieties instead of just two. Sonneborn (p. 335) further cited three evidences that cytoplasmic exchange between mates does in fact occur regularly in *P. bursaria*. (1) Harrison and Fowler (1946) observed transfer of antigens from one mate to the other during conjugation. (2) As mentioned earlier, the abnormalities that occur during conjugation between varieties 2 and 4 appear to be due to interactions between the nucleus of each variety and the cytoplasm of the other, and they occur before nuclei are exchanged (Chen, 1946a). (3) Jennings (1944b) observed that conjugation between an old and a young clone usually kills both mates, abnormalities appearing during conjugation itself. The evidence subsequently obtained by Sonneborn and Schneller (1955) that nuclear abnormalities are induced by old cytoplasm reinforces my interpretation of the dependence of death of the young mate in Jennings' crosses on the action of cytoplasm received from the old mate. Since cytoplasmic exchange is relatively rare in *P. aurelia*, usually the old mate dies and the young one lives. To these evidences may now be added the fact that Chen's numerous figures of conjugation stages indicate the regular occurrence of a wide region of free cytoplasmic continuity between the mates. There can thus be little doubt that massive exchange of cytoplasm between mates is the rule in *P. bursaria*.

This then provides the explanation for the uniformity of mating type in a synclone. Nanney (1956b) has come to the same conclusion and shows in detail how *P. bursaria* fits the group B system. Exposure of the four macronuclear anlagen to a common cytoplasm differentiates all of them to control the same mating type. If my formulation of the operation of the cytoplasmic action in group B is correct, then there should be four, instead of two cytoplasmic differentiators in flux equilibrium in variety 1 of *P. bursaria*. Of the four, one is in high concentration in one mate, another in the other mate; the mixture of these two cytoplasms thus gives high concentrations of two of the four. This accounts readily for the fact that usually over 80% of the synclones produced from a cross are like one or the other parent in mating type, and that, of these, roughly half are usually like the one parent and half like the other parent. Whether temperature can affect these proportions is unknown. The hypothesis also explains some other remarkable observations, as will now appear.

The phenomenon of self-differentiation has this peculiar feature: it always results in the same two types in any one clone, but results in different types in different clones of the same original mating type. That is, a clone of type A may differentiate into A and B; another clone of type A may differentiate into A and C. Yet if either the new A or the new B in the first clone self-differentiates again, the A will always yield B and the B will always yield A. Likewise, in the other clone, the new A and C will always differentiate only into C and A, respectively. All combinations of two types are possible in different clones, but any one clone has only one such combination of possibilities.

The physical basis of these remarkable relations has never been elucidated. Kimball (1943) suggested that each possible combination of alternative mating types producible by self-differentiation was determined by a different genotype. He agreed with Jennings' conclusion that the usual uniformity of mating type in a synclone bespoke genic determination and suggested two reasons for Jennings' failure to find workable genotypic formulas. First, each formula should be not for a particular mating type, but for a particular pair of alternative types. Second, the formulas should take into account Chen's (1940) finding of polyploidy in *P. bursaria*. Kimball's proposals have not thus far led to the further analysis he called for. In my opinion, his suggestion is unsatisfactory, although the genotype probably does play some part in the system. The chief difficulties with Kimball's proposal are that it provides no insight into the basis for the choice between the two alternative types in each synclone or into the mechanism by which either of the alternatives can be inherited through great numbers of fissions.

These difficulties are precisely the ones which are accounted for by the analysis of the group B method of mating type inheritance in *P. aurelia*. In extending this to variety 1 of *P. bursaria*, four instead of two cytoplasmic differentiators are assumed in order to meet the requirements of four instead of two mating types. In a steady state system, these four would stand in a hierarchy of relative concentrations in each mating type, and the one initially in highest concentration would be the one corresponding to the expressed mating type. However, the relative concentrations of the other three might well vary in different synclones of the same mating type. With such a system of cytoplasmic constitutions, the cytoplasmic differentiator present in second highest concentration could be the one which determines the alternative mating type producible by self-differentiation. On this view, there are in each clone very rare occasions on which the two most concentrated cytoplasmic differentiators exchange relative positions in the hierarchy. Then, when age renders the macronuclei susceptible to redifferentiation, those animals which have the changed hierarchy of differentiators change mating type. As in *P. aurelia*, the determination of the specific hierarchy in any synclone or subclone is a resultant of the action of the genotype and of external conditions.

This is perhaps not the appropriate place to develop in detail how the previously unexplained peculiar features of mating type inheritance in *P. bursaria* become intelligible in the light of this approach to the problems; but, to indicate the fruitfulness of the hypothesis, one further example will be cited. Jennings (1942, Table 2) lists five crosses between mating types B and C, with different clones involved in each cross. One of the five crosses stands out as giving unusual results. Instead of the expected result, namely a 1:1 ratio of synclones of type B to synclones of type C, Jennings reported 39 type B and none type C. The clue to this exceptional result, in my opinion, is found in Jennings' paper of 1941 which states that clone Lo 12, which was the type C parent in the cross just mentioned, self-differentiated into types C and B. On my hypothesis, the production of type B to the virtual exclusion of other alternatives is a direct consequence of mixture of cytoplasm between two mates, one of which had the B cytoplasmic differentiator in highest concentration and the other of which had it in second highest concentration, as indicated by the type to which it could differentiate.

In like manner, the occasional production of synclones differing in mating type from either parent can be explained. These new types often correspond to the second most concentrated cytoplasmic differentiator in one or the other or both parent clones, as inferred from the type to which the parent self-differentiates. Although this information is available for only a minority of the parents reported upon by Jennings, in these cases the agreement with expectation is on the whole impressive. However, there are also a few instances of apparent disagreement. The outstanding exception appears in one of the five crosses of mating types B and D. This yielded a 1:1 ratio of A to D instead of B to D. Since the B parent self-differentiated into C and the D parent arose by self-differentiation in a B clone, my hypothesis leads to the expectation of an excess of B progeny, whereas in fact B progeny were totally lacking. However, it is to be noted in this case that the D parent was derived from an inbred ancestor which had for a few generations repeatedly differentiated only into A and D, the B arising as a late exception (Jennings, 1941). Here then is a possible case of a less probable and stable cytoplasmic state reverting to a more probable and stable equilibrium. In view of the obvious complexities of the general system, involving a fourfold steady state system acted upon by genic and environmental factors, I am of the opinion that an apparent exception of the sort just mentioned is not serious enough to invalidate the hypothesis. What is now needed is a number of experimental tests directed specifically toward disentangling the various components of the system as formulated in the hypothesis. Many such tests come at once to mind. Meanwhile, this hypothesis is formally capable both of accounting for most, if not all, of the known kinds of facts and of leading to testable predictions.

One aspect of the breeding results still remains as puzzling in the extreme. This is the contrast between the high viability after selfing and outcrossing and the low viability after crosses between sib clones. I cannot suggest a satisfactory explanation, but attention might be called to one point: selfing occurs between two types which, on my hypothesis, are necessarily as much alike in cytoplasmic constitution as two different mating types can be: the most concentrated cytoplasmic differentiator in each is the same as the second most concentrated differentiator in the other. How this could influence the viability of the resulting synclones is far from clear, but it would be of interest to compare mortality in sib crosses and outcrosses between clones that have this sort of cytoplasmic relation and clones that are more diverse in cytoplasmic constitution.

*Evolution, Speciation and the Breeding System.* The evolutionary relation between the systems of mating type determination in *P. bursaria* and *P. aurelia* are close, and the differences are simple. In principle, the system in *P. bursaria* is the most complex, the one in the group A varieties of *P. aurelia* is the simplest, and the one in the group B varieties is intermediate. The single unique feature in *P. bursaria* is the regularity of massive cytoplasmic exchange between mates. In the group B varieties of *P. aurelia* this feature occurs only in a small minority of the conjugating pairs. In group A, the cytoplasmic role in mating type determination is inapparent because of the lack of evidence of flux equilibrium and the overriding influence of temperature and of a genotypic action which is the same in both mating types. All other features are found developed to varying degrees in both species: caryonidal differences, nuclear differentiation, cytoplasmic differentiators and the genes affecting their concentration, dual and multiple mating type systems, and mating type differentiation sometimes associated with and sometimes independent of a preceding fertilization.

The regular exchange of cytoplasmic differentiators during conjugation in *P. bursaria* is one of its key adaptations to outbreeding. It almost completely prevents the occurrence of mating within a synclone during maturity: only 2.6% of the synclones are exceptions to this rule. This feature, the long period of immaturity, the long life cycle, multiple mating types and the high mortality from sib crosses, assure a highly developed system of outbreeding. Variations from population to population occur, but the populations are prevented from progressive divergence by the forced interbreeding. A common gene pool is thus maintained over a wide range.

As a consequence, speciation occurs on a much smaller scale than in the inbreeders, *P. aurelia* and *P. caudatum*. Only 3 varieties of *P. bursaria* have been found in the United States; they were discovered by 1939, within two years after the discovery of mating types, and no additional ones have been found in the large number of collections subsequently examined. At the same time, the first three varieties of *P. aurelia* were found; but in the following years ten additional varieties were discovered in the United States. Probably few if any additional varieties of *P. bursaria* will be found in this region; but doubtless more, possibly many more, varieties of *P. aurelia* will be discovered, for several new ones turned up within the past year. The difference is accentuated by the existence of karyotypic differences between local populations of *P. aurelia* and their regular production of F2 mortality after crosses between representatives of different populations. Speciation is indeed rich in the inbreeder, poor in the outbreeder.

## *Tetrahymena pyriformis*

*Mating Types and Varieties and Their Distribution.* Like the other Ciliates which have been studied in this connection, *T. pyriformis* consists of a number of sexually isolated varieties. Elliott and Gruchy (1952) discovered variety 1; Gruchy (1955) reported varieties 2 through 8; and variety 9, which seems not to occur in North America, was found by Elliott and Hayes (1955) in Colombia and in the vicinity of the Canal Zone. In the latter places and in Mexico, they also found variety 2, which is widely distributed from Canada to Central America. The other seven varieties are known only from North America. Gruchy (1955) observes that varieties 1, 4, 5, and 8 have been found only in the northern states; that variety 3 seems to occur mainly in running water; and that varieties 1, 5, 6, and 8 have been found only in standing water. In striking contrast to the varieties of *P. aurelia*, of which two or more are often found together in the same body of water, the varieties of *T. pyriformis* usually occur alone, two varieties in the same collection having been found only once (Elliott and Hayes, 1955). Altogether, cultures from about eighty collections in the Americas have been identified as to variety; no collections from other continents have yet been reported upon.

The varieties of *T. pyriformis* are diagnosed primarily on the basis of sexual reactions. If a clone from a new collection mates well with a standard mating type culture, it is assumed to belong to the same variety as the standard. Variations in intensity of the mating reaction appear not to occur under optimal cultural conditions (Gruchy, 1955). In the case of *P. aurelia*, it will be recalled, varieties 4 and 8 interact sexually with almost normal maximal intensity. It is therefore possible, but perhaps not likely, that in *T. pyriformis* some different varieties have been, on the basis of strong intermating, misclassified as belonging to the same variety. Tests of genetic isolation, which have proved important in *P. aurelia* and *P. bursaria*, have not yet been carried out extensively in *T. pyriformis*. (There are practical difficulties here, as will appear.) For this reason, the status of the nine reported varieties, particularly in relation to the problem of gene flow with which we are concerned here, is still dubious. However, it seems clear that there are at least nine varieties among the materials examined, for these are all reported to be isolated by failure to intermate. The study of gene flow within these varieties could serve only to increase their number, if it changes the situation at all.

Another systematic exclusion of possible additional varieties comes from the practice of ignoring all isolates from nature which prove to be selfers. Gruchy (1955) reports that about 5% of the wild strains collected were selfers. If some varieties regularly self, like variety 13 of *P. aurelia*, these would be overlooked.[6] At present there seems to be no satisfactory way to get around this, for it has proved difficult or impossible to discover the variety to which such strains belong. One wonders whether the technique of using the animals left over after a culture selfs and before it is fed again would work, as it does with variety 13 of *P. aurelia*. Nanney (1953) tried another and ingenious approach to the problem in wild selfers of another species of *Tetrahymena*, but without success. These selfers gave 100% death of exconjugants. On the assumption that exconjugant survival might be improved by outcrosses, he mixed selfers derived from different sources and examined the viability of the conjugants isolated from the mixtures. No difference was observed from the results obtained with selfing conjugants from a single source. This should not discourage further use of the method since it is the result expected if the two strains of selfers belonged to different varieties. Perhaps other combinations of selfers would give a different result, indicating that they belong to the same variety. The method could even be profitably extended to mixtures between a selfer and a pure type.

*Varietal Differences.* The nine varieties appear to be alike in many respects, even in some that distinguish varieties of other species of Ciliates. No diurnal mating periodicities have been found, and differences in temperature requirements for mating seem not to exist (Elliott and Hayes, 1953; Gruchy, 1955). The haploid chromosome set consists of one large, one small, and three intermediate chromosomes, all with median or submedian centromeres; and this seems to be the same in all varieties that have been examined, i.e., in all but varieties 3, 7, and 8 (Ray, 1954, 1956; Ray and Elliott, 1954; Gruchy, 1955). There are, however, some differences among the varieties. Differences in distribution and in preference for running or still water were mentioned earlier in this section; others follow.

1. Size, morphology, and cytology. According to Gruchy (1955), animals of varieties 3 and 7 are unusually large, and

those of variety 5 are unusually small. Gruchy further remarks that Corliss found other peculiarities in varieties 3 and 7, the cilia being arranged in fewer rows and part of the meridians appearing heavier and more definite; but earlier, Corliss (1954) mentioned that he could find no differences in structural details among the varieties. Presumably, these differences are statistical and not suitable for identification purposes.

2. Mating types, the mating reaction, and mating type inheritance. The most striking differences among the varieties are in the number of mating types they contain. There are two each in varieties 5, 7, and 8; three each in varieties 4 and 6; five in variety 9; seven in variety 1; eight in variety 3; and eleven in variety 2 (Elliott and Gruchy, 1952; Elliott and Hayes, 1953; Nanney and Caughey, 1953, 1955; Elliott and Hayes, 1955; Gruchy, 1955). The system of possible matings within a variety follows the patterns of *P. aurelia* and *P. bursaria:* no type mates with other clones of the same type; each type mates with all the others of the same variety. Of course, if any of these varieties proves by genetic tests to be more than one variety, that would reduce the number of types in the varieties involved.

On the other hand, there may well be more mating types in some of these varieties than are now known. In other Ciliates, the progeny test has been used to discover whether additional mating types can occur. In *T. pyriformis,* primary reliance has thus far been placed on finding the types in nature, although some effort has been made to use the progeny test. Its importance and also its limitations have been shown for variety 1 by Nanney and Caughey (1953) and Nanney, Caughey, and Tefankjian (1955). From crosses between the two original strains of type I and type II, they obtained among the progeny types III through VII. Only types I, II, and III have been found in nature so far (Elliott and Gruchy, 1952; Elliott and Hayes, 1953; Gruchy, 1955). But Nanney and his co-workers have also shown that clones of different genotypes are restricted as to the mating types they can produce among their sexually derived progeny. One genotype cannot produce type I; another cannot produce types IV or VII. Thus, even the progeny test is limited, and the full number of types in some varieties may not be known until similar studies have been made on crosses among a number of strains of the variety from different sources. The matter is further complicated by the low frequency with which certain types appear among the progeny and the somewhat different frequencies of types under different conditions (Nanney, Caughey, and Tefankjian, 1955). All these difficulties will of course be resolved in time. Meanwhile, Gruchy (1955) has tried to look for new types from crosses and found a new type in variety 6 in this way; but the difficulties of immaturity and viability have made this impossible in some varieties.

Unlike *Paramecium,* the animals in all varieties of *T. pyriformis* fail to agglutinate or react immediately when complementary types are mixed. There is a refractory period of at least an hour or two and, in variety 9, of 24 to 36 hours; then pairing is direct without agglutination (Elliott and Gruchy, 1952; Elliott and Hayes, 1953; Nanney and Caughey, 1953; Ray, 1955). This has raised in several investigators' minds the question of whether there may be sex hormones which induce selfing. Elliott and Hayes (1953) presented evidence that selfing is not induced in this way and proof was provided by Nanney and Caughey (1953). Nevertheless, the latter do not totally exclude the possibility of some sort of sex hormone action; but the former and Gruchy (1955) hold that there is no action of fluid in which one type has lived on animals of the other type. It would be interesting to know whether exposure to such fluids could reduce the refractory period. In view of Ray's (1955) report of a 24-to-36-hour refractory period in variety 9, perhaps this would be the variety of choice for such a test.

The method of mating type inheritance is known only for varieties 1 and 2. The work on variety 2 has just begun, but Hurst (unpublished) has made the important discovery that this variety exhibits the same pattern of mating type inheritance as *P. bursaria;* that is, a synclone is usually uniform with respect to mating type, some synclones from a cross having the mating type of one parent and some having the mating type of the other parent. This indicates that the basic mechanism is the same as in group

B of *P. aurelia*, but with regular exchange of cytoplasm between mates. The existence of 11 mating types in variety 2 would, if our hypothesis is correct, indicate a flux equilibrium among 11 cytoplasmic differentiators. Although this might seem extraordinary and unduly complicated, it should be remembered that equal complexity in flux equilibrium is present in the serotype system of the group B varieties of *P. aurelia*. Nanney (1956b) has pointed out the detailed resemblances and differences between the mating type and serotype systems. Variety 2 of *T. pyriformis* may well prove to be the material of choice for testing the interpretations proposed above for *P. bursaria*.

Mating type inheritance in variety 1 has been analyzed by Nanney and Caughey (1953, 1955), Nanney, Caughey, and Tefankjian (1955), and Nanney (1956b). In contrast with variety 2, variety 1 shows the group A pattern of mating type determination and inheritance. Leaving selfers aside for the moment, each caryonide seems to be independently determined for mating type. Hence, among the four caryonides from a pair of conjugants, there are usually present hereditary differences in mating type; in fact, as expected, even the two caryonides from a single exconjugant are usually different in mating type. With seven possible mating types, a large number of combinations is found among the two caryonides from an exconjugant or among the four from a pair of conjugants.

There are, however, genic restrictions as to which mating type potentialities can be realized. One gene permits the development of any type except I; another gene permits the development of any but IV and VII. These genes behave as alleles. The authors point out some difficulties raised by these discoveries with respect to Metz's scheme of pairs of complementary mating type substances (see earlier discussion under *Paramecium bursaria*), but realize that these difficulties need not be fatal to Metz's hypothesis. The mating type potentialities may be modified by genes at different loci which act in different ways.

3. The life cycle. There appear to be some varietal differences in the length of the immature period, but the literature contains conflicting statements of the details. Part of this is clearly due

to the use of different methods of culture by different investigators or by the same investigator at different times. Sometimes cultures are transferred (fed) once a month, sometimes once a day. This leads to great differences in the observed immature period. With rapid growth, the immature period is very brief, if it exists at all, in varieties 2, 4, and 6 (Ray, Elliott, and Clark, 1955); exconjugant clones of these varieties could mate again within 3 to 5½ days. In contrast, Gruchy (1955) stated that a few exconjugants of variety 6 were not mature after four weeks of growth, although most matured within this period. Variety 1 has been most fully studied in this respect. Elliott and Hayes (1953) give two to four weeks as the immature period; but Nanney and Caughey (1955), growing the animals at a rate of about 10 fissions per day, claimed that 16% of the exconjugants mature after 50 fissions, 43% after 60 fissions, 83% after 70 fissions, and 100% after 80 fissions. If the animals were starved before each transfer (every 12-13 fissions), 100% were mature by 60 fissions. Because of the time involved in starvation (transfers every third day), the time till maturity was prolonged although the number of fissions was reduced. Variety 8, according to Gruchy (1955), matures in nine days of rapid growth, which is about the same as variety 1; but with monthly transfers, they were still immature at six months. Gruchy further reports that exconjugants of varieties 3 and 7 were still immature after eighteen and nine months, respectively. These varieties seem to be difficult to cultivate and grow slowly or erratically; this may be involved in the seemingly very long period of immaturity.

Although there is no question about the existence of an immature period or about its being followed by a period of maturity, the possibility that maturity is followed by a period of senility is just beginning to engage attention. Before inquiring into the existence of varietal differences in the period of maturity, the question of whether maturity ever comes to an end must be dealt with. This is by no means an uncalled-for question. Cultures of *T. pyriformis* have been maintained in axenic medium for decades, and this has created widespread conviction that this species is potentially immortal and hence lacking a life cycle.

Recent discoveries have, however, made it necessary to reevaluate the situation.

The first shock came with the report of Elliott and Nanney (1952) that one of the long-cultivated strains, E, has been amicronucleate for at least 20 years. But the full impact of this discovery becomes clear only in connection with other recent discoveries about *T. pyriformis* and *P. aurelia*. After mating types were discovered in the former species, efforts were made to obtain collections far and wide in the Americas in order to find the mating types and varieties in this region. This search turned up a surprising and, in my opinion, highly significant fact: one-third to more than one-half of the collections of *T. pyriformis* from nature proved to be amicronucleate. Thus 50 of the 154 collections examined by Gruchy (1955) and more than half of the collections of Elliott and Hayes (1955) in Central and South America were amicronucleate. To me these facts signify that the amicronucleate condition is, at least in some varieties, a regular and long-lasting stage of life.

The absence of micronuclei is clearly to be characterized as a senile condition, for two reasons. First, an animal without micronuclei is genetically dead. Its genotype, confined to the macronucleus, could not be transmitted to sexual progeny, since gamete nuclei cannot arise directly or indirectly from a macronucleus. Second, as more and more amicronucleate cultures have been studied, it has become clear that absence of a micronucleus is invariably associated with inability to mate, i.e., even to unite in pairs (Elliott and Nanney, 1952; Elliott and Hayes, 1955; Gruchy, 1955). Hence, amicronucleates have no mating type and, once they get into this condition, the variety to which they belong can no longer be ascertained. However, one can guess that the amicronucleates found in Central and South America belong to variety 2 or 9 or both, because these are the only varieties thus far found in that region. Since maturity by definition is the stage of life during which fruitful mating can occur, amicronucleates are not mature and cannot later become mature. In short, they are senile.

The situation is in part parallel to the one existing in variety

15 of *P. aurelia:* these organisms also seem to lose their micronuclei in old age and to be genetically dead even though the line of descent can continue to reproduce a very long time in this condition, if not indefinitely. It will be recalled that we find Johnson's long-lived axenic culture of variety 15 (known in the literature as *P. multimicronucleatum*) to be amicronucleate, and other strains of this variety show the same condition. However, amicronucleates of *P. aurelia,* unlike those in *T. pyriformis,* are able to mate even though they cannot transmit nuclei to progeny produced by their mating.

If the view developed here is sound, long-continued laboratory observation should reveal the sequence of immaturity, maturity, and senility, i.e., loss of micronuclei, and should further reveal the time at which senility, so defined, begins. Such observations are just beginning to be made. The results already indicate a complication apparently resulting from two different causes of micronuclear loss. On the one hand in some clones considerable variation in the number of micronuclei, accompanied by the production of amicronucleate animals, appears early. Although Elliott and Hayes (1953) found no relation between micronuclear number and age up to 40 days in variety 1, Nanney, Caughey, and Tefankjian (1955) found that amicronucleate animals characteristically arise with high frequency in some clones of the same material, beginning about 40 to 50 fissions after conjugation. Nanney (1956a) has intensively studied these clones, which he calls semi-amicronucleates. Amicronucleate animals seldom arise before the fortieth fission and may not appear until the one hundredth fission or later, i.e., at about ten days. All the amicronucleate animals in these clones die within a short time. This condition is clearly determined by the genotype; it is not by any means an invariable feature of all clones.

On the other hand, according to Nanney (unpublished), normal clones, after reaching 500 to 1500 fissions, produce amicronucleate animals and show other signs of aging. These amicronucleates, like those produced earlier in certain clones, may also be nonviable, for the entire clone is deteriorating. However, some amicronucleates are clearly capable of living and reproducing for a long time, as mentioned above.

Whether senility can be characterized by criteria other than loss of micronuclei is not yet clear. There are indications that selfing may also be a feature of senility in some varieties, as it is in varieties 15 and 16 of *P. aurelia*. My reason for suspecting this is that Gruchy (1955) reports differences as to the viability of exconjugants obtained by selfing in different collections of selfers from nature. It will be recalled that Giese (1957) found the selfed progeny of senile clones in variety 16 of *P. aurelia* to show increasing nonviability with increasing parental age. The differences reported by Gruchy may be of the same sort. Unfortunately, as already mentioned, the varieties to which the selfers belong has not been, and perhaps cannot be, ascertained. Until the phenomenon is discovered to arise with age in clones earlier identified in the laboratory, the occurrence of selfing as a feature of senility will remain uncertain and, if it does occur, the varieties in which it occurs will remain unknown.

At present, the life cycle in *T. pyriformis* appears to be something like this: there is an initial period of immaturity which is followed by a long period of maturity; senility may begin with age-induced selfing, which at first yields viable progeny and progressively yields lesser proportions of survivors until none survive; or senility may begin with loss of micronuclei and last at least for decades. Because of the paucity of available information, little or nothing is known about varietal differences in senility. Some may well become selfers, others may lose micronuclei, some may do first the one and then the other. All this remains to be discovered.

There is also practically no available comparative information on the period of maturity. Detailed studies are virtually confined to variety 1. Many caryonides are characterized by purity for a single mating type throughout maturity, but a considerable fraction begin maturity as selfers according to Nanney and Caughey (1953, 1955). These workers discovered that pure types can often be isolated from such selfing caryonides and that starvation during immaturity is highly effective in stabilizing potential selfers as a pure type. They pointed out that early selfing is a developmental stage characterized by progressive limitation of expressible mating potentialities. Early selfing has also been reported in variety 6 by Gruchy (1955) and may well occur in other varieties.

This early selfing is to be distinguished from age-induced selfing, just as it was in *P. aurelia*. The selfers collected from nature would probably in most cases be past the incompletely determined early stage of development. Moreover, they would probably have been much exposed to the stabilizing action of starvation. In fact, since conjugation is induced by starvation, the exconjugants would at once be exposed to this stabilizing influence and early selfing would be much rarer in nature than in the laboratory where exconjugants are routinely provided with excess food. Finally, the available reports indicate that selfers collected from nature are not stabilized as one type by starvation in the laboratory. A sharp distinction between the two kinds of selfers should therefore be made.

To sum up the situation on varietal differences in periods of the life cycle, very little is yet known about this, except for the existence of differences in length of the immature period. However, from what is known about the marked difference between varieties 1 and 2 in the method of mating type inheritance and in geographical distribution, one may expect them to differ in breeding system and in many features of their lives adapted to this difference. Since the first two varieties to be studied show such marked differences, it is to be expected that others also will. This is already suggested by the observations on selfers and amicronucleates found in nature.

4. Are the varieties of *T. pyriformis* readily identifiable? Present knowledge provides only one way to identify the varieties and that is by their mating reactions. Other differences are scarce and of little or no use in identification. There is, if anything, even less ground for assigning them specific names than there was in *P. bursaria* or *P. aurelia*. The question of whether they should be considered biological species is in principle again the

same. However, the degree to which gene flow occurs or is possible between different populations is not yet well worked out in *T. pyriformis*. Some further facts bearing on this will be brought out in the following section.

*Breeding Systems.* Some information is available with respect to viability and variability of exconjugants after crosses between representatives of a single local population of variety 1 and, with respect to viability alone, after crosses between representatives of different local populations of varieties 5, 6, and 9. Convincing evidence that natural strains of variety 1 are highly heterozygous has been obtained by Nanney, Caughey, and Tefankjian (1955) and by Nanney (1956a and unpublished). Among the sexually produced descendants of two individuals of complementary mating types that had been isolated from the same natural source, two alleles for mating type potentialities segregated. The analysis likewise yielded evidence of polygenic differences controlling the production of the semi-amicronucleate condition, other degenerative conditions, and low viability in general. Two of the segregant families were further inbred separately. In spite of selection, viability of exconjugants in each inbred family occurred in lower and lower frequencies in successive inbred generations; but then selection resulted in improvement, and by the sixth inbred generation it was possible to select a strain which gave 100% viability at conjugation. Usually, crosses between the different inbred lines, at the stage when they were giving poor survival and much abnormality, yielded more normal sets of exconjugants.

These results of Nanney show that variety 1 of *T. pyriformis* differs from variety 1 of *P. bursaria* in the effects of inbreeding. In the latter it seems to lead quickly to extinction, but in *T. pyriformis* completely normal and viable inbreeding lines can eventually be selected.

Crosses between representatives of different local populations were made in varieties 5 and 6 by Gruchy (1955) and in variety 9 by Ray (1955). In the single parental combination tested for each variety, all the exconjugants of varieties 5 and 9 died, but viable exconjugants were obtained from the cross in variety 6.

Viable exconjugants from a cross in variety 3 were also obtained by Gruchy, but I cannot find information as to whether the clones crossed were from the same or different populations. Some comment is in order on the meaning of their statements about obtaining viable conjugants. Apparently this is not intended to mean that all or even a large proportion of the exconjugant clones lived, only that some did. This is inferred because Elliott and Hayes (1953) reported that variety 1 exconjugants were viable, while Nanney and Caughey (1953) and Nanney, Caughey, and Tefankjian (1955) emphasize that crosses of the same strains as those used by Elliott and Hayes yielded a high frequency of abnormalities and few viable clones. In the absence of detailed mortality data on the other varieties, it is difficult or impossible to know the extent to which different natural populations are isolated by genetic barriers to gene flow operating through mortality in the F1 and later generations.

Ray's (1955, 1956) observations on the cytology of conjugation in varieties 1 and 9, however, are very suggestive. In both varieties, he found chromosomal abnormalities at the first meiotic division. They were abundant in variety 9, which gave no viable progeny, and less common in variety 1, which gave some viable progeny. Among the abnormalities noted were clumping of chromosomes, lagging on the spindle, multivalent associations, and chromosome bridges at anaphase. The last two abnormalities of course indicate that the strains were structural heterozygotes. These abnormalities were observed when strains from nature conjugated. The natural strains thus might have been hybrids themselves, but another interpretation to be given in the next paragraph is not excluded. Whether the strains being crossed, which were from different sources, also had different chromosomal arrangements remains unknown.

The interpretation of the preceding observations is rendered difficult if we keep in mind the complications that have been exposed in the fuller studies on other Ciliates. In both *P. aurelia* and *P. bursaria* age is a most important factor influencing survival or death after conjugation. Moreover, chromosomal aberrations are induced by age in *P. aurelia*, and further study may show that they are similarly induced in *P. bursaria*. The important question therefore arises as to the extent to which age may be a factor in the observed exconjugant mortality and in the observed chromosomal aberrations. Until this is cleared up, any conclusions as to the breeding system based on material in this section, except the inbreeding studies on variety 1, would be premature. The alleles for mating type potentiality discovered within a single population are good evidence that some degree of heterozygosity occurs within a population. The observation of inbreeding degeneration followed by selection of fully normal lines also probably indicates genic heterozygosity, though part of these results might be consequences of age-induced chromosomal aberrations in the starting material.

When we now look back over the various features that have been discovered in *T. pyriformis* and ask ourselves whether this organism is an inbreeder or an outbreeder, the situation seems confused. The organism clearly has some outbreeding features, and it shows evidence that they work. Among them are the usual ones: a period of immaturity, multiple mating types (in most varieties), a very long life cycle, and low viability after inbreeding. The finding of heterozygosity for genes within a single local population is good evidence that outbreeding does in fact take place. On the other hand, inbreeding features also seem to be present. Among them, the most outstanding are (1) the occurrence of selfing clones in considerable proportions and (2) the method of mating type inheritance in variety 1, which assures the presence in most clones of animals capable of mating with each other.

However, these various aspects of the biology of the organism, as discovered in the laboratory, must be considered in the light of conditions in nature. Three factors tend to reduce the importance of the inbreeding features. First, the likely recurrence of starvation in nature would tend to reduce, if not prevent, the development of selfing clones during early maturity. Second, although mating type inheritance provides potential mates among close relatives in variety 1, the immature period permits time to spread and so reduces the chance of mating with closely related individuals. Third, inbreeding nonviability tends to eliminate those which do inbreed; and, if there are present simultaneously outbred competitors, these would have an advantage in subsequent population increase.

Looking at the picture as a whole, therefore, I am inclined to interpret *T. pyriformis* as an outbreeder, but as one that has an already highly developed potential for evolving inbreeding descendants. The fact that Nanney has been able in the laboratory to select inbred lines of variety 1 that give virtually no exconjugant death shows that the organism could do the same in nature, if conditions were propitious for it. The fact that some selfing clones which give viable sexual progeny are found in nature could mean that the same sort of selection in nature preceded their origin; but I am rather of the opinion that this represents a last resort for old clones which have failed to outcross. Finally, it is of much interest that Corliss (1952) has in fact discovered a species of Tetrahymena (*T. rostrata*) which regularly responds to starvation by going 100% into autogamy—the closest type of inbreeding. *T. pyriformis* is in fact of great interest in our present discussion because it shows how an outbreeder is preadapted, as it were, to become an inbreeder on demand. This may be considered as strengthening our view concerning the evolutionary derivation of inbreeders from outbreeders. The mechanisms needed for an inbreeder are present, but counterbalanced so that they operate relatively poorly; remove the restraints and an inbreeder results. The restraints actually seem to be developed to different degrees in different varieties, as indicated, for example, by variations in the length of the immature period and in the mode of mating type inheritance.

As the varieties of *T. pyriformis* are compared in more detail, it may well be discovered that the different mechanisms for inbreeding and outbreeding are quite differently developed and balanced among them, as among the varieties of *P. aurelia*. In this connection, comparison between variety 1 and variety 2 is of special interest. The latter is clearly more of an outbreeder than the former. It shows synclonal uniformity in mating type, while variety 1 does not. For this reason, the report that its immature period may be a few days shorter is of less significance. Even without an immature period, the method of mating type inheritance inhibits the closest form of conjugation inbreeding. Variety 2 is also distinguished by having more known mating types than any other variety and, as we have seen, multiplicity of mating types is of adaptive value to an outbreeder. Finally, variety 2 has the widest known latitudinal distribution in nature, and we have previously seen how this too is correlated with outbreeding. The fact that the first two varieties to be studied in detail show such marked differences in features associated with different breeding systems leads to the supposition that the species *T. pyriformis*, like *P. aurelia*, consists of varieties with a large array of breeding patterns, perhaps from rather extreme outbreeders to extreme inbreeders.

## *Euplotes*

*Species, Varieties and Mating Types.* Until Kimball (1939, 1942) discovered mating types in *Euplotes patella* and worked out the breeding relations among many collections of organisms that closely resembled or were identical with *E. patella*, there was great confusion in the literature concerning those species of *Euplotes* that look much like *E. patella*. Sorting his collections into those that would or would not interbreed, Kimball came out with at least five groups. One group did not conjugate. Each of the other four included a number of collections that could conjugate in mixtures with each other, but no strain of one group would conjugate with any strain of another group. Pierson (1943) made a careful morphological study of Kimball's five groups of strains and of some additional strains belonging to these groups. On the whole (exception noted below), the groups were found to be morphologically distinct. The one which was not observed to conjugate was identified as *E. woodruffi*. One proved to be a new species, *E. aediculatus*. One was *E. eurystomus*. Two, this is the exception, could not be distinguished morphologically; both were *E. patella*. There may have been more than one group in some of the other species already mentioned, but this is not clear from the published account. There is also some vagueness as to whether there was a third group of *E. patella*. This vagueness is apparently due to the fact that the only group extensively studied as to breeding relations was one of the groups of *E. patella*.

The groups of strains of *E. patella*, although they are sexually isolated, were not assigned specific names by Pierson because she agreed with the views of Sonneborn (1938) and Kimball (1943) that this should not be done unless the species could be readily identified. They correspond exactly to the varieties of *Paramecium*. Kimball (1943) and Pierson (1943) agree that sexual isolation probably preceded morphological divergence in the speciation of *Euplotes* (and other Ciliates as well). This is based on the plausible assumption that the varieties of *E. patella*, which are morphologically identical so far as known, are more closely related than the previously mentioned species, which nevertheless are much alike in morphology and well marked off from other species of *Euplotes*. The very similar species are therefore considered to be intermediate in evolutionary divergence and relationship between morphologically very different species and morphologically indistinguishable varieties.

The preceding work established the existence of mating systems based on mating types in *E. patella*, *E. eurystomus*, and *E. aediculatus*. Katashima (1952) reported mating types in *E. harpa*, which is apparently not a member of the group of species that is most closely related to *E. patella*. Thus far, no survey has been made which would indicate the number of varieties in any species of *Euplotes*. I suspect that it may be small. The full number of mating types in a variety is also at present unknown. Katashima (1952) reports upon only two cultures collected in Hiroshima City; these were of different mating types. Kimball (1939, 1942) also confined his studies to the progeny of two individuals of complementary mating type obtained from neighboring ponds, but he included clones derived sexually from the two original cultures. In this way, he discovered four more mating types. It might be supposed that this procedure would reveal all the mating types of the species, as it appears to do in *P. bursaria*. Kimball (1942) points out, however, that his discoveries of the genetic basis of mating type specificity (see below) do not at all preclude the existence of more types. One might actually expect a very large number of types in nature, as will appear.

*Mating Relations and Inheritance of Mating Type.* The mating relations reported by Kimball for the six mating types he found seem superficially to be like those in *P. bursaria,* but they probably differ in several respects. Conjugation occurs when any two of the six types are mixed together. However, by using visibly marked animals, Kimball proved that the conjugant pairs are not always crosses between the two types that were mixed together. In some combinations of cultures, one finds both crosses and selfing. In fact, if one adds animals of one mating type to fluid in which animals of certain other types have lived, the one and only type introduced into the fluid selfs. The animals clearly secrete "hormones" that induce conjugation in animals of certain other types. Sonneborn (1947) concluded from a detailed examination of the data of Kimball (1942) and Powers (1943) that crosses occur only when *each* type secretes into the fluid a hormone which induces the other type to conjugate. As Kimball showed, in such mixtures three kinds of conjugant pairs are formed: crosses between the two types and selfing of each type.

This system has a simple genetic basis and remarkable consequences. Altogether, among the six mating types, only three hormones are produced. Each hormone is determined by a single gene, and the three genes are codominant alleles. There are thus three homozygotes, each producing one and a different hormone; and three heterozygotes, each producing two of these same three hormones in a different combination. These are the six mating types. Because mating type is directly controlled by the genotype and each mating type has a different genotype, each synclone is necessarily uniform for mating type because conjugation has rendered it uniform in genotype. Without going into further details, the system works out so that crosses between different synclones take place (along with double selfing) only in three of the five possible combinations of any one mating type with the others. In the other two combinations one member selfs, the other does not, and no crosses occur.

*The Breeding System.* Superficially the mating relations seem like a very fumbling and inefficient arrangement to encourage outbreeding, inefficient because not all combinations of diverse types can mate, and fumbling because it can induce selfing as well. However, there appear to be good reasons to suppose the system is not really as bad as it seems. In fact, when one takes into account all the information discovered in the laboratory, but makes allowance for the differences between the laboratory conditions used in analysis and those that probably prevail in nature, the system appears to be well suited for outbreeding and neither fumbling nor inefficient. Actually, it seems if anything to be a more nearly perfectly developed outbreeding system than that of any multiple type system in *Paramecium* or *Tetrahymena.* I shall now try to make this clear.

First, the six known types were derived from but two neighboring individuals. Both were heterozygotes for the hormone genes —in itself a little indication that we are dealing with an outbreeder that does not prevailingly self in nature. Three of the four genes at the hormone locus differed in these two individuals. These are the three alleles the combinations of which yield the six known mating types. Is it reasonable to suppose that by chance these two individuals possessed all the hormone alleles that exist in nature in this species? Decidedly not, in my opinion. It seems far more likely that a wide sampling of various natural populations would have revealed a large number of such alleles. The situation might even be comparable to the one known for incompatibility alleles in Basidiomycetes and higher plants, in some of which a hundred or more alleles have been estimated to occur. Certainly it would be a miracle if all the alleles that exist in this variety of *E. patella* were the three found in the only two individuals studied. With many multiple alleles, there would be many mating types; and multiple mating types are, as has been repeatedly emphasized in this paper, an adaptation for outbreeding.

Second, Kimball reports the regular existence of an immature period of a month or more following conjugation and Katashima (1952) found a comparable immature period (three weeks or more) in *E. harpa*. The existence of a considerable immature period is hardly to be considered fortuitous. As set forth above, it exists also in *T. pyriformis*, in *P. bursaria*, and in outbreeding varieties of *P. aurelia*. Its meaning appears to be to discourage inbreeding by making mating impossible when the organism is most likely to be near its relatives and by giving time to spread and meet a stranger that would make a suitable mate. Of other stages in the life cycle, nothing is known except that there is a long mature period and a long life cycle, if there is a terminating cycle at all, as I believe there is. All that is known is characteristic of outbreeders.

Third, Kimball mentioned that there was high mortality in his crosses, all of which involved inbreeding. As we have seen in the accounts of *Paramecium* and *Tetrahymena*, inbreeding mortality is a sign of an outbreeder.

All these strong indications that *E. patella* is an outbreeder make me conclude that the superficially fumbling, inefficient, and contradictory features may be misleading us. The major difficulty seems to be the fact that mating types which can cross also induce each other to self. This is no way for an outbreeder to behave. Perhaps it is largely a laboratory phenomenon and rare or nonexistent in nature. In the laboratory these observations were made in depression slides, i.e., in about 1 ml. of fluid containing large numbers of euplotes. Contrast this with a pond or river. The contrast is important, for we deal with extracellular hormones acting through the surrounding fluid medium. Surely, in a pond the dilution of the hormones would be enormous and rapid. There would be a sharp concentration gradient away from the position of the secreting animal. Further, the concentration of animals producing hormones would be expected to be very much less than in the laboratory. Consequently, the hormones could probably work only for short distances, possibly only on two euplotes that were close together, if not in actual contact. I can see little likelihood of the hormonal mechanism leading in nature to anything like the amount of selfing observed in the laboratory.

However, there may well be some selfing in nature as well. This is indicated by two facts. First, selfing was observed repeatedly within single clones. Second, viable inbred progeny were obtained in the laboratory. Although as mentioned, mortality was high among the inbred progeny of the two original isolates, Kimball and Powers were able to pursue genetic analysis with them. Whether it was possible to select in this way combinations relatively free from further inbreeding mortality is not ascertainable from their papers, so far as I can discover. It is also not clear whether continued inbreeding led to progressive deterioration, but Kimball eventually abandoned the organism because of difficulties in maintaining cultures. Either inbreeding degeneration or aging or both may have played a part in this. To the extent that inbreeding yields viable progeny and to the extent that selfing occurs in nature, E. *patella* may be considered as a facultative or potential inbreeder, although provided with life features that make outbreeding the usual practice. Here, as in *Tetrahymena,* we may see traces of how an organism which is customarily an outbreeder contains the seeds of an inbreeding evolutionary descendant.

**Other Ciliates**

Mating types and something about the breeding system are known for relatively few Ciliates other than those already dealt with. All can be quickly dealt with except *Oxytricha bifaria* about which much of interest is known. A system of three mating types which show the *P. bursaria* relations (each type mating with both of the others) was found by Sonneborn (1938, 1939) in *P. trichium.* Diller (1948) and Wichterman (1937) have found strains of *P. trichium* which self very readily. I have also encountered such strains. On the other hand, those in which I reported the three mating types did not self, and Wichterman (1953) has had one strain under cultivation for twelve years without finding selfing. Whether the selfing and nonselfing strains belong to different varieties is unknown. If they do, the former would probably be found to be an inbreeder, the latter an outbreeder. Sonneborn (1938) reported a pair of mating types in *P. calkinsi.* Wichterman (1950, 1951) found a second variety which also contained two mating types. I have found multiple interbreeding mating types in an organism kindly identified for me as *Colpidium truncatum* by Corliss. This organism, which is closely related to *Tetrahymena*, likewise has its breeding system much complicated by selfing. Downs (1952) reported a system of five interbreeding mating types in *Stylonychia putrina* collected from a single source and he (1956) now finds two varieties with eleven types in one variety and fifteen in another. In *S. putrina*, selfing is rare; only one clone of one mating type selfed among 41 clones of the first five types studied.

In *Oxytricha bifaria*, Siegel (1956) reported a system of nine interbreeding mating types. He investigated the possibility that *O. bifaria* might manifest relations like those in *Euplotes* because these organisms are both hypotrichous Ciliates and so are more closely related to each other than they are to other (holotrichous) Ciliates in which mating types have been found. Siegel looked for action of sex hormones, but could find no evidence of their existence. He also devised means of discovering whether the conjugating pairs in mixtures of two mating types were true crosses or selfers and showed that they were regularly true crosses. Indeed, selfing appears to be exceptional in his material. Only one mating type (III) selfed, and this frequently showed about 1% of the animals conjugating even when not mixed with another type. The two members of selfing pairs were separated before becoming firmly united. Both gave rise to cultures like the parent, that is, they were predominantly or entirely type III, but selfed slightly.

Crosses of certain mating types gave types among the progeny which sometimes differed from both parents; but the genetic details were not reported. Whether, as in *Euplotes*, mating types are determined by a series of multiple alleles is still unknown. The large number of mating types in *O. bifaria* and in *S. putrina* suggests that such a basis may well exist. On the other hand, Siegel found the same mating type in several natural sources, and this indicates that the number of existing types may not be great.

The life cycle of *O. bifaria* begins with a period of immaturity lasting for variable periods, from less than one month to over two years. Although Kay (1946) reported endomixis (autogamy?) in certain strains of this species, Siegel did not find it and was unable to induce it. Perhaps Kay's strains belonged to a different variety. In the absence of conjugation, Siegel's variety of *O. bifaria* slowly weakens and eventually dies, the whole cycle usually lasting more than two years.

Some information is available on the survival of exconjugant clones. Those derived from the selfing of type III almost always proved nonviable. Some crosses (types I by II, III by V, and V by VIII) gave high percentages of viable progeny; other crosses (types I by III and II by V) yielded a majority of nonviable progeny. It is to be noted that the type III culture, which yielded few survivors when it selfed, gave high survival when crossed to type V. This type III culture was one collected from nature, not one produced from crossing in the laboratory. Its age was therefore unknown. The possibility that selfing of this culture and the death which is its consequence were due to age seems excluded, for Siegel found no stage of selfing in other cultures which were followed through senility to death. Had this information not been available, one would have been tempted to suggest that old-age selfing is a regular feature of the life cycle. This should serve as a warning of the insecure basis of some of the conjectures that I have made about life cycles in *Tetrahymena* for which full laboratory studies of the whole cycle are not available.

In discussing the significance of the facts about *O. bifaria*, Siegel stresses those features of life that facilitate the occurrence of mating, without differentiating between inbreeding and outbreeding. He is particularly impressed by the occurrence of autogamy and the existence of only two mating types in some species of Ciliates and by the absence of autogamy and the existence of multiple mating types in other species. Although recognizing certain exceptions, he attributes this correlation to the necessity for fertilization in preventing senescence and death by starting new clones. Animals in species with multiple types have more opportunity to meet an appropriate mate; since the probability of doing so is less in two-type varieties, autogamy may serve as a safety device to cover the risk of failure to meet a mate.

Sonneborn (1955) agreed with some of these views and to some extent also does so in the present paper. Yet, taking into account other facts and more comprehensive correlations, I have here developed in detail the thesis that all of them have their primary significance as mechanisms adapted to different systems of breeding, either inbreeding or outbreeding. The association of a considerable period of immaturity with multiple mating types is not interpretable on Siegel's hypothesis. These two features seem to act at cross purposes: one favors mating, the other prevents it. Their association is, however, adapted to outbreeding. In varieties that have these features, it is important to find a stranger, not just to undergo fertilization indiscriminately. The absence of autogamy and the long life cycle in such varieties have the same meaning, as do the other features discussed throughout this paper. As implied here and developed in detail above, the length of the life cycle is highly variable from variety to variety; there is no need for quick fertilization in an outbreeder and a short life cycle is adequate for inbreeders with their various devices: little or no immaturity, much selfing, early autogamy, and various genetic means of assuring presence of both mating types. Actually, the situation seems to be just the reverse of the one postulated by Siegel: in two-type systems, the whole biology of the organism is designed to assure early meeting of related mates; whereas, in multiple type systems, usually the meeting of mates is prevented during early life so as to increase the chance of mating with a stranger.

So far as *O. bifaria* is concerned, the variety examined by Siegel seems clearly to be an outbreeder. It rarely selfs; it has no autogamy; it has multiple mating types; it has a considerable period of immaturity, long maturity, and a long life cycle. Only the questions of inbreeding degeneration and outbreeding viability remain unsettled. The high death rate after selfing fits the picture, but the variable results of outcrosses need clarification.

That other varieties may exist with different breeding systems is perhaps suggested by Kay's report of what might be autogamy.

## Obligatory Inbreeding and Asexual Protozoa

*The Transition from Sexual to Asexual Reproduction in the Flagellates of Wood Roaches and Termites.* The Ciliates illustrated gradations from extreme outbreeders to nearly the extreme of inbreeding. The highly evolved and morphologically complex Flagellates that live in the gut of wood roaches and termites illustrate the transition from obligatory close inbreeding to asexual reproduction. Cleveland (1934-54) has reported extensively on the Flagellates of wood roaches. They undergo sexual reproduction at or near the time the host moults. Only at this time do the roaches pass fecal pellets laden with the Flagellates; and at the same time the eggs, which are laid earlier, hatch. Sexual processes in the Flagellates are thus timed to coincide with transmission to new individual hosts.

That the two processes should be timed in this way is hardly without meaning. Obviously this would make possible selection of those genetic variations in the symbionts which happen to be adapted to variations of the individual infected host. Yet, as I shall at once show, there is little opportunity for either host or symbiont variability. It would therefore seem more likely that the present correlation in time of symbiont sexual processes and transmission to new hosts is a relic of earlier conditions when both symbiont and host were more variable than they are today.

The wood roach lives in small isolated colonies and migration seems to be very limited. Under such conditions, the amount of genetic variability expected among the individuals of a single roach colony would be small. Correspondingly there would be little need for genetic variability among the Flagellate symbionts living in so constant a milieu.

In agreement with this, Cleveland's accounts of the sexual processes in the Flagellates indicate little opportunity for genetic recombination. Although some of the Flagellates, such as *Euco-*

*monympha,* show complete two-division meiosis, Cleveland describes remarkable abbreviations of, and variations on, the standard sexual processes. According to him, *Oxymonas, Saccinobaculus, Notlia, Leptospironympha,* and *Urinympha* undergo a remarkable one-division meiosis. Neither centromeres nor chromosomes duplicate, chiasmata are not observed, and homologous chromosomes pass intact to opposite poles. This would clearly reduce the possibilities for genetic recombination. Cleveland reports for other Flagellates processes which would place even greater restrictions on genetic variability. In *Rhynconympha,* the first meiotic division is accompanied by cell division, but the second is not, and the sister nuclei thus produced reunite in autogamy. In *Urinympha* he maintains that the diploid nucleus undergoes meiosis in one division and that the two reduced nuclei reunite in autogamy. In the haploid *Barbulanympha,* fusion of sister nuclei is said also to occur sometimes. Some of these Flagellates thus appear to have no possibility of outbreeding. Organisms reach the ultimate limit of inbreeding when fertilization *always* takes the form of autogamy. These are the considerations concerning host and symbiont variability which led me to conclude that present conditions are not adequate to explain the coincidence in time between infection and fertilization, and that earlier conditions of greater host and symbiont variability probably provided the selective pressure necessary for its evolution.

A corollary of this conclusion is that the aberrant and abbreviated sexual processes now observed in the symbionts represent stages in the atrophy and loss of meiosis and fertilization. Cleveland (1951b) on the contrary rejected this possibility and concluded that they were stages in the origin and evolution of sexuality. To him this seemed the more reasonable interpretation of what he recognized as the comparative genetic uselessness of the cytological processes he described. The major support for his view comes from the juxtaposition of two facts: (1) these symbionts are among the morphologically most complex and highly evolved Flagellates; and (2) sexual processes are not satisfactorily established as occurring in the simpler animal

Flagellates (see review by Wenrich, 1954). Whether the second point is to be ascribed to incompleteness of investigation or to actual absence of sexuality in the lower animal Flagellates remains unknown. However, many examples of perfectly standard meiotic and fertilization processes are well known among the plant Flagellates. One could hardly assume therefore that sexuality in Flagellates first arose in the complex symbionts of wood roaches. Rather one needs to inquire why so many of the simpler Flagellates lost it, if indeed they do lack it. That the Flagellates of wood roaches show stages in the loss, not the origin, of meiosis and fertilization is further indicated by a comparison of the situation in wood roaches and termites.

The wood roaches are primitive members of the group that gave rise to termites. Hence, the later evolutionary stages of the symbionts might be expected in the termites. Closely related Flagellates are indeed also found as symbionts in the termites. But without exception these show no sexual reproduction whatever. They reproduce asexually only. It is true that Cleveland's evidence correlates the sexual cycles in the Flagellates of wood roaches with the moulting hormones of the roach, and exactly the same hormonal conditions would not be expected in termites. But any organism which is under selective pressure to maintain the hard won evolution of sexuality should not find such a change of hormone pattern beyond its capacity to cope with. More likely, the change in the mode of life of the evolving host provided no more pressure for genetic variability on the symbionts, and possibly less.

Each colony of termites is isolated underground. The extent to which the brief nuptial flight preceding the founding of new colonies permits outbreeding of the host is not, so far as I am aware, known. It seems that there is a considerable degree of synchronization in swarming, however, so that interbreeding between neighboring colonies might well occur. On the other hand, the enormous number of eggs laid by a single female is a powerful factor for assuring a high degree of uniformity among the members of a colony and its descendant colonies. The relatively small degree of host variability and presumably very slow host changes during the course of evolution have obviously been coped with adequately by the symbionts without recourse to sexual reproduction (Kirby, 1949). Thus, under conditions of relaxed selection for sexual reproduction, it seems that it degenerated step by step from normal meiosis and crossing, to one-step meiosis, to autogamy, and finally to asexual reproduction.

*The Problems.* In the series of transitional stages from non-obligatory but preferred inbreeding to exclusively asexual reproduction, changes take place in the species problem and in the problem of adjusting the features of life to assure maintenance of an unbroken line of descent through evolutionary time. I am not sufficiently familiar with the intimate details of the lives of the Flagellates to attempt to work out for them, as I did for the Ciliates, a full picture of how the major life features are adapted to the mode of reproduction. A few of the more important and obvious features will be mentioned below.

Those who have considered the species problem in relation to types of reproduction often make a sharp contrast between species in outbreeding organisms and in obligatorily inbreeding and asexual organisms. For example, Dobzhansky (1937, 1941) draws a sharp line between the two, maintaining that there is little in common save the word "species," which has very different meanings in the two cases. He further states, "in asexual groups either every biotype is to be called a species or else species do not exist there at all." These comments are of course based primarily on the concept of a species as a common gene pool, which Dobzhansky has strongly urged for 20 years. The nature of species in this special sense does not suddenly change, but undergoes continuous, progressive change in a single direction as one passes through the various degrees of outbreeding and inbreeding to asexual reproduction. One might therefore expect to find equivalent levels of biological organization in different systems of breeding and reproduction. I shall explore that possibility after taking up first the problem of asexual species in the routine taxonomic sense.

Hoare (1952) maintains that only morphological differences

should be used as species differentials. Although his discussion is primarily about parasites, he clearly intends to include a broader domain. In essence, this amounts to denying any difference in principle between criteria of taxonomic species in sexual and asexual organisms. In fact, Hoare includes both sexual and asexual parasites in his discussion. Physiologic differences he rejects in general as alone insufficient to mark off species. He particularly lists among such inadmissible traits differences in drug resistance, virulence, disease produced, serotypes, and host specificity. I shall add specific objections to most or all of these traits, but I shall also bring out objections to certain morphological or visible traits. My chief point in this connection will be that the *genetic* difference between species in asexual organisms should be as nearly as possible of the same kind and magnitude as in sexual organisms. To achieve this equivalence some genetic information gained from the study of sexual organisms will have to be used in guiding the choice of differential traits. I shall illustrate my point of view by presenting and then discussing certain observations on Trypanosomes and Rhizopods.

*Trypanosomes.* The Trypanosomes are of special interest in relation to the soundness of serological traits as species differentials. Since the time of Ehrlich, the complexity and variability of serologic traits in Trypanosomes has been known, and this has prevented their abuse as species differentials. Recently, Inoki *et al.* (1952a,b) have shown particularly well that the multiple alternative serotypes which can arise in a single clone are not selected gene mutations, but alternative expressions of an array of potentialities which were from the start potentially expressible. In so far as the analysis has been carried, it has yielded results which are comparable to those obtained in *P. aurelia*. As set forth above, the genetic analysis in *P. aurelia* shows that each clone possesses at a series of loci genes for the various serotypic potentialities and that each individual and subclone expresses serotypically genes at only one locus of the series. The parallel phenomena in the Trypanosomes point to the same sort of genetic basis though breeding analysis is not possible.

In view of this parallel, it seems likely that similar situations occur in other asexual organisms, including some in which serologic traits are used to define species. Failure to appreciate that organisms with the same genotype can express serologically unrelated antigens could easily lead to assigning organisms with the same genotype to different species. Only by knowing the full array of potentially expressible antigens in a clone and by taking into account the grades of difference in cross reactions between corresponding serotypes in the arrays producible in different clones could a sensible beginning be made in establishing serotypic criteria for distinguishing species. Lederberg (1955) points out other weaknesses of the use of serologic species criteria in bacterial taxonomy based upon his work on gene transduction.

Physiological differences such as virulence and host specificity are used as species differentials in Trypanosomes and other parasitic asexual organisms. Both of these criteria may be objected to on the ground that comparable traits in other organisms (e.g., bacteria and viruses) are known to arise as single mutational steps and to be inherited as single gene traits. So simple a genetic difference would surely not be acceptable as a species differential in sexual organisms, and there is no good apparent reason why this should be done in asexual organisms. It is sometimes argued that confinement to different hosts is equivalent to geographic isolation. But this is only true if the adaptation to different hosts is outside the range of simple genetic differences. Otherwise the mutational array of host specificities within each single line of descent will yield a common array of potential hosts.

Restriction of species differentials to visible differences does not necessarily avoid this difficulty. Some visible differences also are of course simple in genetic basis. For example, certain species of Trypanosomes are known to differ only with respect to the presence or absence of a kinetoplast. This is a self-reproducing cytoplasmic granule. Under the action of certain chemicals, the kinetoplast can be irreversibly lost. Sublines of the same clone can differ in this way. Yet they must be assigned to different species if this is accepted as a valid species differential. The situation is closely parallel to the existence of killer and sensitive paramecia in the same clone by reason of loss of kappa from some individuals. And it is no more sensible to use the one or the other as a species differential. These cases are of additional interest in showing that visible hereditary differences need not necessarily be preceded by invisible physiological genetic differences, as many have maintained.

*Rhizopods and the Results of Selection.* Jennings (1916) and his students were able to select strikingly different morphological types within a clone in various Rhizopods. The different types thus obtained tended to perpetuate themselves in the absence of further selection. For example, Jennings obtained the most varied shell (test) types in *Difflugia:* spineless and many spined, with few teeth around the mouth and with many, and so on. Contrariwise, it was possible to start with two diverse forms and select the same type from both. Here again there is no indication that physiologic genetic difference precedes morphologic difference in the divergence of two forms. Even persistence of the difference among unselected progeny is no assurance that the two types belong to different clones, and similarity of type is no guarantee of belonging to the same clone.

*Criteria of Species in Asexual Organisms.* From the foregoing examples and discussion a few useful general principles seem to emerge. Species differences in sexual organisms are based upon complex, not simple, genetic differences. To reduce species differences to a single trait dependent upon a single gene difference is to equate species differences in asexual organisms to the level of individual differences in sexual organisms. From this point of view, a single, genetically simple visible or morphological trait is in no better case than a single, genetically simple physiological trait. If asexual species are to represent a level of evolutionary divergence comparable to the one used in taxonomy of sexual species, the distinctions would have to be not only morphological but genetically complex. To employ differentials which can be wiped out by one or even a few mutational steps is in my opinion indefensible.

*Variation and Speciation in Asexual Organisms.* The conditions of life for an asexual organism are likely to be less varied than those encountered by most sexual organisms. A large body of water or a particular host, for example, acts as a buffer against radical variation in the milieu. But some variations in conditions are inevitable and the organism has to be prepared to adapt to it or run the risk of extinction. Lacking the rapid and wide-ranging variability obtainable by genetic recombination, other sufficient mechanisms for yielding adaptive variations have been selected. One of these is the combination of haploidy and large, rapidly produced populations with mutation. This has apparently been adequate for many asexual organisms.

Another and faster-operating mechanism is one which is less widely appreciated. It is the one illustrated by the serotype systems of Trypanosomes and *Paramecium*. This mechanism involves (1) including in the common genotype of the species sets of loci for alternative and mutually exclusive phenotypes and (2) incorporating mechanisms for shifting phenotypic expression from one to another on demand. This is clearly a very satisfactory adaptation to the need for rapid variation during asexual reproduction. As such, it is found not only in organisms that lack sexual processes, but also in those that have sexual processes along with a well-developed asexual phase, such as the Ciliates. Further, there is some evidence that it plays an important role in developmental differentiation, the asexual phase of higher organisms. For example, genes for both sexes are present in hermaphrodites, but come to expression in different cells; and other aspects of differentiation probably have a comparable genetic basis. Thus, at every level of asexual reproduction, this mechanism of variability seems to be in operation.

Speciation in both asexual and sexual organisms is probably fundamentally the same. In both cases, the basic event is genetic adaptation to different conditions. Perhaps asexual organisms selectively live under relatively constant or very slowly changing conditions. This selection of constant environments would tend to retard conspicuous speciation, but beneath the visible uniformity there might well be much accumulated genic and chromosomal change, such as exists among local populations of strongly inbreeding varieties of *P. aurelia*. On the other hand, selected variants can be rapidly multiplied and spread by asexual reproduction. The individual wastage of recombination and the swamping of variations by crosses are avoided. As in any system which has proved itself in nature, what is lost by one feature of the system is compensated by others. But a short-term advantage is of course no guarantee of long-term persistence. Very often temporarily successful asexual organisms may have been wiped out on occasions of stress. Those which have adapted to constant or but slowly changing environments may, however, succeed even on the geological time scale, as studies on the Flagellates of roaches and termites have shown (Kirby, 1949).

The nonexistence of species in asexual organisms is asserted only by those who define species as syngens. For reasons given earlier, I have rejected this definition and have accepted the necessity for readily recognizable distinctive features in a species. In the present section, emphasis has been placed upon the complex genetic basis of species differences in sexual organisms, and genetic differences of the same kind and magnitude were demanded for species differentials in asexual organisms. On the whole, the taxonomy of the Protozoa does in fact utilize such species differentials. The chief infractions of the rule are found among the parasitic and medically important Protozoa which have come to be known in great detail. This intimate knowledge has led the "splitters" to use such genetically simple criteria for species differentials that their species become the equivalent of individuals or families in sexual organisms. "Lumping" is now needed until the discontinuities between groups become genetically complex as well as readily discernible.

Before proceeding to a consideration of the more important question of the possible existence in asexual organisms of the equivalent of the syngen in sexual organisms, some further comments on the usage of the term species in general and its parallelism in sexual and asexual organisms are in order. Difficulties in the application of the term species arise from the attempt to make it do double duty in serving both as designating an evolutionary unit, the one which shows minimal irreversible discontinuity, and as designating a readily recognizable group. For most of the last hundred years, it has been tacitly assumed that these two aspects of species coincide and that, since the evolutionary unit often represents a real level of biological organization, species exist as real entities, not as human constructs made for the convenience of biologists. Modern researches, on the contrary, have shown that the evolutionary unit is not always readily recognizable. In such cases, the proponents of the modern biological species concept admit the dual implications of the term species by distinguishing named species, those which are readily recognizable, from sibling species (Mayr, 1948), those which are often not named because they are not readily recognizable. This leads to the totally illogical procedure of referring to a group of species as a species, i.e., a group of unnamed sibling species has a single species name. Moreover, by reserving species names for readily recognized groups, this procedure tacitly admits that the convenience of biologists must first of all be served by the term species. Agreeing with this, I propose, in the interest of logic and clarity, to use a different term, syngen, for the evolutionary unit. The same group of organisms would be both a species and a syngen if both the criteria of ready recognition and minimal irreversible evolutionary divergence were met. Otherwise, not. The reluctance of proponents of the modern biological species concept to take this logical step seems to be based on their love of the word species. Its early connection with evolution theory, its wide usage, and general familiarity with it make it a great prize to capture and invest with absolute existence as a definite level of biological organization of the greatest evolutionary significance. This cannot be done without sacrifice of logic and introduction of confusion.

Once the brave but hopeless effort of the proponents of the modern biological species concept is abandoned, it will again be generally recognized that species in sexual and asexual organisms have been and can be defined on essentially the same principle. The principle is simply minimal irreversible evolutionary divergence that yields readily recognizable difference. With this principle guiding him, no greater task is placed upon the skill of the taxonomist in classifying asexual organisms into species than in doing the same with sexual organisms. He has in fact already done it and, on the whole, has done it just as satisfactorily in the asexual Protozoa as in sexual Protozoa or higher organisms, with the readily corrected exceptions pointed out above.

The question of whether asexual equivalents of syngens exist and can be recognized is more difficult, but not hopeless. The key to progress in this direction is to recognize in the syngens of sexual organisms a distinction between the means of ascertainment and that which is ascertained. Breeding methods are the means of ascertainment. The syngen which is thereby ascertained proves to be a discrete group, the discontinuity of which with other discrete groups is based upon complex genetic differences. The discontinuity and the complexity of its genetic basis are the essential features of the difference between closely related syngens. This is what prevents the flow of genes between them and that is why the test of gene flow is a means of ascertainment. The definition of a syngen on the basis of gene flow is an admirable operational definition. But conceptually the syngen could also be defined on the basis of discontinuity and its complex genetic basis. The same basis can be used for a concept of the asexual equivalent of the syngen. Indeed, the same term, syngen, might be used in both cases for it implies not only a common pool of genes but also generation by a common group. Although another term might be preferred for asexual organisms, the same one will tentatively be employed here.

The obvious difficulty with the concept of the asexual syngen just set forth is the lack of a simple, neat operational definition comparable to the one that serves for the sexual syngen. This statement of the difficulty of course assumes that there is no gene flow among different lines of descent in organisms not known to have sexual processes. In the light of modern researches on transformation and transduction in bacteria, this assumption might be questioned. If such mechanisms or others come to be discovered in other asexual organisms, so much the better; but this cannot be counted upon and may not occur. For the present at least, other approaches must be sought. The problem then is how to recognize irreversible evolutionary divergence in the absence of a breeding test. In sexual organisms such divergence is associated with complex and discontinuous genetic differences. In asexual organisms also, such differences may be used as the sign of irreversible divergence.

To discover whether such complex discontinuities exist within a taxonomic species of asexual organisms, one would have to make detailed comparative studies of many strains of the species obtained from throughout its range. The comparisons would have to include studies of the life cycle, cytology, morphology, physiology, and ecology. Many features would have to be examined experimentally under the same conditions to distinguish phenotypic from genotypic variation. As with "sibling species," syngens discovered within an asexual species would not differ in readily identifiable traits. If there were closely related sexual species, the genetics of traits that differ in the asexual strains should be studied in the sexual relatives. In short, every possible means of arriving at a sound judgment about the complexity of the genetic differences and their discontinuities would have to be utilized. This would be a great labor; but great labor is also involved in the analysis of a sexual species into its syngens. The results of such studies on the species of *Paramecium*, summarized in this paper, represent the work of 20 years by an increasing group of investigators and the work is far from complete.

The concept of an asexual syngen as having passed the threshold of irreversible evolutionary divergence, but not having reached readily recognized differentiation, is thus the exact equivalent of the sexual syngen and refers to the same level of biological organization. In both cases, this level is the one which embraces all the individuals that can potentially contribute to the further evolution of the group. Thoday (1953) has presented a comparable point of view. The evolutionary unit is as distinct and isolated in asexual organisms as it is in sexual organisms, no more and no less so. The syngen is sharp and definite in many outbreeders; it becomes fuzzier and fuzzier the more inbreeding is adopted. Similar differences in its distinctness may be anticipated among asexual organisms. The difference between syngens in the two cases is not in the concept or in its correspondence with reality, but in the method of ascertainment which is the inevitable result of the different methods of reproduction.

### General Comments and Conclusions

Before passing to a general statement of the major conclusions, two outstanding limitations of the present paper should be pointed out. First, the restriction of the treatment of Ciliates to those in which mating types have been found may well introduce important biases. Such a discovery is likely to be made only in organisms that are not too specialized in their conditions of life, for the investigator has to be able to cultivate them in the laboratory. They also must conjugate under the conditions provided. This excludes entirely asexual Ciliates, if such exist, and also those with undiscovered fastidious requirements for mating. This might also exclude organisms with the longest immature periods or shortest mature periods. Further, even if conjugation occurs, the demonstration of mating types is difficult or impossible if there is excessive selfing. This would exclude some extreme inbreeders. In fact, most work on conjugation and life cycles in Ciliates prior to studies on mating types were performed on selfers. In this category belong the classic studies on *Uroleptus mobilis* by Calkins (1919, 1920) and on *Spathidium spathula* by Woodruff and Spencer (1924). These and other studies are rich in information on life cycles. There is much need to bring them into relation to the modern work and particularly to make investigations on breeding systems and speciation in these organisms.

The second general limitation of the present paper is due to the twin vices of ignorance and haste. As a newcomer to the fields of taxonomy, speciation and evolution, I have only the most superficial acquaintance with the large and impressive literature in these fields. As a result, I have doubtless failed in my obligations to make appropriate references to this literature. My only defense is that strong pressure from editor and publisher allowed me no time to correct my ignorance.

The major conclusions to be drawn from the present study concern three general topics: (1) the problem of taxonomic species and the modern biological species concept; (2) the relation of breeding systems and methods of reproduction to the species problem; and (3) the significance of differences in the major features of life. These topics have been touched upon repeatedly throughout this paper in connection with the treatment of the various organisms discussed. The following résumé brings together the relevant points on each topic.

*The Problem of Taxonomic Species and the Modern Biological Species Concept.* The modern biological species concept of a potentially common gene pool is in principle limited to a fraction of existing organisms and can in practice be applied to but a small fraction of these. It excludes all purely asexual organisms and all obligatory self-fertilizers. Into this category fall perhaps half or more of the Protozoa and a large number of plants in all the major groups, as well as many invertebrates. Among the organisms to which the biological species concept applies in principle, relatively few can ever be sorted with assurance into such species because the task of accomplishing this in any one organism is so great.

The fact that closely related biological species sometimes cannot be readily identified leads to the logical inconsistency of grouping a number of species into a species. Thus the species *Paramecium aurelia* consists of at least 16 biological species (referred to as varieties in this paper). These can be identified only under special conditions which are not generally available or only with great labor and prolonged research. Biological species of this sort are referred to as sibling species (Mayr, 1948) and may remain unnamed. A single species name may be used for the whole group of them. Regardless of the mental reservations about what the "real" species are in such cases, this procedure is tantamount to admitting that identifiability cannot be ignored in a workable species concept. Beginning with logical inconsistency we end in terminological inconsistency, using the word species to denote three different things: taxonomic species not analyzed as to gene flow, biological species, and groups of closely related and practically unidentifiable sibling species.

The enthusiasm for the biological species concept in spite of such limitations and confusions is perhaps to be explained in part by the aura which surrounds the word species and in part by the justifiable satisfaction of having delimited a biological level of organization of great evolutionary importance. Darwin forged a close connection between the term species and evolution theory. Everyone knows the term. To confiscate it for an objectively defined evolutionary unit is tempting. In the process, certain broader considerations were put aside.

Classification into species preceded evolution theory and had as its prime functions ordering and identification. It still has, and for *all* kinds of organisms, not for outbreeders only. In spite of the reproductive implications of the biblical account of creation, with each kind of organism reproducing true to type, and in spite of the early scientific usage of species as referring to the type-maintaining original creations, it can hardly be denied that the primary function of the word species was to designate a unit of identification. This long usage has become deeply ingrained. The universal need for it with respect to all organisms is recognized by practically all biologists. However distasteful it may be to some geneticists, there seems to be no possibility that biologists in general will ever agree to restrict the term species to the carriers of a common pool of genes.

Even the proponents of that concept of species would probably agree with most of my statements in the immediately preceding paragraphs. Dobzhansky (1941) himself makes most of these points. He says that the systematist's task must be primarily to pigeon-hole, that the term species is used in more than one sense by being applied both to taxonomic species and common gene pools, that species are discrete by reason of complex genetic differences, and so on. But he is willing to accept this situation along with its logical inconsistency, terminological inconsistency, and confusion.

Only deep attachment to the word species seems to prevent recognition that the needs of biologists would best be served by having a separate term, such as syngen, for the unit of evolution when it is known not to be the same as the unit of identification. When they are known to be the same, then they are at once both a species and a syngen. The reverence now accorded, often unjustly, to the term species would in time be transferred to the term syngen, or whatever term may be adopted, because of its greater evolutionary significance. The taxonomist and the geneticist would both be free to perform their different tasks with a logical and consistent terminology, neither being frustrated by conflict with the other. The major remaining problem is whether the two terms can be generalized.

Generalization of the term species for the unit of identification offers no great difficulty in principle. It simply requires a verbal statement which embodies current sound taxonomic practice. The first requirement is that the species be readily and visibly identifiable as a group of organisms manifesting discontinuity in morphology with all other groups. Secondly, the level of visible discontinuity ideally is the simplest one judged to be untransgressible by genetic means, either recombination or mutation. Groups of organisms that are judged to show minimal, irreversible, visible divergence thus are assigned to different species.

The generalization of syngen is more difficult. Failure to recognize the equivalent of the syngen in organisms that cannot outbreed is due to the identification of the method of ascertainment with the thing ascertained. The discreteness and reality of the latter is based upon the genetic complexity of the features that mark off a syngen from closely related syngens. This is what prevents gene flow between them, and it is the reason that the test of gene flow serves for ascertainment. Conceptually therefore a syngen can be defined by the complexity of its genetic distinctness from other syngens. A syngen, like a species, has thus passed the threshold of irreversible evolutionary divergence; but, unlike a species, it need not show readily recognized visible differentiation. This level of biological organization is difficult to discover in any kind of organism, but it can be done with any kind. With outbreeders, the method is to test for gene flow. With asexual organisms, the method is to compare different strains of a species from every possible point of view and, with the fullest possible array of facts, to arrive at judgments on discontinuities and the probable complexity of their genetic basis. In spite of the difference in method imposed by the difference in reproduction, corresponding levels of biological organization are thereby defined. In both cases, the level of organization includes the group of individuals that can potentially contribute to the further evolution of the group. Differences in the sharpness of delimitation of the asexual syngen are to be expected. Similar differences are found in sexual organisms, as will now appear.

*The Relation of Breeding Systems and Methods of Reproduction to the Species Problem.* The preceding attempt to generalize the biological species or syngen runs counter to the view of proponents of the biological species concept (e.g., Dobzhansky, 1937, 1941) that biological species do not exist among obligatory inbreeders or asexual organisms. This denial, as indicated above, is based upon an operational definition of biological species. Since the operation, testing gene flow, is impossible in asexual organisms, they deny the existence in them of the thing this operation discovers in sexual organisms, i.e., the biological species or syngen. Their statement of the situation thus implies an abrupt change in the organization of nature and in the units of evolutionary divergence correlated with an abrupt change from outbreeding to obligatory inbreeding and asexual reproduction. By subordinating the method of ascertainment to the thing ascertained and by seeking methods of ascertainment in asexual reproduction, the concept of biological species or syngens was generalized. This implies the absence of an abrupt change in the organization of nature and in the units of evolutionary divergence with changes in breeding system or method of reproduction.

No such abrupt change is in fact found in the present review of conditions in the Protozoa. On the contrary, this review shows a progressive series of changes in the organization of nature and in the units of evolution running parallel to the sequence of breeding systems from extreme outbreeding through various degrees of out- and inbreeding to extreme inbreeding. The passage from inbreeding to asexual reproduction is accompanied by the final gradual change in the series. Nowhere does an abrupt, marked change occur. The continuity of the whole series stresses the unity of the problems of species and syngens in both sorts of reproduction and justifies the attempt to generalize these concepts.

The progessive series of changes accompanying the sequence of breeding systems in the Ciliates includes chiefly the following: the number of closely related syngens per species increases; the range of distribution of a syngen, especially in latitude, becomes more and more restricted; the number of local populations per syngen becomes smaller and smaller; the genetic divergence of each local population becomes greater and greater. Thus, the outbreeding species *Paramecium bursaria* includes only three syngens in North America, while the inbreeding species *P. caudatum* includes 12, and others include more, perhaps many more. Outbreeding syngens, like variety 1 of *P. bursaria*, variety 15 of *P. aurelia*, and variety 2 of *Tetrahymena pyriformis*, cover a range from cold to very warm latitudes; while inbreeding syngens, such as variety 10 or 14 of *P. aurelia*, may be restricted to a small number of populations near the Gulf of Mexico. An outbreeding syngen, such as variety 1 of *P. bursaria*, is genetically so much alike throughout its range that crosses between widely separated populations yield far more viable progenies than inbreeding, and no differences in chromosome morphology or size are found throughout its range. Different populations of an inbreeding syngen, such as variety 4 of *P. aurelia*, are able to interbreed in the laboratory, but the F2 shows high nonviability, sometimes approaching 100%, in contrast to the virtual absence of nonviability after inbreeding within a population. Further, each of its local populations that has been examined has a unique chromosome number in an aneuploid series and shows unique chromosome morphology and size. Syngens with intermediate breeding systems, such as many varieties of *P. aurelia*, are intermediate in range and in the genetic divergence of their populations. The extreme narrowness of range and divergence of populations may be shown by the *P. caudatum* of Europe; each local population gives more or less F1 nonviability when crossed with other local populations. But that situation is not yet entirely clear.

Actual, as contrasted with potential, gene flow thus becomes progressively more limited with the passage through various grades of breeding systems from extreme outbreeding to extreme inbreeding. As the extreme of inbreeding is approached, each of the many local populations of a Ciliate actually approaches in its isolation and genetic differentiation the status of a syngen. Although evolutionary divergence has not yet passed the point of no potential for return, the actual level of biological organization is as different as possible from that observed in outbreeding syngens. Yet all gradations are found from one extreme to the other. Although little information is available on asexual Protozoa, the degree of isolation and genetic differentiation of local populations supposed to exist in them could be at most but little more than actually occurs near the inbreeding extreme of the series in sexual organisms. Both the potential and the actual group of interbreeding populations contracts progressively with progressive changes in the breeding system and the final change in method of reproduction. The significance of the syngen as an objective and real level of biological organization correspondingly decreases as the local population progressively increases in such significance.

*The Significance of Differences in the Major Features of Life.* Among the Ciliates, differences in the major features of life appear chiefly as quantitative variations in the duration of the various stages of a fundamentally similar life cycle, as variations in the details of a fundamentally similar mechanism of mating type determination, and as variations in the form of fertilization characteristic of the senile stage of life. Previously these variations were known as mere brute facts. In this paper they have been shown to be of the greatest significance as adaptations to the breeding system by which the syngen perpetuates itself.

Variations in the duration of stages in the life cycle are found with respect to periods of immaturity, maturity, senility, and total life. These are all long in outbreeders, short in inbreeders, and intermediate in those with intermediate breeding systems. The same sort of general relation has been pointed out in certain plants by Stebbins (1950): self-fertilizers are chiefly annuals and outbreeders are chiefly perennials. In Ciliates, long immaturity serves to give outbreeders an opportunity to migrate away from close relatives before they can mate; short or no immaturity permits inbreeders to mate when they are still nearest to their close relatives. Long maturity gives outbreeders time to find suitable mates; short maturity in inbreeders restricts the opportunity to mate with others. Long senility, during which mating is still possible but less likely to yield viable progeny, occurs in outbreeders and extends the opportunity to find suitable mates; the relatively short senility of inbreeders is characterized by inability to mate at all, or by greatly reduced ability to mate. During senility in inbreeders, the form taken by fertilization is autogamy, the closest form of inbreeding; and this stage of life is by far the longest, immaturity being contracted virtually to nothingness and maturity lasting only a matter of some days. Outbreeders do not undergo autogamy. If they fail to find a suitable unrelated mate, eventually complementary mating types arise within a single line of descent and conjugation occurs within the line. This serves to rejuvenate, initiating a new life cycle, and at the same time serves to maintain heterozygosity, which is not accomplished by autogamy.

Outbreeders commonly have systems of multiple interbreeding mating types, while inbreeders commonly have but two. Multiple types serve to increase the possibility of mating when an outbreeder meets a stranger. The method of mating type determination in outbreeders results in uniformity of mating type of the synclone (the two clones from a pair of conjugants). This prevents the closest forms of inbreeding. Inbreeders, however, regularly yield more than one mating type in a synclone; this serves to make the closest relatives capable of mating with each other. The several systems of mating type inheritance (except the one in *Euplotes*) are based on a single fundamental system of mating type determination, with relatively minor differences in detail.

In asexual Protozoa, the most important differences in life features are with respect to the rapidity of reproduction, the size of populations, the ploidy level, and the constancy or variability of the conditions of life. Haploids that reproduce rapidly into large populations can probably depend upon mutations to provide sufficient genetic variability for adaptation to widely varied conditions. They are thus comparable to the sexual outbreeders. On the other hand, asexual diploids are comparable to sexual inbreeders. They cannot achieve so readily genetic adaptation to varied conditions. They are thus confined largely to relatively constant or very slowly changing conditions, such as life in a host or in a considerable body of water which acts as a buffer against rapid change. A different sort of genetic basis for variability is found in them: the possession of multiple gene loci for alternative and mutually exclusive traits which are readily transformed one into another in response to different environmental conditions. Various intermediate types of Protozoa are found to manifest various combinations of the mechanisms and features just mentioned, e.g., haploid asexuals with small populations and slow reproduction share some of the diploid asexual features, as does the well-developed asexual phase of some sexual Ciliates.

All systems of breeding and both methods of reproduction have proved themselves by existing and persisting in nature. Each must be well adapted to the conditions under which it is found. What one system or method lacks must be compensated by some other feature it possesses. This does not mean that all systems are equally likely to succeed on the evolutionary time scale. Radical changes in conditions may periodically purge inbreeders and asexuals which flourish under relatively constant conditions. Their specialization to such conditions may give them a short-term advantage over outbreeders. But those inbreeders and asexuals that select the constant environments of hosts have also persisted long on the evolutionary time scale, as is clearly shown by the Flagellate symbionts of roaches and termites.

The survey of the Protozoa indicates that there are relatively few fundamental alternatives open to such organisms, perhaps to any organisms. On the one hand, they can evolve systems of genetic variability that permit adaptation to a wide range of conditions of life, as is done by sexual outbreeders and asexual haploids that rapidly produce large populations. On the other hand, they can evolve highly specialized adaptations to relatively constant and restricted conditions of life, as is done by sexual inbreeders and asexual diploids. With the choice of one or the other fundamental alternative go a host of further adaptations to the system of breeding or reproduction, as set forth above in detail for the Ciliates. The superficially chaotic differences in the features of life of different closely related syngens acquire deep significance as systematically selected means to the great end of successful survival. The correlated problems of speciation that tax the taxonomist and generate concepts from the geneticist are trivial man-made difficulties that divert attention from the fundamental beauty and simplicity of nature's ways of meeting nature's problems.

**Summary**

*Statement of Problems and General Background.* (1) Problems. Can biological species, i.e., the common gene pools, of Ciliates be readily identified? Should they be assigned species names? Should they be considered as species? Is it possible to arrive at concepts of species and asexual equivalents of "biological species" which would be generally applicable to inbreeders and asexual organisms as well as to outbreeders? What is the bearing of breeding systems and methods of reproduction on species and evolutionary problems? What is the relation of the varied life features of Ciliates to their breeding systems?

2. Background. Inbreeders and outbreeders among the Ciliates were early recognized. Since the discovery of mating types (1937), each taxonomic species has been found to include a number of "biological species" or varieties. Each variety includes a number of populations which are potentially able to interbreed; but the varieties are sexually or genetically isolated. Information is fullest on several taxonomic species of Paramecium. The chief differential characters of these species are given.

*The Specificity of Mating Types and Isolation of Varieties in P. aurelia.* Of the 16 known varieties, all have two mating types except variety 16 which has four.[3] Although flow of genes can occur in the laboratory between different populations of the same variety, the F2 generation shows in some varieties considerable mortality, approaching 100% in certain crosses. Parallel and corresponding mating types occur in different varieties, but each mating type is uniquely specified by its pattern of sexual reactions. In most cases there is no attempt of different varieties to interbreed or the attempt is unsuccessful; sexual isolation is complete. In the few combinations of varieties that can interbreed, the F1 or F2 is almost completely nonviable, and survivors are always weak. The latter are incapable of yielding normal vigorous progeny either by inbreeding or by backcrosses. Genetic isolation is virtually complete.

Only two varieties are known to have worldwide longitudinal and latitudinal distributions; two have wide, but not so wide, distributions. The rest are restricted to a single continent, sometimes to but a small part of it. The varieties are distributed in temperature clines, many overlapping; not uncommonly more than one variety is found in a small sample of a single body of water.

*Differences among the Varieties of P. aurelia and the Problem of Identification.* Differences of the following sorts are found among the varieties: mating type specificity; geographical distribution; maximal and minimal tolerated temperature (these not always being completely correlated with geographic distribution); fission rates at the same temperature; body length; number of micronuclei in early life; the occurrence of killer and resistant strains; specificity of serotypes; system of determination and inheritance of serotypes; system of mating type inheritance and determination; the existence of and nature of diurnal periodicities in sexual reactivity; the temperature optima for mating; other conditions for mating; the occurrence of and duration of a period of immaturity after conjugation (immaturity never occurs after autogamy); the duration of the period of sexual maturity; the frequency of occurrence of lines that self-conjugate during maturity; the occurrence of autogamy or selfing during senility; the period of survival after loss of micronuclei during senility.

Each of these many differences between varieties was described in full. No one of them, except mating type and serotype specificities, can be used to define or identify all the varieties, nor can any combination of the others serve this purpose. Most of the traits just distinguish a few *groups* of varieties.

Only mating type specificity is sufficient to identify all varieties. Serologic specificity may, with fuller knowledge, eventually serve the same purpose. Great labor, technical knowledge, skill, and elaborate materials are required to use either mating types or serotypes for identification. Either *living standard* cultures of all mating types or hundreds of specific antisera are required. The difficulties and hazards involved in identification clearly preclude assigning species names to the varieties.

Objections are also raised against considering the varieties to be species. The term "syngen" is proposed to replace "biological species."

*Breeding Systems and Evolution in P. aurelia.* Fertilization serves two functions: genetic recombination and initiating new life cycles. Death (genetic, somatic, or both) eventually occurs in the absence of fertilization. Fertilization occurs either as conjugation or as autogamy. Varieties that undergo autogamy do so at a later stage of life (senility) than the one (maturity) in which they are ripe for conjugation. Varieties that lack autogamy are adapted only to cross conjugation during maturity, but to cross conjugation and self-conjugation during senility. Natural selection has thus given prior choice to conjugation and cross conjugation. The relative adaptation to cross conjugation and to selfing or autogamy differs greatly among the varieties. Few are strong outbreeders, many are intermediate, and many are strong inbreeders. The different breeding systems of different varieties are supported by differences in their life features. Outbreeders have long periods of immaturity, maturity, senility, and total life. This gives opportunity to migrate away from relatives before mating and to have much time to find a suitable mate. Inbreeders have short lives, with no immature period and a very brief mature period. They must mate quickly with the close relatives near at hand or undergo autogamy. Other adaptations to the breeding system, fully explained in the text, concern the number of mating types (two in inbreeders, more in outbreeders), the fission rate (low in outbreeders, high in inbreeders), the form of fertilization during senility (selfing in outbreeders, autogamy in inbreeders), and the method of mating type inheritance (synclonal uniformity in outbreeders, its absence in inbreeders). The breeding system actually determined depends on the particular *combination* of these features; when an inbreeding and an outbreeding feature occur together, one may negate the other. Thus the combination of a period of immaturity with a very brief mature period followed by autogamy leads to inbreeding in spite of the usual association of immaturity with outbreeding. Other examples are also cited.

Each local population examined in one inbreeding variety has a uniquely different chromosome set. This seems to be the basis of F2 mortality following crosses between populations. Strongly inbreeding varieties are thus a loose assemblage of highly isolated local populations, each of which has a status that just falls short of being comparable to that of a whole outbreeding variety. An outbreeding variety represents a clear-cut level of biological organization; inbreeding varieties do not. In the latter, the local population, not the variety, is the significant unit of biological organization.

Data are given to indicate the low population densities even at the season of maximal populations; and to show that, within a population of certain varieties, mating is not random; inbreeding prevails. Other observations indicate that migration into and out of a single body of water is a very rare event: the composition of a population remains constant over the years. Speculations on means of migration and its consequences under various conditions are presented.

The varieties are divisible into two groups, A and B, on the basis of differences in mating type inheritance. This is paralleled by differences in the serotype system and the presence or absence of killer strains.[4] It seems to represent an early evolutionary cleavage in *P. aurelia*. Similarities in serotypic and mating type specificities are greater within group A and within group B than between groups. Geographical distribution of the varieties indicates that within each group a widely distributed variety has given rise near the temperature limits of its range to narrowly confined varieties. The group A and group B mating type systems are found in other species of Paramecium and in *Tetrahymena pyriformis*, suggesting repeated independent evolutionary divergence along these lines. A general formulation of the two systems is given from which it appears that one or a few changes in details could shift from one system to the other. The major isolating mechanisms in *P. aurelia* are listed.

*Paramecium caudatum.* This species consists of 16 known varieties, with two mating types each. Only four varieties occur on two continents, the rest on one only. The varietal differences include: mating type specificity, geographical distribution, animal size and shape, micronuclear size and shape, optimal temperatures for mating, and resistance to a killer of *P. aurelia*. Only mating type specificity serves to distinguish all varieties. None of the other differences and no combination of them can, so far as present knowledge goes, serve to identify the varieties. Assigning species names would be futile.

With respect to gene flow, the varieties fall into two groups, 1 and 2. The 9 varieties of group 1 are absolutely isolated sexually from all other varieties. The 7 varieties of group 2 show a limited system of interbreeding, but the pattern of mating reactions of each mating type is unique. The possibility of gene flow among them has not yet been excluded. European workers are unable to sort their collections into varieties. They find fair viability after conjugation within a local population; but variable, though usually less, viability after conjugation between different populations. This refers to the F1 only. Little is known about F2 results in *P. caudatum* after crosses between populations of the same or different varieties. The F1 results on the European material suggest that each local population may be more or less isolated from all others, but the matter is in an unsettled condition.

Most varieties, if not all, of *P. caudatum* appear to be inbreeders. Immature periods are unknown and, if they exist, must be relatively short. Selfing is common. The group B method of mating type inheritance assures the usual existence of both mating types in a synclone. Some selfing is induced in mixtures of mating types. Autogamy seems to occur in some varieties, but not in most. Selfing appears to take its place. All these features, plus outbreeding F1 nonviability, point toward inbreeding. There may be one or more outbreeding varieties, but the evidence on this is still not clear.

*Paramecium bursaria.* Five varieties of *P. bursaria* are well known. One of these has two mating types, three have four mating types, and one has eight mating types. The sexual isolation of these varieties is complete except that one of the two mating types of one variety conjugates with four of the eight types of another variety. This exception is fruitless: all the conjugants die. Fruitful conjugation can occur between local populations of the same variety. Each variety thus constitutes a potentially common gene pool which is completely cut off from all others. Two of the varieties are confined to North America, two to Europe; one is common to Asia and North America.

Varieties differ in the following traits: number of mating types, specificity of mating types, diurnal periodicities in sexual reactivity; speed of conjugation; length and thickness of chromosomes; and geographical distribution. Identification of varieties is simpler than in the other species, partly because not more than three varieties are known from any land mass. These three could be readily distinguished if all the mating types were available, even without living mating type standards. Since the task of obtaining all mating types of a variety could be very laborious and time-consuming, this is not a satisfactory basis for a species description.

In variety 1, the only one very extensively studied, the life features mark it clearly as a strong outbreeder. The immature period is 3 to 5 months with continuously favorable conditions, but even a short unfavorable period triples the duration of immaturity. In the laboratory, maturity lasts about three years, senility about four to five years. In nature, all these stages of life are probably much longer than in the laboratory. Unlike the previous species, *P. bursaria* can mate readily throughout senility; but during this period the fertility, i.e., survival after conjugation, progressively decreases. This variety thus has remarkably long periods for dispersal before mating can occur and much longer periods (at least 7 years) during which mating can occur, as befits an outbreeder. The selection of such long mating periods suggests that the achievement of fruitful cross-mating is relatively rare. Other adaptations to, and correlates of, outbreeding are the systems of multiple mating types, the almost invariable synclonal uniformity in mating type, the total absence of selfers of the sort found in other species, the very much higher death rate after inbreeding (sib mating) as compared with outbreeding, the rapid rise in death rate through successive inbreedings. A puzzling exception is associated with the very rare production of a complementary mating type in a clone. When this mates with the original type in the clone, survival is as good as in outbreeding. Whatever the basis of this exceptional inbreeding, the result is highly adaptive, since, like autogamy and selfing in senility in *P. aurelia*, it both rejuvenates and permits the retention of heterozygosity.

The synclonal uniformity in mating type is based upon the *P. aurelia* group B system of mating type determination plus the regular massive exchange of cytoplasm between mates. The rare exceptions (2.6% of the synclones) which yield more than one mating type per synclone reveal the basic B system. A hypothesis is presented to account for all the remarkable kinds of observations on the inheritance of mating type which have hitherto remained unexplained. The system of mating type determination in *P. bursaria* is the most complex of those discussed. Loss of the mechanism assuring regular cytoplasmic exchange between mates transforms it to the *P. aurelia* form of the group B system. Loss of one or two other features (explained in the text) converts the B into the A system. The *P. bursaria* system is a strong adaptation to outbreeding by almost completely blocking mating within a synclone.

The highly developed complex of adaptations to outbreeding in *P. bursaria* inhibits progressive divergence of local populations. As a result of the wide distribution of potentially common gene pools, speciation is poor in comparison with that of inbreeders. In the United States there are probably only three varieties of *P. bursaria* as compared with at least 14 (probably more) varieties of *P. aurelia*.

*Tetrahymena pyriformis.* At least nine varieties of this species have been found in a survey confined to North, Central, and northern South America. There may well be more varieties in this region, even among the materials already collected, because a large proportion of the collections could not be classified for technical reasons (selfers and amicronucleates) and the basis of classification is solely the capacity to mate, genetic isolation not yet having been systematically studied. All but two of the varieties are confined to North America, four of them being concentrated in the northern latitudes of the United States. Of the other two, one (variety 2) ranges from Canada to northern South America, and the other (variety 9) is limited to the southernmost part of this range. All varieties have five chromosomes in the haploid set, and their distinctive morphology is the same in all varieties. The number of known mating types is two in three varieties, three in two varieties, and five, seven, eight, and eleven in one variety each. These are minimal figures based chiefly on collections from nature; breeding analysis in the laboratory has revealed some not yet found in nature and may reveal more.

Varietal differences include the following: number of mating types, system of mating type determination, geographical distribution, preference for running or still water, size, possibly ciliary patterns, and length of immature period. Little is yet known about possible differences in the characteristics and duration of the periods of maturity and senility. In some varieties, senility is marked by loss of micronuclei and capacity to mate; but the organisms may live at least 20 years in this condition. Variety 1 may begin to reach this stage at ages 500 to 1500 fissions. Senility may be marked by the onset of selfing with decreasing survival accompanying increasing parental age in some varieties; but this is not yet certain. Among the varietal differences, at present only mating type specificity can be used to identify all the varieties. This requires live standard cultures of at least two mating types of each variety. Since identification by routine taxonomic procedures is impossible, assigning specific names to the varieties is not practicable.

Knowledge of the breeding system is greatest for varieties 1 and 2. Variety 1 with seven mating types shows the *P. aurelia* group A system of mating type determination. A pair of alleles limiting mating type potentialities has been found. This variety has an immature period and is heterozygous in nature. It shows hybrid vigor and inbreeding degeneration, but normal fertile inbred lines can be selected eventually. During meiosis in crosses of wild strains evidences of chromosomal aberrations are apparent. The possible relation of these and the nonviability after conjugation to age-induced chromosomal aberrations has not been studied. Much selfing occurs during maturity in variety 1 when cultivated with excess food, but this is greatly reduced by periodic starvation. In nature it therefore is probably not very common. Taking all these features into account, variety 1 appears to have an intermediate breeding system. It has many outbreeding stigmata: immaturity, multiple mating types, natural heterozygosity, and inbreeding degeneration. But it also has some inbreeding features: distribution is concentrated in a narrow range of latitude; synclones almost always include more than one mating type; normal fertile inbred lines are obtainable; selfing occurs; and structural chromosomal differences between natural strains lead to abnormality and death when they are crossed. Some of these inbreeding features are to a degree compensated by outbreeding features: synclonal nonuniformity of mating type by immaturity and selfing by the stabilizing effect of starvation. Others, such as chromosomal aberrations may be peculiarities of aging. On the whole, therefore, variety 1 seems to be primarily an outbreeder, but less extremely so than *P. bursaria*. It has latent mechanisms for inbreeding, which it would appear could easily give rise to inbreeding descendants.

Variety 2 is better adapted to outbreeding and shows more stigmata of that system of breeding. It has an immature period, the largest number of mating types (11), the widest latitudinal range, and synclonal uniformity of mating type due to the *P. bursaria* form of the *P. aurelia* group B mechanism of mating type determination. Little is known about the breeding systems in the other varieties. Crosses between different populations within variety 5 and within variety 9 yielded no survivors; the crosses of variety 9 showed much chromosomal abnormality at meiosis. Some survivors were obtained from crosses of populations of variety 6. Again the role of aging in these results is unknown. Further work may well reveal other marked differences among the varieties in their breeding systems.

*Euplotes.* There is a group of morphologically distinct species which are, however, much like *E. patella*. In three of these species (*E. aediculatus, E. eurystomus, E. patella*) and in the less closely related *E. harpa*, mating types forming one or more varieties are known, but there has been no effort to discover the full number of varieties in any region in any of these species. Knowledge is chiefly confined to one variety of *E. patella*, indeed to the progeny of two wild individuals from neighboring ponds. These were of different mating type and yielded four other mating types among their sexual progeny. Doubtless more, possibly many more, mating types of this variety exist in nature.

The method of mating type determination is entirely different from the ones thus far described. The six types are determined by the three homozygous and three heterozygous combinations of three codominant alleles. Because of the genotypic identity of the two exconjugants of a pair, each synclone is uniform in mating type. Mating is normally controlled by the action of three hormones secreted into the medium, each hormone being under the control of one allele. An animal is activated, or put into mating condition, by exposure to a hormone which it cannot itself produce. Animals thus activated can mate with each other regardless of whether their mating types are alike or different. Thus, "mating type" has a somewhat different meaning in this material.

*E. patella* is an outbreeder in spite of some superficial indications to the contrary. Since any two activated animals can mate with each other, mixtures of two mating types lead to much selfing. But in nature, the lesser population densities and the large volume of diluent probably restrict hormone action to contiguous pairs, and these could conjugate only if each activated the other, a result which is possible only if they are of certain different mating types. This is then actually an adaptation to outbreeding, not inbreeding. Other indications of outbreeding are: the system of multiple (probably many) mating types, a regular immature period, and high mortality after inbreeding. On the other hand, some selfing may occur, as this was observed even within a clone. Its possible relation to age is unknown. Eventually work on *E. patella* was abandoned because the animals became difficult to culture. Whether this was due partly to aging and inbreeding is unknown.

*Other Ciliates.* In *Paramecium trichium* three interbreeding pure mating types are known. Selfers occur, but their relation to the former is unknown. *P. calkinsi* has at least two varieties with two mating types in each. *Colpidium truncatum* has multiple mating types and much selfing in the one variety observed. *Stylonychia putrina* includes at least two varieties with at least 11 and 15 mating types, respectively, and selfing is very rare.

One variety with nine mating types has been found in *Oxytricha bifaria*. Some were collected in nature, others arose among their sexual progeny. Selfing is very rare, no hormones can be demonstrated, and conjugation is regularly between unlike mating types. This variety is an outbreeder. In addition to multiple mating types and the rarity of selfing, other indices of outbreeding are a considerable immature period and a long mature period. The one culture which selfed a little usually failed to survive selfing, but usually survived when crossed to one of the two other types reported upon. The system of survival and death in relation to inbreeding and outbreeding is still unknown.

*Obligatory Inbreeding and Asexual Protozoa.* The Flagellates which are symbiotic in wood roaches and termites seem to show an evolutionary series from sexual to asexual reproduction. The sexual species are found in wood roaches; the related asexual species occur in termites. The coincidence in time between sexual processes in the Flagellates and transmission to new roach hosts seems like a mechanism for providing genetic variability in the symbiont at the time it might be needed for adaptation to varied hosts. But the biology of the roaches indicates little variation among the individuals of a colony, and the sexual processes in the symbionts are such as to yield little or no variability in them.

Hence, these sexual processes are viewed as vestiges of more nearly perfect recombination processes of the ancestors of the symbionts when the latter were more variable. This is consistent with the absence of sexuality in the closely related symbionts of termites, which are more highly evolved than the roaches, and with the fuller development of sexuality in some of the simpler plant Flagellates.

There are two major species problems in asexual Protozoa. One grows out of the gene pool concept of species, for there are no common gene poles in such organisms and therefore no species in that sense. The other grows out of the routine taxonomic usage of the term species, for little attention has been given to establishing similar usage in sexual and asexual organisms. In the latter, physiological differences alone have been used in some cases, though morphological differences are sometimes held to be essential. In my opinion, similarity of usage requires recognition of genetic differences of the same kind and magnitude in both sexual and asexual organisms.

The situation in Trypanosomes illustrates the inadequacy of both physiological and some morphological bases for distinguishing species. Serologic differences in Trypanosomes can be independent of genic differences, as in *Paramecium aurelia,* and similar conditions may be expected in other asexual organisms. Ignorance of the genetic identity of lines showing very different serotypes could lead to assigning members of the same clone to different species. Other physiologic differences, such as virulence and host specificity, are used to distinguish species of Trypanosomes; but in some organisms these are known to be single gene differences. Confinement to different hosts is *not* equivalent to geographic isolation when the adaptation to different hosts depends on single mutational steps. Simple morphological differences, such as the presence or absence of a kinetoplast, are also used to distinguish species. These types also arise at a single step by a mechanism comparable to loss of kappa in *Paramecium*. This too is indefensible, for it could assign two products of one fission to different species. The extent to which considerable morphological variations can quickly arise within a clone is well shown by the rapid selection of very diverse types within a clone in Rhizopods and the selection of similar types from initially different clones. Thus genetically simple morphological differences as well as genetically simple physiological differences are equally inadmissible as species differentials if species are to represent equivalent evolutionary divergences in sexual and asexual organisms. Many criteria now used in asexual organisms to distinguish species are comparable to individual differences in sexual organisms.

Two major mechanisms of variation are recognizable in asexual organisms. One is based upon mutation alone. It operates successfully in haploids that rapidly produce large populations. The other works equally well in diploids. This involves having in the genotype multiple loci for alternative, mutually exclusive, and rapidly transformable traits. It is illustrated by the serotype system in *Paramecium* and Trypanosomes. This mechanism operates in purely asexual organisms, in sexual organisms with a well-developed asexual phase, and in the asexual phase, that is, the phase of development, in higher organisms.

Speciation appears to be fundamentally the same in sexual and asexual organisms. In both it is based upon genetic adaptation to changing conditions. It may proceed more slowly and less abundantly in some asexual organisms because of their choice of relatively constant conditions of life, such as in a host organism or in a considerable body of water. The asexual method of reproduction has proved its adequacy if not its advantage under certain conditions, by its widespread occurrence in nature. It has means of rapidly spreading and multiplying advantageous variants, and it avoids the wastage of recombination and the swamping of variations which results from crosses. However, short-term advantage is no guarantee of long-term persistence. While some asexual organisms may be wiped out in times of stress, others have succeeded on the evolutionary time scale by adapting themselves to a constant or very slowly changing environment, such as life in a host.

It is possible to adopt principles of classification into species and syngens which are equally applicable to sexual and asexual organisms. This requires reducing species to one of its current meanings. By restricting it to designate a readily identifiable group which shows the minimal irreversible discontinuity with other groups, species can be set up just as well in asexual as in sexual organisms. It is primarily designed for the convenience of biologists and is not intended to represent a biological reality. The latter is reserved for the term syngen. This designates the group which is potentially able to contribute to the further evolution of their descendants, i.e., a group within which irreversible evolutionary divergence has not yet occurred. When this cannot be defined by breeding methods as in asexual organisms, it is in principle possible to do so by intensive comparison of every aspect of many strains and judging by every means available the complexity of their genetic differences and the limits of untransgressible discontinuities. This is precisely what is discovered by breeding methods in sexual organisms. In both cases the syngen is independent of readily recognized differences and depends only upon complex genetic deviation associated with untransgressible discontinuity.

*Conclusions.* The conclusions on the three major subjects dealt with in this paper are given in full in the section and will not be repeated here.

**Addenda**

1. In the six-month interim between writing and proofreading, new discoveries about *P. aurelia* make necessary the following changes.

2. The *P. aurelia-multimicronucleatum* complex is not divisible into well-marked species on the basis of currently used criteria: body size and micronuclear number; but at least three distinct types are distinguishable by size of micronuclei and structure of their central chromatic area. The micronuclear diameters are 2.6 $\mu$, 3 $\mu$ and more than 4 $\mu$; their chromatic areas are solid, doughnut shaped, and spongy, respectively. The first combination is typical *multimicronucleatum* (varieties 13, 15, and new 16, see addendum 3) and it is usually associated with more than two micronuclei;

the second is typical *aurelia;* the third is characteristic of variety 12. (See pp. 161, 162, 169, 178, 179, 198, 232.)

3. My variety 15 and Giese's variety 16 are almost certainly identical; both will hereafter be called variety 15. It has two constant mating types (XXIX and XXX) with diurnal periodicity stated in the text. It also has two other types (XXXI and XXXII). XXXI behaves as if it was XXIX at night, XXX during the day; XXXII behaves in the reverse way. There are thus four mating types based upon two mating specificities. Not all strains require tube culture in order to conjugate. Another variety (from Mexico) is now called variety 16; it has the *aurelia* body size, but *multimicronucleatum* micronuclear size, number, and structure (see addendum 2). (See pp. 163, 166, 169, 171, 178, 179, 185, 189, 190, 194, 198, 207, 301.)

4. Beale (in press) has found a mate killer in variety 1 of group A. (See pp. 179, 226, 234, 304.)

5. The short immature periods after these variety 15 crosses appear to be due to macronuclear regeneration. (See pp. 191, 207.)

6. Newly found strains of variety 13 contain caryonides some of which stay pure for mating type XXV during the period between fertilizations. Autogamy has now been found in variety 13. (See pp. 188, 193, 198, 214, 258, and Table III.)

## REFERENCES

Austin, M. L. 1951. Sensitivity to paramecin in *Paramecium aurelia* in relation to stock, serotype, and mating type. *Physiol. Zool.*, **24**, 196.

Austin, M. L., D. Widmayer, and L. M. Walker. 1956. Antigenic transformation as adaptive response of *Paramecium aurelia* to patulin; relation to cell division. *Physiol. Zool.*, **29**, 261.

Beale, G. H. 1952. Antigen variation in *Paramecium aurelia*, variety 1. *Genetics*, **37**, 62.

Beale, G. H. 1954. *The Genetics of Paramecium aurelia*. Cambridge University Press, Cambridge, England.

Beale, G. H., and M. Schneller. 1954. A ninth variety of *Paramecium aurelia*. *J. Gen. Microbiol.*, **11**, 57.

Butzel, H. M., Jr. 1953. *Genetic Studies on Paramecium aurelia*. Dissertation, Indiana University, Bloomington, Indiana.

Butzel, H. M., Jr. 1955. Mating type mutations in variety 1 of *Paramecium aurelia*, and their bearing upon the problem of mating type determination. *Genetics*, **40**, 321.

Calkins, G. N. 1919. *Uroleptus mobilis* Engelm. II. Renewal of vitality through conjugation. *J. Exptl. Zool.*, **29**, 121.

Calkins, G. N. 1920. III. A study in vitality. *J. Exptl. Zool.*, **31**, 287.

Chen, T. T. 1940. Polyploidy and its origin in Paramecium. *J. Heredity*, **31**, 175.

Chen, T. T. 1946a. Conjugation in *Paramecium bursaria*. II. Nuclear phenomena in lethal conjugation between varieties. *J. Morphol.*, **79**, 125.

Chen, T. T. 1946b. Varieties and mating types in *Paramecium bursaria*. I. New variety and types, from England, Ireland, and Czechoslovakia. *Proc. Natl. Acad. Sci. U.S.*, **32**, 173.

Chen, T. T. 1956. Varieties and mating types in *Paramecium bursaria*. II. Variety and mating types found in China. *J. Exptl. Zool.*, **132**, 255.

Chen, Y. T. 1944. Mating types in *Paramecium caudatum*. *Am. Naturalist*, **78**, 334.

Cleveland, L. R. 1950a. Hormone-induced sexual cycles of flagellates. II. Gametogenesis, fertilization, and one-division meiosis in *Oxymonas*. *J. Morphol.*, **86**, 185.

Cleveland, L. R. 1950b. Hormone-induced sexual cycles of flagellates. III. Gametogenesis, fertilization, and one-division meiosis in *Saccinobaculus*. *J. Morphol.*, **86**, 215.

Cleveland, L. R. 1950c. Hormone-induced sexual cycles of flagellates. IV. Meiosis after syngamy and before nuclear fusion in *Notila*. *J. Morphol.*, **87**, 317.

Cleveland, L. R. 1950d. Hormone-induced sexual cycles of flagellates. V. Fertilization in *Eucomonympha*. *J. Morphol.*, **87**, 349.

Cleveland, L. R. 1951a. Hormone-induced sexual cycles of flagellates. VI. Gametogenesis, fertilization, meiosis, oöcysts, and gametocysts in *Leptospironympha*. *J. Morphol.*, **88**, 199.

Cleveland, L. R. 1951b. Hormone-induced sexual cycles of flagellates. VII. One-division meiosis and autogamy without cell division in *Urinympha*. *J. Morphol.*, **88**, 385.

Cleveland, L. R. 1952. Hormone-induced sexual cycles of flagellates. VIII. Meiosis in *Rhynchonympha* in one cytoplasmic and two nuclear divisions followed by autogamy. *J. Morphol.*, **91**, 269.

Cleveland, L. R. 1953. Hormone-induced sexual cycles of flagellates. IX. Haploid gametogenesis and fertilization in *Barbulanympha*. *J. Morphol.*, **93**, 371.

Cleveland, L. R. 1954. Hormone-induced sexual cycles of flagellates. X. Autogamy and endomitosis in *Barbulanympha* resulting from interruption of haploid gametogenesis. *J. Morphol.*, **95**, 189.

Cleveland, L. R., S. R. Hall, E. P. Sanders, and J. Collier. 1934. The wood-feeding roach *Cryptocercus*, its Protozoa, and the symbiosis between Protozoa and Roach. *Mem. Am. Acad. Arts Sci.*, **17**, 185.

Corliss, J. O. 1952. Le cycle autogamique de *Tetrahymena rostrata*. *Compt. rend.*, **235**, 399.

Corliss, J. O. 1954. The literature on *Tetrahymena*: Its history, growth, and recent trends. *J. Protozool.*, **1**, 156.

De Garis, C. F. 1935a. Lethal effects of conjugation between *Paramecium aurelia* and double monsters of *Paramecium caudatum*. *Am. Naturalist*, **69**, 87.

De Garis, C. F. 1935b. Heritable effects of conjugation between free individuals and double monsters in diverse races of *Paramecium caudatum*. *J. Exptl. Zool.*, **71**, 209.

Diller, W. F. 1948. Nuclear behavior of *Paramecium trichium* during conjugation. *J. Morphol.*, **82**, 1.

Dippell, R. V. 1954. A preliminary report on the chromosomal constitution of certain variety 4 races of *Paramecium aurelia*. *Caryologia*, suppl. **1954**, 1109.

Dippell, R. V. 1955. Some cytological aspects of aging in variety 4 of *Paramecium aurelia*. *J. Protozool.*, **2** (suppl.), 7.

Dobzhansky, T. 1937. What is a species? *Scientia*, **61**, 280.

Dobzhansky, T. 1941. *Genetics and the Origin of Species*. Columbia University Press, New York, N.Y.

Downs, L. E. 1952. Mating types in *Stylonychia putrina*. *Proc. Soc. Exptl. Biol. Med.*, **81**, 605.

Downs, L. E. 1956. Mating types and varieties in *Stylonychia putrina*. *Proc. Soc. Exptl. Biol. Med.*, **93**, 586.

Ehret, C. F. 1953. An analysis of the role of electromagnetic radiations in the mating reaction of *Paramecium bursaria*. *Physiol. Zool.*, **26**, 274.

Elliott, A. M., and D. F. Gruchy. 1952. The occurrence of mating types in *Tetrahymena*. *Biol. Bull.*, **103**, 301.

Elliott, A. M., and R. E. Hayes. 1953. Mating types in *Tetrahymena*. *Biol. Bull.*, **105**, 269.

Elliott, A. M., and R. E. Hayes. 1955. *Tetrahymena* from Mexico, Panama, and Colombia, with special reference to sexuality. *J. Protozool.*, 2, 75.

Elliott, A. M., and D. L. Nanney. 1952. Conjugation in *Tetrahymena*. *Science*, 116, 33.

Erdmann, R. 1927. Endomixis bei *Paramecium bursaria*. *Sitzber. Ges. naturforsch. Freunde Berlin* (1925), 24.

Erdmann, R., and L. L. Woodruff. 1916. The periodic reorganization process in *Paramecium caudatum*. *J. Exptl. Zool.*, 20, 59.

Finger, I. 1956. Immobilizing and precipitating antigens of *Paramecium*. *Biol. Bull.*, 111, 358.

Finger, I. 1957a. Immunological studies on immobilizing antigens of *Paramecium aurelia*. *J. Gen. Microbiol.* (in press).

Finger, I. 1957b. Inheritance of immobilization antigens of *Paramecium aurelia* variety 2. *J. Gen. Microbiol.*, 16, 350.

Giese, A. C. 1941. Mating types in *Paramecium multimicronucleatum*. *Anat. Record*, 81 (suppl.), 131.

Giese, A. C. 1957. Mating types in *Paramecium multimicronucleatum*. *J. Protozool.*, 4, 120.

Giese, A. C., and M. A. Arkoosh. 1939. Tests for sexual differentiation in *Paramecium multimicronucleatum* and *Paramecium caudatum*. *Physiol. Zool.*, 12, 70.

Gilman, L. C. 1939. Mating types in *Paramecium caudatum*. *Am. Naturalist*, 73, 445.

Gilman, L. C. 1941. Mating types in diverse races of *Paramecium caudatum*. *Biol. Bull.*, 80, 384.

Gilman, L. C. 1949. Intervarietal mating reactions in *Paramecium caudatum*. *Biol. Bull.*, 97, 239.

Gilman, L. C. 1950. The position of Japanese varieties of *Paramecium caudatum* with respect to American varieties. *Biol. Bull.*, 99, 348.

Gilman, L. C. 1954. Occurrence and distribution of mating type varieties in *Paramecium caudatum*. *J. Protozool.*, 1 (suppl.), 6.

Gilman, L. C. 1956a. Distribution of the varieties of *Paramecium caudatum*. *J. Protozool.*, 3 (suppl.), 4.

Gilman, L. C. 1956b. Size differences among twelve varieties of *Paramecium caudatum*. *J. Protozool.*, 3 (suppl.), 4.

Gruchy, D. F. 1955. The breeding system and distribution of *Tetrahymena pyriformis*. *J. Protozool.*, 2, 178.

Harrison, J. A., and E. H. Fowler. 1946. A serologic study of conjugation in *Paramecium bursaria*. *J. Exptl. Zool.*, 101, 425.

Hiwatashi, K. 1949. Studies on the conjugation of *Paramecium cau-*

*datum.* I. Mating types and groups in the races obtained in Japan. *Science Repts. Tôhoku Imp. Univ., Biol.,* **18,** 137.

Hiwatashi, K. 1951. Studies on the conjugation of *Paramecium caudatum.* IV. Conjugating behavior of individuals of two mating types marked by a vital staining method. *Science Repts. Tôhoku Imp. Univ., Biol.,* **19,** 95.

Hoare, C. A. 1952. The taxonomic status of biological races in parasitic Protozoa. *Proc. Linnean Soc. London,* **163,** 44.

Inoki, S., T. Kitaura, Y. Kurogochi, H. Osaki, and T. Nakabayasi. 1952a. Genetical studies on the antigenic variation in *Trypanosoma gambiense. Japan. J. Genetics,* **27,** 85.

Inoki, S., T. Kitaura, T. Nakabayasi, and H. Kurogochi. 1952b. Studies on the immunological variations in *Trypanosoma gambiense.* 1. A new variation system and a new experimental method. *Med. J. Osaka Univ.,* **3,** 357.

Jennings, H. S. 1910. What conditions induce conjugation in *Paramecium? J. Exptl. Zool.,* **9,** 279.

Jennings, H. S. 1911. Assortative mating, variability and inheritance of size in the conjugation of *Paramecium, J. Exptl. Zool.,* **11,** 1.

Jennings, H. S. 1916. Heredity, variation and the results of selection in the uniparental reproduction of *Difflugia corona. Genetics,* **1,** 407.

Jennings, H. S. 1938. Sex reaction types and their interrelations in *Paramecium bursaria.* I and II. *Proc. Natl. Acad. Sci. U.S.,* **24,** 112.

Jennings, H. S. 1939. Genetics of *Paramecium bursaria.* I. Mating types and groups, their interrelations and distribution: mating behavior and self sterility. *Genetics,* **24,** 202.

Jennings, H. S. 1941. Genetics of *Paramecium bursaria.* II. Self-differentiation and self-fertilization of clones. *Proc. Am. Phil. Soc.,* **85,** 25.

Jennings, H. S. 1942. Genetics of *Paramecium bursaria.* III. Inheritance of mating type, in crosses and clonal self fertilizations. *Genetics,* **27,** 193.

Jennings, H. S. 1944a. *Paramecium bursaria:* Life history. I. Immaturity, maturity and age. *Biol. Bull.,* **86,** 131.

Jennings, H. S. 1944b. *Paramecium bursaria:* Life history. II. Age and death of clones in relation to the results of conjugation. *J. Exptl. Zool.,* **96,** 17.

Jennings, H. S. 1944c. *Paramecium bursaria:* Life history. III. Repeated conjugations in the same stock at different ages, with and without inbreeding, in relation to mortality at conjugation. *J. Exptl. Zool.,* **96,** 243.

Jennings, H. S. 1944d. *Paramecium bursaria:* Life history. IV. Relation of inbreeding to mortality of exconjugant clones. *J. Exptl. Zool.,* **97,** 165.

Jennings, H. S. 1945. *Paramecium bursaria:* Life history. V. Some relations of external conditions, past or present, to ageing and to mortality of exconjugants, with summary of conclusions on age and death. *J. Exptl. Zool.,* **99,** 15.

Jennings, H. S., and P. Opitz. 1944. Genetics of *Paramecium bursaria.* IV. A fourth variety from Russia. Lethal crosses with an American variety. *Genetics,* **29,** 576.

Johnson, M. 1955. *Viability of Inbred and Outbred Paramecium caudatum.* Thesis, University of Durham, England.

Johnson, W. H. 1952. Further studies on the sterile culture of Paramecium. *Physiol. Zool.,* **25,** 10.

Jollos, V. 1916. Die Fortpflanzung der Infusorien und die potentielle Unsterblichkeit der Einzelligen. *Biol. Zentr.,* **36,** 497.

Katashima, R. 1952. Studies on *Euplotes.* I. Conjugation and cytogamy induced by split pair method in *Euplotes harpa. J. Sci. Hiroshima Univ. Ser. B, Div. 1,* **13,** 111.

Kay, M. W. 1946. Studies on *Oxytricha bifaria.* III. Conjugation. *Trans. Am. Microscop. Soc.,* **65,** 131.

Kimball, R. F. 1939. Mating types in *Euplotes. Am. Naturalist,* **73,** 451.

Kimball, R. F. 1942. The nature and inheritance of mating types in *Euplotes patella. Genetics,* **27,** 269.

Kimball, R. F. 1943. Mating types in the ciliate Protozoa. *Quart. Rev. Biol.,* **18,** 30.

Kirby, H. 1949. Systematic differentiation and evolution of Flagellates in Termites. *Rev. soc. mex. hist. nat.,* **10,** 57.

Lederberg, J. 1955. Genetics and microbiology. In *Perspectives and Horizons in Microbiology,* S. A. Waksman, editor. Rutgers University Press, New Brunswick, New Jersey, p. 24.

Lee, H. 1949. Change of mating type in *Paramecium bursaria* following exposure to X-rays. *J. Exptl. Zool.,* **112,** 125.

Levine, M. 1953. The interaction of nucleus and cytoplasm in the isolation and evolution of species of *Paramecium. Evolution,* **7,** 366.

Margolin, P. 1956a. An exception to mutual exclusion of the ciliary antigens in *Paramecium aurelia. Genetics,* **41,** 685.

Margolin, P. 1956b. The ciliary antigens of stock 172, *Paramecium aurelia,* variety 4. *J. Exptl. Zool.,* **133,** 345.

Maupas, E. 1888. Récherches expérimentales sur la multiplication des infusoires ciliés. *Arch. zool. exptl. gén.*, **6**, 165.

Maupas, E. 1889. La rajeunissement karyogamique chez les ciliés. *Arch. zool. exptl. gén.* **7**, 149.

Mayr, E. 1948. The bearing of the new systematics on genetical problems. The nature of species. *Advances in Genetics*, **2**, 205.

Melechen, N. 1955. *Antigenic Stability and Variability in Paramecium aurelia, Variety 3*. Thesis, University of Pennsylvania, Philadelphia, Pa.

Melvin, J. B. 1949. *Hybrid Mortality in Intervarietal Crosses in Paramecium aurelia*. Thesis, Wellesley College, Wellesley, Massachusetts.

Metz, C. B. 1954. Mating substances and the physiology of fertilization in Ciliates. In *Sex in Microorganisms*, American Association for the Advancement of Science, Washington, D.C., p. 284.

Müller, W. 1932. Cytologische und vergleichend-physiologische Untersuchungen über *Paramecium multimicronucleatum* und *Paramecium caudatum*, zugleich ein Versuch zur Kreuzung beider Arten. *Arch. Protistenk.*, **78**, 361.

Nanney, D. L. 1953. Nucleo-cytoplasmic interaction during conjugation in *Tetrahymena*. *Biol. Bull.*, **105**, 133.

Nanney, D. L. 1954. Mating type determination in *Paramecium aurelia*. A study in cellular heredity. In *Sex in Microorganisms*, American Association for the Advancement of Science, Washington, D.C., p. 266.

Nanney, D. L. 1956a. Inbreeding deterioration in *Tetrahymena pyriformis*. *Genetics*, **41**, 655.

Nanney, D. L. 1956b. Caryonidal inheritance and nuclear differentiation. *Am. Naturalist*, **90**, 291.

Nanney, D. L., and P. A. Caughey. 1953. Mating type determination in *Tetrahymena pyriformis*. *Proc. Natl. Acad. Sci. U.S.*, **39**, 1057.

Nanney, D. L., and P. A. Caughey. 1955. An unstable nuclear condition in *Tetrahymena pyriformis*. *Genetics*, **40**, 388.

Nanney, D. L., P. A. Caughey, and A. Tefankjian. 1955. The genetic control of mating type potentialities in *Tetrahymena pyriformis*. *Genetics*, **40**, 688.

Pierson, B. F. 1943. A comparative morphological study of several species of *Euplotes* closely related to *Euplotes patella*. *J. Morphol.*, **72**, 125.

Powelson, E. E. 1956. Differences in the silver-line system and various measurements in individuals, stocks and varieties of *Paramecium aurelia*. *J. Protozool.*, **3** (suppl.), 9.

Powers, E. L., Jr. 1943. The mating types of double animals in *Euplotes patella*. *Am. Midland Naturalist*, **30**, 175.

Preer, J. R., Jr. 1950. The role of genes, cytoplasm and environment in the determination of resistance and sensitivity. *Year Book Am. Phil. Soc.*, p. 161.

Pringle, C. R. 1955. Thesis, Univ. Edinburgh, Edinburgh, Scotland.

Pringle, C. R. 1956. Antigenic variation in *Paramecium aurelia*, variety 9. *Z. Abst. Vererbsl.*, **87**, 421.

Ray, C., Jr. 1954. Chromosome behavior during conjugation of mating types I and II of variety 1 of *Tetrahymena*. *Biol. Bull.*, **107**, 318.

Ray, C., Jr. 1955. Irregularities during meiosis in variety 9 of *Tetrahymena pyriformis*. *Biol. Bull.*, **109**, 367.

Ray, C., Jr. 1956. Meiosis and nuclear behavior in *Tetrahymena pyriformis*. *J. Protozool.*, **3**, 88.

Ray, C., Jr., and A. M. Elliott. 1954. Chromosome number of four varieties of *Tetrahymena*. *Anat. Record*, **120**, 812.

Ray, C., Jr., A. M. Elliott, and G. M. Clark. 1955. The immaturity period in three varieties of *Tetrahymena pyriformis*. *Biol. Bull.*, **109**, 367.

Siegel, R. W. 1954. Mate-killing in *Paramecium aurelia*, variety 8. *Physiol. Zool.*, **27**, 89.

Siegel, R. W. 1956. Mating types in *Oxytricha* and the significance of mating systems in Ciliates. *Biol. Bull.*, **110**, 352.

Skaar, P. D. 1956. Past history and pattern of serotype transformation in *Paramecium aurelia*. *Exptl. Cell Research*, **10**, 646.

Sonneborn, T. M. 1937. Sex, sex inheritance and sex determination in *Paramecium aurelia*. *Proc. Natl. Acad. Sci. U.S.*, **23**, 378.

Sonneborn, T. M. 1938. Mating types in *Paramecium aurelia*: Diverse conditions for mating in different stocks; occurrence, number and interrelations of the types. *Proc. Am. Phil. Soc.*, **79**, 411.

Sonneborn, T. M. 1939. *Paramecium aurelia*: Mating types and groups; lethal interactions; determination and inheritance. *Am. Naturalist*, **73**, 390.

Sonneborn, T. M. 1941. The effect of temperature on mating reactivity in *Paramecium aurelia*, variety 1. *Anat. Record*, **81** (suppl.), 131.

Sonneborn, T. M. 1942a. Inheritance in Ciliate Protozoa. *Am. Naturalist*, **76**, 46.

Sonneborn, T. M. 1942b. More mating types and varieties in *Paramecium aurelia*. *Anat. Record*, **84**, 92.

Sonneborn, T. M. 1943. Gene and cytoplasm. *Proc. Natl. Acad. Sci. U.S.*, **29**, 329.

Sonneborn, T. M. 1947. Recent advances in the genetics of *Paramecium* and *Euplotes*. *Advances in Genetics*, **1**, 263.

Sonneborn, T. M. 1950. Methods in the general biology and genetics of *Paramecium aurelia*. *J. Exptl. Zool.*, **113**, 87.

Sonneborn, T. M. 1954a. The relation of autogamy to senescence and rejuvenescence in *Paramecium aurelia*. *J. Protozool.*, **1**, 38.

Sonneborn, T. M. 1954b. Patterns of nucleocytoplasmic integration in *Paramecium*. *Caryologia* (**suppl. 1**), 307.

Sonneborn, T. M. 1955. Heredity, development and evolution in *Paramecium*. *Nature*, **175**, 1100.

Sonneborn, T. M. 1956a. New varieties and mating types of *Paramecium aurelia*. *Anat. Record*, **125**, 566.

Sonneborn, T. M. 1956b. Distribution of the varieties of *Paramecium aurelia*. *Anat. Record*, **125**, 567.

Sonneborn, T. M. 1956c. The distribution of killers among the varieties of *Paramecium aurelia*. *Anat. Record*, **125**, 567.

Sonneborn, T. M., and R. V. Dippell. 1943. Sexual isolation, mating types, and sexual responses to diverse conditions in variety 4, *Paramecium aurelia*. *Biol. Bull.*, **85**, 36.

Sonneborn, T. M., and R. V. Dippell. 1946. The significance of race 31 as a link between group A and B varieties of *P. aurelia*. *Anat. Record*, **96**, 19.

Sonneborn, T. M., and R. V. Dippell. 1956. *Giant Paramecium aurelia* (?). *J. Protozool.*, 3 (**suppl.**), 9.

Sonneborn, T. M., and M. V. Schneller. 1955. Genetic consequences of aging in variety 4 of *Paramecium aurelia*. *Genetics*, **40**, 596.

Stebbins, G. L., Jr. 1950. *Variation and Evolution in Plants*. Columbia University Press, New York, N.Y.

Thoday, J. M. 1953. Components of fitness. *Symposia Soc. Exptl. Biol.*, **7**, 96.

Vivier, E. 1955. Contribution à l'étude de la conjugaison chez *Paramecium caudatum*. *Bull. soc. zool. France*, **80**, 163.

Wenrich, D. H. 1954. Sex in Protozoa: a comparative review. In *Sex in Microorganisms*, American Association for the Advancement of Science, Washington, D.C., p. 134.

Wichterman, R. 1937. Conjugation in *Paramecium trichium* Stokes

(Protozoa, Ciliata) with special reference to nuclear phenomena. *Biol. Bull.*, **73**, 397.

Wichterman, R. 1948. The time schedule of mating and nuclear events in the conjugation of *Paramecium bursaria*. *Turtox News*, 26.

Wichterman, R. 1950. The occurrence of a new variety containing two opposite mating types of *Paramecium calkinsi* as found in sea water of high salinity content. *Biol. Bull.*, **99**, 366.

Wichterman, R. 1951. The ecology, cultivation, structural characteristics and mating types of *Paramecium calkinsi*. *Proc. Penn. Acad. Sci.*, **25**, 51.

Wichterman, R. 1953. *The Biology of Paramecium*. McGraw-Hill-Blakiston Co., New York, N.Y.

Wood, H. K. 1953. *Some Factors Affecting Delayed Separation of Conjugants in Paramecium aurelia*. Thesis, Indiana University, Bloomington, Indiana.

Woodruff, L. L., and H. Spencer. 1924. Studies on *Spathidium spathula*. II. The significance of conjugation. *J. Exptl. Zool.*, **39**, 133.

Wright, S. 1940. Breeding structure of populations in relation to speciation. *Am. Naturalist*, **74**, 232.

# AN EMBRYOLOGIST'S VIEW OF THE SPECIES CONCEPT

JOHN A. MOORE: DEPARTMENTS OF ZOOLOGY, BARNARD COLLEGE AND COLUMBIA UNIVERSITY, NEW YORK

> Nature never made species mutually sterile by selection, nor will men.—Darwin to T. H. Huxley, Jan. 7, 1867, in *More Letters,* vol. 1, p. 277, Letter 197.

If an embryologist is asked to contribute his views to a symposium on "The Species Concept" it could be for any of several reasons, such as his interest in the origin of things, in this case species, or the possibility that information on gametes and early embryos can aid our understanding of this taxonomic category.

The data that I will present have a bearing on the origin of species and on the complex intergroup and intragroup relations in natural populations. Most of these data have appeared previously (Moore, 1946, 1947, 1949a,b, 1950, 1954, 1955) but they have not been used for the specific problem of this symposium.

## Current Views on the Origin of Species

The work of systematists in the last part of the nineteenth century and in the twentieth century has made a strong case that one pattern of speciation involves divergence in geographically isolated populations. The assumed events may be outlined in a simple fashion as follows: A single species becomes distributed over a large area, which is divided by barriers into a number of smaller areas. The barriers are such that gene exchange between the populations of the geographically isolated areas is greatly reduced or almost absent. With the passage of time each population evolves in an essentially different phyletic line. The course of these separate evolutions is assumed to involve the selection of genomes that best fit the isolated populations for survival in their own environments. Eventually some of the isolated populations may diverge from others to such an extent that one would call them different species.

Mayr (1948) has expressed these notions as follows:

Wherever extrinsic factors cause a retardation or interruption of gene flow between portions of a species (geographical or spatial isolation), these subdivisions of the species tend to drift apart genetically. The rate of this change is different in different species.

If the isolation is sufficiently complete and lasts sufficiently long, it will permit the evolution of isolating mechanisms, which will inhibit the interbreeding of the two daughter species after the elimination of the extrinsic isolating factors.

This is, of course, the now familiar concept of allopatric speciation. Many have regarded it as an adequate explanation of the common pattern of evolution in terrestrial vertebrates and probably in many other organisms as well. It is this position that I will attempt to defend. On the other hand, Dobzhansky and, to a lesser extent, a few others have thought it necessary to invoke an additional phase in the formation of species from geographic races. They do not believe that the species level of divergence can be reached while the population remains separated and that a culminating sympatric stage is necessary for the development of full species differences. The problem for them is whether it is possible to obtain fully effective interpopulation isolating mechanisms as a consequence of evolution in allopatric groups.

Thus, Dobzhansky (1940) develops his argument after first asking the question ". . . whether isolating mechanisms develop as a necessary consequence of the accumulation of genetic differences in general, or whether they represent a separate category of genetic changes which appear and become established only under certain special conditions." He believes that the second possibility is the correct one, namely, ". . . that the origin of isolation is a process separate from that of the origin of other species differences." His reasons for so believing are as follows:

This theory starts with the premise that each species, genus and probably each geographical race is an adaptive complex which fits into an ecological niche somewhat distinct from those occupied by other species, genera and races. The adaptive value of such a complex is determined not by a single or a few genes, but is a property of the genotype as a whole. Furthermore, the adaptive complex is attuned to its environment only so long as its historically evolved pattern remains, within limits, intact. It is true that interbreeding of different adaptive complexes may sometimes result in emergence of new genotypes which fit into unoccupied or sparsely settled ecological niches—hence the evolutionary role of hybridization. Nevertheless, hybridization usually leads to the formation of disharmonious recombinations.

Considerations such as these have prompted some writers (Dobzhansky, 1937a,b; Sturtevant, 1938; cf. Fisher, 1930) to assume that occurrence of hybridization between races and species constitutes a challenge to which they may respond by developing or strengthening isolating mechanisms that would make hybridization difficult or impossible. Where hybridization jeopardizes the integrity of two or more adaptive complexes, genetic factors which would decrease the frequency or prevent the interbreeding would thereby acquire a positive selective value, even though these factors by themselves might be neutral. Race formation is essentially the development of genetic patterns which are adapted to a definite environment. Speciation is a process resulting in fixation of these patterns through the development of physiological isolating mechanisms. Clearly, raciation and speciation should not be conceived of as entirely independent processes, but the development of physiological isolating mechanisms must nevertheless be supposed to intervene only after the divergence of the adaptive complexes had been initiated. If races are to become species, isolating mechanisms must arise when the distinct adaptive complexes are exposed to the risk of disintegration due to interbreeding.

Dobzhansky has modified his view to some extent in more recent publications. The quotation can be taken, however, as an expression of a theory that is fairly widely subscribed to in biology.

It is my feeling that in the majority of situations it is possible for fully diverged species, adequately equipped with isolating mechanisms, to evolve in allopatric populations. Furthermore, I feel that it is unnecessary to assume a culminating sympatric phase *involving selection against hybridization* as has been done by Dobzhansky. In developing my reasons questioning these views, I shall first cite a sequence of cases that can serve as a model for the origin of fully effective isolating mechanisms in allopatric populations.

### Stages in the Evolution of Frogs

The processes involved in the evolution of one species into one or more different species requires a very long time. It has been suggested that one million years might be taken as a rough average. A period of this magnitude is equivalent to at least 20,000 lifetimes of work for a student of evolution. Clearly one individual could not expect to reach any significant conclusions during his own lifetime by measuring the changes in a single species. Instead he must study a variety of organisms in different stages of speciation and then reconstruct the probable course of events. It must always be remembered that this procedure is indirect, and that the answer depends to a considerable extent on the choice of examples.

We will now list some data obtained from several frogs in a manner suggesting a pattern of evolution.

There are five common species of the genus *Rana* in the New York area. All are "good" species and no taxonomist questions their distinctness. Each one has its own characteristic habitat, morphological features, behavior pattern, and geographic distribution. Some years ago, I presented data showing the close relation between the breeding behavior, geographic distribution, and temperature adaptations of the embryos of these species (Moore, 1949b). There is no question of the importance of these adaptations. *Rana sylvatica*, which is an early spring breeder, has embryos with a cold-adapted physiology that could not allow survival during the summer when *R. catesbeiana* breeds. Neither could the heat-adapted *R. catesbeiana* embryos survive in the cold water of ponds in early spring.

In all but one species the temperature adaptations correlated so well with geographic distribution that it seemed clear there must be a causal connection. The exceptional species was *R. pipiens*, which had a geographic distribution much greater than could be anticipated on the basis of the temperature adaptations of individuals from the New England area.

Studies of *R. pipiens* from south of their predicted geographic boundary showed that these southern forms differed from their northern counterparts in their temperature adaptations. This situation is by no means surprising. One would anticipate that a widely distributed species would be broken into local populations, each adapted to its immediate environment. *R. pipiens* embryos adapted for early spring breeding in the cold-temperate environment of Quebec and northern New England would not be expected to survive under the temperature conditions encountered in the semitropical zones of southern Florida or eastern Mexico.

For our present purposes it is of considerable interest to note that developmental (and presumably genetic) incompatibility is associated with adaptation to different temperatures. If individuals from similar temperature-adapted populations are crossed, the embryos are normal in their development. On the other hand, if individuals from different temperature-adapted populations are crossed, the embryos are very abnormal and may die early in development.

The extreme results are obtained in crosses of Vermont x Florida, Wisconsin x Texas, and Vermont x eastern Mexico individuals. In the last cross most of the embryos die as abnormal gastrulae or neurulae. Thus, these southern *pipiens* and northern *pipiens* are behaving as different species so far as their potential ability to exchange genes is concerned. These are phenomena associated with the crossing of individuals from widely separated latitudes. Adjacent populations, however, can be crossed, and they give entirely normal offspring.

We could consider the populations of *R. pipiens* to be one species, two species, or to be one species in a crucial stage in the process of splitting into different species. I believe the third alternative is the most meaningful way to consider the data. The point to be emphasized, however, is that the northern and southern forms have developed isolating mechanisms, in reference to each other, to such a degree that they could coexist and remain distinct. Since these differences that serve as isolating mechanisms have developed in widely separated places, it is not necessary to invoke Dobzhansky's hypothesis of a culminating sympatric phase of speciation.

What is the basis of the differences between the southern and northern frogs? Are they the incidental and fortuitous concomitants of divergence? I prefer to think not for the following reasons. The northern and southern embryos differ greatly in their embryonic temperature adaptations. They differ in their range of temperature tolerance, temperature coefficient for development, and rate of development. Clearly they are quite dissimilar in their physiologies, which presumably are genetically determined. When we cross individuals with these dissimilar physiologies, the offspring are highly abnormal. It is my belief that the developmental abnormalities in the "hybrids" are due to the inharmonious association of two genetic systems that differ in their modes of action. If this is the correct interpretation, the genetic differences that are the basis of the embryonic adaptations are likewise the basis of an isolating mechanism, namely, hybrid inviability. Two sets of data indirectly support this possibility:

1. The magnitude of the differences in embryonic temperature adaptations between any two populations parallels the degree of abnormality shown by their "hybrids." This suggests that the same genes might be the basis for both phenomena.

2. The genes that are responsible for the embryonic temperature adaptations and the genes that interact to produce the hybrid malformations have their effects at precisely the same time in the life cycle. This also suggests that the genetic basis for the embryonic temperature adaptations and hybrid inviability is the same.

If this interpretation is correct, the very process of adaptation to diverse climatic conditions has resulted in genetic differences that would also be completely effective isolating mechanisms.

This brief synopsis of the situation in *R. pipiens* will serve as the first stage in our model of geographic speciation. It is a widespread species occupying a diversity of habitats. So far as the data go, it appears that it is a continuous group of interbreeding populations. Some of the widely separated populations, however, have diverged to such an extent that they behave as different genetic species. This divergence could not have been promoted by selection against hybrids, according to Dobzhansky's hypothesis, since there is no possibility of hybrids being formed.

Our question now is this: Where does the species go from here? Some have regarded divergence in a continuous population, such as *R. pipiens,* as a blind alley in evolution and, unless some extrinsic factors can divide the species, it can never be more than a group of interbreeding allopatric populations. I favor this view, but the evidence for it is so incomplete that there is little point in one being counted in adherence or in opposition.

There is no difficulty in imagining the formation of two species in *R. pipiens* if we invoke some extrinsic factors which would obliterate the populations in the middle portions of the species' range or if there arose several well-placed barriers to gene flow. A variety of physical and biological factors could serve this purpose.

The next stage in our model of speciation will be a species that at one time undoubtedly had a continuous distribution but it is now found in two widely separated regions.

*Crinia signifera* occurs in eastern and in western Australia but not in a wide band of unsuitable country that extends from the Great Australian Bight to the northwest coast. There are no detectable morphological differences between the eastern and western individuals and no one thought they were other than the same species until cross fertilization experiments were performed. When one crosses eastern and western individuals, however, profound hybrid defects and generally so little survival are observed that it is necessary to consider them separate species.

The fact that the eastern and western forms are morphologically identical is best interpreted in terms of their descent from a common ancestor. During the Pleistocene the climate was such that there could have been a continuous population of these frogs across Australia. There is good evidence for periods during which most of the continent was moist. These humid times were followed by dry intervals, some of which were drier than the present-day climate.

We can imagine, therefore, that at one time in the past the frogs ancestral to *C. signifera* occupied a continuous zone across Australia. Somewhat later the continuous distribution was broken by the development of arid central regions. The east and west populations are now different species. There is no way of telling whether they became different after being isolated or whether, as in *R. pipiens*, the differences developed while the range was continuous.

With either interpretation, or some combination of them, it is clear that there is no need to invoke the concept of a culminating sympatric stage in speciation. The differences that now serve as isolating mechanisms arose in allopatric populations.

## Other Arguments against the Hypothesis

We should not conclude from these two examples in frogs that the hypothesis of the development of isolating mechanisms by selection, in zones of overlap, against hybridization is incorrect. For most cases, however, I do think it is unnecessary. The following are some of the reasons.

Undoubtedly there have been many instances where a species population has become divided and has remained so for long periods of time. When the land connection between North America and South America was reestablished in recent geological times, the continuous ranges of many marine organisms was broken. At the present we find many pairs of related species, the two members of each pair being separated by the Central American isthmus. The most probable interpretation of this situation is that divergence was a consequence of evolution in isolation. Since the two diverging groups never came together there would have been no possibility of their developing or reinforcing isolating mechanisms through the selection against hybridization.

There must be numerous similar instances both for terrestrial and aquatic organisms. The chance dispersal of organisms from a continent to a remote island, the breaking of connections between continents or bodies of water, the extermination by biological or physical causes of the individuals in portions of a species' range are some situations that can divide an originally continuous population. In many instances the isolated groups will evolve with no further relation to each other.

The situations just described exclude the possibility of hybridization, and we will now consider those instances where once separated populations become continuous.

When two once continuous and then separate populations reestablish contact, a variety of consequences can ensue. If the period of separation has been short, or if genetic divergence has been slight, the two groups would be expected to become parts of the same breeding population in the zone of overlap. If the period of separation has been long and if genetic divergence has been sufficient for effective isolating mechanisms to develop, the two overlapping populations could maintain their discreteness.

The two situations just considered are extremes and there is little doubt about what will occur. It is with the intermediate situation that we shall now give our attention, for it is here that there is doubt as to the course of events. The situation will be this: Two portions of a population become geographically isolated and during their period of isolation an appreciable amount of genetic divergence occurs. The amount of divergence will be defined as sufficient to result in a hybrid less well adapted than either parent population. It is in this situation that Dobzhansky has postulated that natural selection will promote the development of isolating mechanisms that will decrease the wastage of gametes due to their combining in poorly adapted hybrids.

When the two populations reestablish contact, under the conditions stipulated, they come together as differently adapted groups. In their period of separation each evolved along paths leading to better adaptation to their separate environments. The very fact that the genomes are no longer able to collaborate normally may be taken as indicative of considerable genetic divergence and probably of considerable adaptational divergence as well. Thus, in any zone of overlap the two groups may be assumed to have somewhat different ecological niches. This may take the form of differences in breeding time, habitat, food preferences, and so on. The net effect will be to limit the possibility of forming hybrids.

The previously evolved adaptive differences will tend to restrict competition between the two populations in the zone of overlap. Where competition does occur, however, there are some conditions under which further divergence of the populations will be promoted. Let us designate the two populations A and B and assume two habitats alpha and beta. Let us further assume that A was previously isolated in a zone where alpha was the commonest habitat and that the course of A's evolution was for adaptation in alpha although it also occupied beta. We will assume that B was found in an area where beta was the prevailing habitat and B became better adapted to beta in spite of the fact that it occupied the available alpha to a certain extent. If the two populations became sympatric in a region with alpha and beta, we would anticipate that each species would become restricted to a single habitat. A being better adapted to alpha would displace B from this habitat. Similarly, we would anticipate that B would displace A from beta. In the zone of overlap, any genetic difference would have a selective advantage if it tended to restrict each species to the habitat to which it was better adapted. *Thus, an isolating mechanism (habitat preference) would be enhanced, but this would be a consequence of natural selection reducing competition and not of natural selection reducing hybridization.*

In the situation described, there seems no reason to believe that the gene differences that promote habitat isolation in the zone of overlap will spread throughout the population. The gene differences will tend to be restricted to the zone where they have an adaptive significance.

Much the same line of reasoning would lead us to question the ability of selection against hybridization to produce isolating mechanisms that characterize the entire population. Let us again postulate two populations, C and D, which were isolated for some time during which they diverged genetically to such an extent that any hybrids they might form would show reduced viability. Let us further assume that C and D reestablish contact and in the zone of overlap they hybridize. Since we have postulated that the hybrids have reduced viability, let us assume with Dobzhansky, that natural selection would promote the development of isolating mechanisms. *These isolating mechanisms would have selective advantage in the zone of overlap but not elsewhere.* Consequently they could not become characteristic of the species as a whole, for the following reason: If we assume with Dobzhansky that "Isolating mechanisms encountered in nature appear to be *ad hoc* contrivances which prevent the exchange of genes between nascent species," and that they are formed by selection in zones of overlap, we can assume that they have adaptive value only in the zone of overlap. If they were of adaptive value to the two species in areas where they were not sympatric, they would be selected but *not for their function as isolating mechanisms*. Genetic differences that are *ad hoc* isolating mechanisms between two species can be selected for only in areas where the two species occur sympatrically. These isolating mechanisms can develop as a consequence of selection for genes that prevent the formation of hybrids or by selection for genes that prevent competition (irrespective of whether or not hybrids are formed).

Outside of the zone of overlap any gene differences that serve as *ad hoc* isolating mechanisms would probably be either neutral or somewhat deleterious. (If they were advantageous they would be selected for any way, but with no reference to any pleiotropic functions as isolating mechanisms.) It is difficult to understand, therefore, how they could become established as isolating mechanisms throughout the ranges of both species unless the ranges were coextensive. Except on small islands, this last possibility would be highly unlikely.

Several additional reasons can be cited for believing that the origin of isolating mechanisms by selecting against hybrid formation cannot be of universal occurrence.

First, there are some species pairs that are able to form apparently normal hybrids, and yet they remain distinct even when sympatric. In this situation there would be no adaptive advantage to restricting hybridization, and hence the isolating mechanisms that do prevent gene interchange could not have been perfected by the natural selection of genes that prevent hybridization.

Second, in areas of overlap where inferior hybrids are formed, we must remember that hybridization must be frequent if natural selection is to increase the frequencies of gene differences that reduce crossing. Consequently, as isolating mechanisms become more effective, the ability of selection to augment them will steadily decrease.

Third, we must not assume that the wastage of gametes is always a disadvantage to the species. There are some special situations where the wastage of gametes might be a real advantage to the population as a whole. As a hypothetical case let us consider a species whose numbers are held in check by predators and available food. If the source of predation is then removed, there will be greater competition for the available food. In some species severe competition may result in 100 per cent mortality. In this, an unusual situation, some wastage of gametes would be an advantage.

There are some observations that have been cited in support of Dobzhansky's thesis, such as those of Koopman (1950). He kept *Drosophila pseudoobscura* and *persimilis* together in a population cage and found that it was possible to augment the existing isolating mechanisms by removing all the hybrids that formed. In other words he was able to select for mating preference. This is not an unexpected result. It is highly probable that all genetic differences that serve as isolating mechanisms are consequences of natural selection (though in most instances I feel they are not selected for as isolating mechanisms per se). The fact that one can select against hybrid formation does not prove that an analogous situation will occur in nature. It is possible to select in artificial situations for all sorts of things that are not selected for under natural conditions.

There are now a number of reports of species pairs which overlap in their ranges and show greater genetic isolation from one another where they are sympatric than elsewhere. In no case that I know of is there good evidence to indicate that selection against the formation of ill-adapted hybrids is involved. The greater differences between the species in the area of overlap might be due to greater genetic divergence resulting from competition (as has been discussed before) or to still other factors.

There are also some reports of overlapping and closely related species showing less genetic isolation where they are sympatric than elsewhere. Volpe (1955) has described a case involving *Bufo americanus* and *B. fowleri*, and he refers to similar cases in other organisms.

**Summary**

The pattern of evolution that involves the formation of new species from geographical races is discussed. The conclusion is reached that the most important aspect of this process is genetic divergence associated with adaptation to the local environment. It is felt that this divergence will involve differences that will also serve as isolating mechanisms. There is no critical evidence that isolating mechanisms develop as *ad hoc* contrivances that prevent hybridization between incipient species. Furthermore, there are some theoretical considerations that make it improbable that the genetic differences that serve as isolating mechanisms commonly arise in this manner.

REFERENCES

Dobzhansky, T. 1940. Speciation as a stage in evolutionary divergence. Am. Naturalist, 74, 312-21.

Koopman, K. F. 1950. Natural selection for reproductive isolation between *Drosophila pseudoobscura* and *Drosophila persimilis*. Evolution, 4, 135-48.

Mayr, E. 1948. The bearing of the new systematics on genetical problems. The nature of species. Advances in Genetics, 2, 205-37.

Moore, J. A. 1946. Incipient intraspecific isolating mechanisms in *Rana pipiens*. Genetics, 31, 304-26.

Moore, J. A. 1947. Hybridization between *Rana pipiens* from Vermont and eastern Mexico. *Proc. Natl. Acad. Sci. U.S.*, 33, 72-75.

Moore, J. A. 1949a. Geographic variation of adaptive characters in *Rana pipiens* Schreber. *Evolution*, 3, 1-24.

Moore, J. A. 1949b. Patterns of evolution in the genus *Rana*. In *Genetics, Paleontology, and Evolution*. G. L. Jespen, G. G. Simpson, and E. Mayr, editors. Princeton University Press, Princeton, New Jersey.

Moore, J. A. 1950. Further studies on *Rana pipiens* racial hybrids. *Am. Naturalist*, 84, 247-54.

Moore, J. A. 1954. Geographic and genetic isolation in Australian amphibia. *Am. Naturalist*, 88, 65-74.

Moore, J. A. 1955. Abnormal combinations of nuclear and cytoplasmic systems in frogs and toads. *Advances in Genetics*, 7, 139-82.

Volpe, G. P. 1955. Intensity of reproductive isolation between sympatric and allopatric populations of *Bufo americanus* and *Bufo fowleri*. *Am. Naturalist*, 89, 303-18.

# THE SPECIES PROBLEM FROM THE VIEWPOINT OF A PHYSIOLOGIST

C. LADD PROSSER: PHYSIOLOGY DEPARTMENT, UNIVERSITY OF ILLINOIS, URBANA, ILLINOIS

The principal task of physiologists is to describe life functions as they occur in well-known organisms; physiologists are less concerned with how organisms came to be than with how they now are. Comparative physiologists attempt to ascertain the ways in which different organisms solve their life problems. By using *kind* of organism as one experimental variable when studying a function, the comparative physiologist can arrive at different kinds of generalizations from those obtained by the more intensive study of single organisms. As part of the effort to describe life processes, some physiologists investigate adaptations to environmental stresses; these adaptations constitute the functional bases for animal distribution, and consideration of them leads inevitably to the species problem. Success of a species at its environmental limits depends on adaptive capacity in structure and function, but the morphology which is related to such essential function is often at quite a different level from that used by the systematist; it may be molecular morphology. At the same time, the morphological characters of the systematist's key, whether adaptive or not, are generally assumed to be associated genetically with physiological characters; the genetic basis for adaptive physiological characters is usually multifactor and complex.

Most of the physiologists and biochemists who have considered phylogeny have compared higher taxonomic units—phyla and classes—and they have generally supported the conclusions reached previously by comparative embryologists and classical phylogenists. Relatively few physiological comparisons have been made of closely related species or of populations living at the opposite range limits of one species. Most physiologists seek homogeneous material, whereas in population studies, controlled heterogeneity is desired. When differences are observed in similar animals, the differences may be among individuals within a genetically similar population, they may be in mutants carried at some level in the population, they may be racial between ecotypes of the same species, or they may be between the total genetic systems of true species. Our operative assumption has been that speciation progresses from individual variation through strains and races to species. Since the complete sequence cannot be observed in any single species but is inferred, it is necessary to piece together as much indirect evidence as possible to test the basic assumption.

**Physiological Criteria of Variation**

Before examining hypotheses and experimental data it is important to consider what physiological criteria can best be used in laboratory comparisons of animals drawn from different populations. We may list seven general categories of criteria of physiological variation.

*Survival.* Tolerance of sudden environmental extremes is the easiest and most widely used criterion of physiological variation. Percentage survival at different times after application of a stress is plotted for different intensities of stress, and $LD_{50}$ (median lethal dose) values for the stress are obtained. More important is the establishment of the maximum stress tolerated indefinitely. The causes of death within short times after exposure to extreme stresses are usually different from those after longer times, and it is important that prior history of the animals be known because acclimation may strongly influence survival. For example, Fry, Brett, and Clawson (1942) and Brett (1944) constructed useful polygons that gave the upper and lower limits of temperature tolerated by fish which had been differently acclimated, and they found in goldfish a 1°C. change in lethal temperatures for each 3°C. acclimation difference. When a stress is applied gradually, the acclimation process increases the degree of stress which can be tolerated, i.e., the rate of application of the stress affects the tolerated limits. Most survival tests are made with adults rather than with young, a procedure which is easier but perhaps less meaningful for speciation, since embryos are frequently more sensitive to stresses. Despite the fact that survival data must be qualified according to acclimation history, time of observation, and rate of application of the stress, the *extremes* to which animals are subjected in nature are more important than are *average* environmental values.

*Reproduction.* Ability to reproduce measures survival over

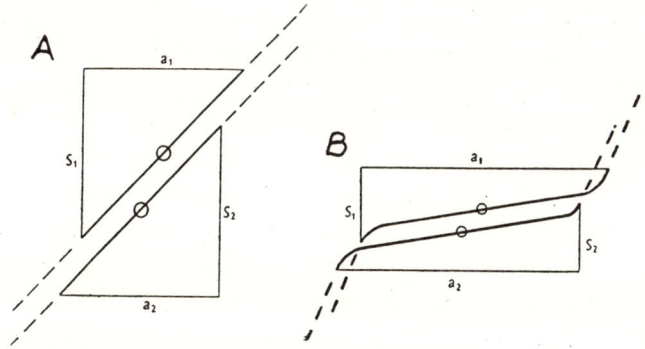

Fig. 1. Diagrams representing relation between an internal state I and an environmental stress O. $a_1$ and $a_2$ represent environmetal limits of two acclimation levels corresponding to survival limits $s_1$ and $s_2$ of internal variation. Broken lines represent regions where only brief survival is possible. (A) Diagram for a conforming (adjusting) animal. (B) Diagram for a regulating animal. From Prosser (1955).

a complete life cycle and approaches the natural situation better than survival of adults only. Unfortunately, rearing animals in the laboratory is possible for relatively few species, and the best comparisons are of reproduction in populations occurring in different ecological situations and acclimated by transplanting to different field conditions. More significant than ability to reproduce is rate of growth of populations, and this can be measured readily in the laboratory only for a few species of small animals, particularly protozoans and insects.

*Measurement of an Internal State as a Function of External Stress.* There are two basic patterns of physiological adaptation to a given environmental change (Prosser, 1955): (*a*) an animal may alter itself so that it conforms internally with the environment (adjusts) (Fig. 1A), or (*b*) it may regulate its internal state, and thus maintain relative constancy in an altered environment (Fig. 1B). Familiar examples of these two patterns are found in poikilotherms, in which the internal temperature changes in conformity with the environment, and homeotherms, which maintain relatively constant body temperature. The tolerated range of internal variation of a given character is greater in a conforming than in a regulating animal (Fig. 1A, B). Acclimation can shift the range of internal variation in a conformer, and it can shift the limits of regulation in a regulator; both patterns of response are adaptive in that they permit tolerance of an altered environment. The genotype fixes the point beyond which acclimation is no longer effective. Conformity to the environment and regulation can be analyzed with respect to temperature, osmotic concentration, ionic balance, oxygen partial pressure, and other aspects of the internal state. Some animals may conform on one side of the tolerated range and regulate on the other side; for example, some shore crabs are weak osmoregulators in a dilute medium and osmoconformers in more concentrated media. Different kinds of animals and different populations of the same species may profitably be compared with respect to the pattern of response of internal state.

*Mechanisms and Limits of Regulation.* In a regulator any environmental change starts feedback mechanisms which tend to compensate for the change. At some limits, the feedback mechanisms fail and survival is no longer possible. The mechanisms at the two limits of stress are not necessarily related and may call upon different organ systems. More physiological studies are concerned with mechanisms of regulation than with any other aspect of the interaction between organism and environment, probably because man is such a good regulator. Yet there are few systematic comparisons of regulating mechanisms in different races and species.

*Recovery from a Deviated State.* Every animal, whether it is a conformer (adjuster) or regulator, can withstand some brief deviation from its "normal" internal state, and under natural conditions every animal operates close to some "optimum." For example, a homeotherm can be given either a fever or hypothermia and then, when the stress is removed, it returns to its "normal" temperature. An animal can have its normal water load (water content) altered by dehydration or by excessive hydration after which in its usual environment it will gain or lose water until it arrives at its "normal" water load. One of the most obvious and yet least understood generalizations in physiology is that life processes go best within limited ranges of osmoconcentration, at certain oxygen pressures, within narrow limits of temperature and of concentration of specific ions; the conditions which are most favorable differ for different animals, but in a given animal there is always a tendency to return to the "optimum" state after deviation. Numerous patterns of return exist, some with an overshoot (Adolph, 1943), and comparison of recovery patterns is a useful way to differentiate different kinds of animals. Most of the comparisons thus far made have been between distantly related forms.

*Rate Functions.* Probably the most subtle and quantitative method of differentiating physiological types is to compare reaction rates within narrow and natural environmental ranges. Rates of oxygen consumption, rates of heartbeat and breathing, rates of growth, are affected by many factors, some more critical than others. The most familiar use of rate functions is in respect to temperature effects, and numerous natural populations have been compared with respect to temperature characteristics ($Q_{10}$, the rate increase calculated per 10°C.) for various functions. There is need to extend rate measurements from organ systems to enzyme systems in order to learn to what extent physiological variation depends on integration in the intact animal and to what extent it may occur at the cellular or subcellular level. Rate functions have meaning for speciation only if related to a critical environmental stress.

*Behavior.* Specific differences in reproductive behavior, particularly as concerns mating and care of young, have been useful for differentiating certain insects and birds. In addition, taxic and orienting behavior, particularly in selection of an optimum region in a gradient, for example of humidity or temperature, may be of use in distinguishing races. Food selection, based on taste "preferences," is partly established habit by "imprinting" and partly based on genetic difference.

In general, the best examples of physiological differences from population to population come from genera which have a wide ecological distribution and which live in transitional environments. Care must be taken in selecting appropriate clines or circles and in separating ecotypes from sibling species. Also there may be interaction among several environmental factors, and it must be remembered that animals react to a total environment although certain factors may be more important than others for a species.

## Allopatric Speciation

The commonly assumed sequence of allopatric speciation of sexually reproducing animals may be summarized as follows:

1. Phenotypic variation, both morphological and physiological, occurs within populations of genetically similar individuals. Part of this variation follows the normal distribution curve of a population in an environment; part of the phenotypic variation is the result of acclimation.

2. Primary adaptive variation becomes genetically fixed by natural selection of randomly occurring genetic changes. Mutations or chromosomal rearrangements are carried in an interbreeding population and genetic strains become established.

3. In a restricted environment or clinal situation ecotypes or races become established. There may be gene flow between races, but in one ecotype a given gene frequency may exceed that in another ecotype.

4. Sufficient isolation, restrictive with respect to gene flow, provides for the fixation of a selected genotype; such isolation may occur between ends of a cline or in insular populations, i.e., isolated by water or by unfavorable terrain surrounding breeding sites. During isolation other differences become established, differences which result in reproductive separation of the popula-

tions. Reproductive isolation may result from chromosomal arrangement or mutation, from gross morphological, physiological, or psychological incompatibility. Speciation is now effectively complete.

5. Further changes resulting in habitat or ecological niche selection may occur so that if the two species come to overlap in range they may be separated ecologically as well as reproductively. The initial or primary adaptive difference may have lost its effectiveness.

Physiological variation is important at steps 1, 4, and 5.

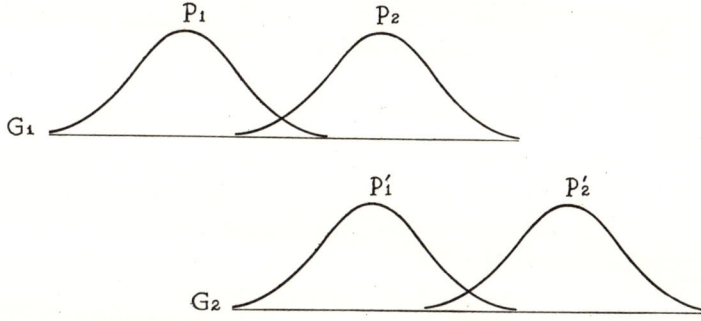

Fig. 2. Schematic representation of distribution of a character in two genotypes $G_1$ and $G_2$ and in two populations of each genotype indicated by phenotypes $P_1$ and $P_2$ for $G_1$ and $P'_1$ and $P'_2$ for $G_2$.

Figure 2 indicates two similar genotypes, each limiting the range within which phenotypic variation can occur with respect to some specified character. Two populations of the same genotype ($G_1$) living in two different environments, for example at two latitudes, may show phenotypic patterns indicated by $P_1$ and $P_2$. By acclimation, individuals of one of these populations can shift their phenotypic expressions to correspond to those of the other population. For example, the minimum lethal temperatures can be altered for cold and warm water populations of fish. In the other genotype or race ($G_2$), one population $P'_1$ is similar to $P_2$ in respect to the measured character, but by acclimation it is found that $P'_1$ cannot be converted to $P_1$ but that it can go beyond either $P_1$ or $P_2$ toward a new type $P'_2$. According to the Baldwin effect, if a genetic change occurs in the direction favored by selection, e.g., from $G_1$ toward $G_2$ in a population already living near its limit with respect to that character, as in $P_2$, it is more likely to become fixed (step 2 above) than if such a change should occur in a population far from this limit, as $P_1$.* Conversely the genotype may become more limited. For example, $P_2$ may become genetically incompetent of acclimation to $P_1$, and thus if these two populations are isolated by some means they may become true species. There are many physiological examples of $P_1$'s and $P_2$'s but few physiological examples of $G_1$'s and $G_2$'s within single species. Distinction as to whether a given difference between two populations is genetic or not can be made only by acclimation or breeding experiments.

## Inadequacies of the Simplified Scheme of Speciation

When one examines with care some of the physiological analyses of populations and when one attempts field studies of so-called races and related species and examines data from population genetics, he begins to question the simple picture of speciation which has been outlined above. None of the following considerations is in itself decisive but in the aggregate they raise serious doubts.

1. Every serious systematist is impressed more by the diversity than by the uniformity of his species. Mutants of many sorts are carried at certain levels in most breeding populations. We readily recognize individual diversity in man and domestic animals without knowing much about similar diversity in natural populations. There is pronounced heterozygosity in virile strains, and balanced polymorphism permits the maintenance of several phenotypes in one population. It is unlikely that genetic factors related to susceptibility to disease and to longevity are found in man only.

* This does not imply any causal relation between phenotype and genetic change. Rather, individuals of $P_1$ would not be likely to move to the geographic range limit where the genetic change would be favored; phenotypically adapted individuals of $P_2$ would tend to survive, hence reproduce better in respect to the critical stress, and if selection pressure is strong, $G_2$ would permit migration to new range limits.

2. The notion that new characters are retained only if they have adaptive advantage for survival is unrealistic, and there are numerous examples of establishment of neutral morphological and ethological patterns. Neutral *characters* may be retained if they are linked or pleiotropic with advantageous characters; according to the theory of genetic drift, neutral *genes* may become established in small populations. As more neutral characters are closely scrutinized, they appear to be associated genetically with selective characters and hence they are present in ratios different from those expected by random assortment (Sheppard, 1953). For example, human blood types are apparently correlated with susceptibility to certain chronic diseases; type A is more frequent and type O less frequent in people with cancer of the stomach than in the general population (Aird *et al.*, 1953, 1954). The real problem of the physiologist is to recognize the critical functional characters. In any case, the characters used by taxonomists to distinguish species are often not adaptive ones.

3. While the range of a species may be very great, actual breeding populations are often very small. Individuals in migrant species tend to return to delimited areas. This restricts the probability of outbreeding. Territoriality is now recognized not only among vertebrates—mammals, birds, and fish—but in some crustaceans and insects, and it probably exists very widely. The concept of gene flow throughout the range of a species is apparently an oversimplification of the actual situation. In general, breeding populations are more susceptible of isolation in fresh water and terrestrial habitats than in marine environments, particularly if in the latter they have pelagic larvae.

4. Many of the so-called physiological races turn out to be true species. When two populations differ enough to be physiologically distinct, they usually have developed sufficient differences to become reproductively isolated. The presence of a few hybrids in zones of overlap does not invalidate the basic species distinction, and in the laboratory there may occur interspecific crosses which would be rare or nonexistent in nature. In clines the terminal populations are often clearly different species, but intermediate populations present difficulties for the taxonomist,

and the cline as a whole may show progressive functional adaptedness. Usually sibling or morphologically similar species are first considered as physiological races, but as they are studied more intensively, they turn out to be natural species. Careful observation usually reveals minor neutral morphological differences by which sibling species may be distinguished.

5. Physiological characters tend to be more sensitive to the environment in their expression than do structural ones, and the potential lability of individual animals is very great. The changes which can be brought about by acclimation are extensive, and as more populations are subjected to acclimation tests, more and more do their differences turn out to be nongenetic. Many so-called races are considered such merely because of their ecology and they have not been subjected to adequate acclimation tests.

6. Most physiological characters which have been analyzed in animals are polygenic. In fact, a well-adapted animal is suited to its entire environment, not merely to a single physical factor, and it is adapted not through the action of one gene but by its whole integrated genotype. A gene will often be expressed differently if the accompanying genes are changed. On this basis it is perhaps futile to expect adaptive physiological characters to be as susceptible of analysis as distinctive morphological ones.

7. It may be argued that the physiological criteria which have been applied are too gross, that measurement of oxygen consumption by an animal is no more precise than stating the number of segments on an appendage, that what is needed are more sensitive tests, measures of single enzymes. However, the one gene-one enzyme hypothesis as developed so beautifully for *Neurospora* may be of limited applicability since it concerns only defects in ability to synthesize relatively small molecules. Very rarely is any animal with complex dietary requirements dependent on any single enzyme pathway for either a specific synthesis or degradation. There are numerous alternate pathways by which energy can be obtained from basic substrates. It is true that interference with cytochrome oxidase would be disastrous for most animals, yet this enzyme must be under multiple control,

and many animals do function without it at restricted energy levels. Also there are several alternate oxidative paths which can bypass the cytochrome system. In other words, a genic change interfering with single enzymes would not be readily detected in complex animals, it would not likely become fixed because it would be negative rather than positive or even neutral. However, when enzyme changes have occurred in the direction of changes in optima for a species, these have been selected; parasitic worms can rely on anaerobic pathways, herbivorous insects tend to have more active carbohydrases, and carnivorous ones more proteases. However, there is no doubt that more sensitive functional tests are needed. Also, it is not always easy to find the most critical or important functional characters.

In summary, physiological adaptedness to the environment may exist in the absence of clear morphological speciation (*a*) in balanced polymorphic populations (these include advantageous heterozygotes and genic expression differing with genetic environment), (*b*) in clines, (*c*) in sibling species, and (*d*) in the extensive phenotypic lability in response of individuals to environmental stress. Physiological characters may not conform to "good" specific characters because (*a*) taxonomic characters are often nonadaptive, (*b*) physiological characters are complex genetically (multifactor), (*c*) adaptive characters usually have large safety factors and parallel or alternate paths, and functional characters are either (*d*) quantitatively so sensitive to the environment that phenotypic variation exceeds genotypic, or (*e*) qualitatively so complex that reproductive isolation has already been established. If evolution is considered as the development of adapted organisms and if adaptation is basically functional, it is of prime importance to analyze populations in terms of physiological characters. However, the origin of species per se may not coincide with the development of adapted organisms.

## Examples of Variation

We may next consider examples of populations which have been examined for physiological variation with respect to various limiting environmental factors. We shall consider first for each stress examples of phenotypic differences, then reported races, and finally species adaptations.

*Water and Ions.* The transitions between sea and fresh water and between regions of low and high humidity on land permit much physiological variation in water economy. Closely related to this is ion balance, although specific elements have scarcely been considered as limiting factors in the distribution of animals. Numerous examples of what appear to be phenotypic osmotic differences between populations are known. North Sea starfish (*Asterias rubens*) are much more tolerant of dilution than populations from the Kiel Canal (Schlieper, 1929). Gills from *Mytilus* from the North Sea (salinity 30°/oo) consumed 80 ml. $O_2$ per gram per hour, whereas after four weeks at a salinity of 15°/oo the corresponding value was 144 ml. $O_2$ per gram per hour; initial $O_2$ consumption by gills from *Mytilus* living in the Baltic at 15°/oo was 141 and after four weeks at 30°/oo it was 84 ml. $O_2$ per gram per hour (Schlieper, 1953, 1955). Blue crabs (*Callinectes sapidus*) collected in the upper parts of an estuary not only tolerate lower salinities but regulate their internal osmoconcentrations down to lower environmental limits than do crabs from the mouth of the river, but after a week in normal sea water, the crabs from the upper river approached in tolerance and regulation those from the river mouth (Anderson and Prosser, 1953).

Several examples of possible osmotic races may be cited. *Nereis diversicolor* in Denmark penetrate into nearly fresh water and tolerate sudden transfer to fresh water better than do *N. diversicolor* from England. These differences have been interpreted as indicating two races. However, tests of the regulation of internal chloride as a function of environmental chloride show similar limits of regulation for populations from Denmark, Scotland, and Finland (Smith, 1955a,b). The limiting salinity may not be the summer value, which has been mostly studied, but rather the spring dilution when the embryos are developing. Perhaps some measure other than chloride regulation would reveal racial differences, but present evidence suggests only nongenetic variation. Two brackish water gammarids, formerly considered one species, are now recognized as two species, *Gam-*

*marus zaddachi,* which tolerates near-fresh water, and *G. salinus,* which lives in the more saline regions of estuaries. The blood concentration curves of the two forms have not been obtained, but they are intersterile, and hence are effectively separate species (Spooner, 1947; Kinne, 1954).

Genetically distinct clones of *Euplotes vassus* differ in their ability to withstand transfer to high salinity corresponding to saline lakes. Populations of *Euplotes* in salt lakes appear to have become selected for salt-resistance (Gause, 1947). By analogy with varieties of *Paramecium* (see below) these strains of *Euplotes* may well be natural species.

The European stickleback, *Gasterosteus aculeatus,* occurs in two subspecies (Heuts, 1947, 1949). One subspecies (*gymnurus*) is predominantly a freshwater fish whereas the other (*trachurus*) predominates in a brackish to marine habitat. The reproduction (hatching) of *gymnurus* is better in fresh water, and that of *trachurus* in salt water. Linked with salt-water tolerance is low number of vertebrae, larger body size, and larger number of lateral plates. Also the interaction between the effects of temperature and salinity in affecting the number of dorsal fin rays differs in the two forms. The physiological mechanisms of differences in salinity have not been examined, but this fish appears to offer an example of the complex interaction of two environmental stresses (salinity and temperature) and the linkage with gross morphological characters. The two subspecies differ in many ways, and despite the fact that they produce fertile hybrids, natural selection favors the two extremes; they appear to be well on the way to becoming natural species.

A similar example, now recognized as sibling species, is the freshwater *Anopheles gambiae* and the morphologically similar *A. melas* which normally breeds in brackish water and can develop in 150% sea water. Hybrids are sterile, and the eggs are morphologically distinguishable (Ribbands, 1944). The tropical forest mosquitoes, *Anopheles bellator* and *A. homunculus,* occupy canopy levels which overlap on one side of each range; *bellator* rises in the forest heights in the evening and descends less far in the day because of its greater tolerance of low humidity (Pittendrigh, 1950). This difference is secondarily derived; that is, the species were originally separated by some other factor.

*Temperature.* More different physiological criteria have been used with temperature than with other environmental stresses. Compensation for cold climates by poikilotherms has been reviewed and population differences have been described recently (Bullock, 1955). The low lethal temperature of some temperate zone marine animals may be above the high lethal temperature of closely related arctic animals. However, lethal temperatures can be altered markedly by acclimation, and familiar seasonal variations in lethal temperatures reflect acclimation. Statements about lethal temperatures of local populations need to include information regarding previous temperature experience.

Many rate functions have been compared for animals from different latitudes measured over a range of temperatures (Bullock, 1955). The rate of oxygen consumption by cold-acclimated poikilothermic animals is usually higher than for warm-acclimated ones, whether the acclimation is seasonal (*Fundulus*, Wells, 1935; crustaceans, Edwards and Irving, 1943, as discussed by Rao and Bullock, 1954), or latitudinal (*Pandalus*, Fox and Wingfield, 1937), or altitudinal (planaria, Bläsing, 1953). Absence of metabolic compensation for temperature change has been reported for other animals (hot springs fish, Sumner and Lanham, 1942; certain terrestrial insects, Scholander *et al.*, 1953; an Alaskan pond gammarid, Krug 1954; and *Rana pipiens*, Fromm and Johnson, 1955). Other rate functions such as heart rate, breathing rate, and cruising speed indicate some compensation by animals from cold environments. For example, *Mytilus californianus* from latitude 48°21′ pumped water at the same rate at 6.5°C. as those from 38°31′ at 10°C. and those from 34°0′ at 12°C. (Fig. 3) (Rao, 1953). Also the animals from colder waters are less affected by temperature, i.e., their $Q_{10}$ is lower than for animals from warmer regions. Adaptation by alteration in the $Q_{10}$, i.e., shift in slope as well as position of the rate-temperature curve, occurs in some animals but not in others (Rao and Bullock, 1954; Scholander *et al.*, 1953; Bullock, 1955). In southern California, limpets (*Acmaea limatula*) from the high intertidal zone have consistently lower heart rates at a given temperature than those from the low intertidal zone (Segal *et al.*, 1953). *Mytilus* show a similar vertical variation in pumping rate; 75 cm. vertically are equivalent in pumping to 330 miles latitudinally when tested at 16°C. (Rao, 1953). By transplantation of limpets and *Mytilus* in the vertical gradient, complete acclimation was

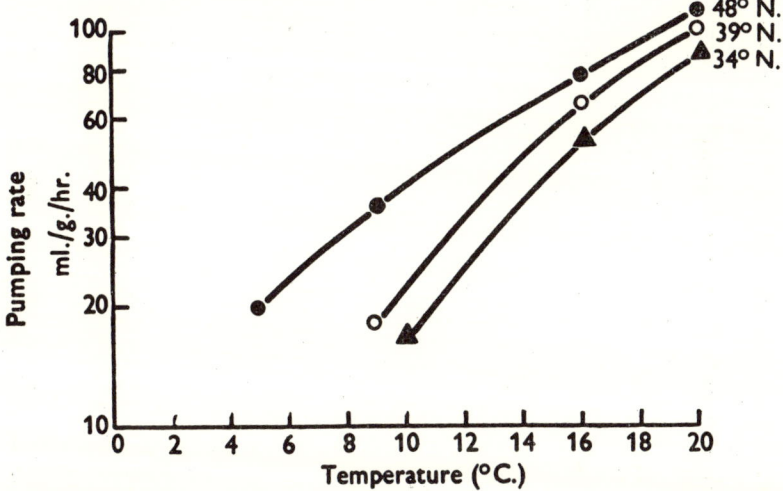

Fig. 3. Rate of pumping of water by 50-g. specimens of *Mytilus californianus* as a function of temperature for populations collected at three latitudes of western North America. From Bullock (1955); redrawn from Rao (1953).

demonstrated within a few weeks. Latitudinal transplantation experiments have not been reported, but it is likely that many of the latitudinal effects with respect to temperature are phenotypic.

Racial differences have been invoked to explain latitudinal differences in reproductive rates; poikilothermic animals from cold waters generally develop more rapidly than those from warm waters when tested at the same temperature, especially in the lower range. A low-temperature race of *Daphnia atkinsoni* is found in pools in England where the breeding season is restricted by summer evaporation (Johnson, 1952). Oysters (*Crassostrea virginica*) from Virginia, where normal spawning occurs at 25°C., failed to spawn during two years in Long Island

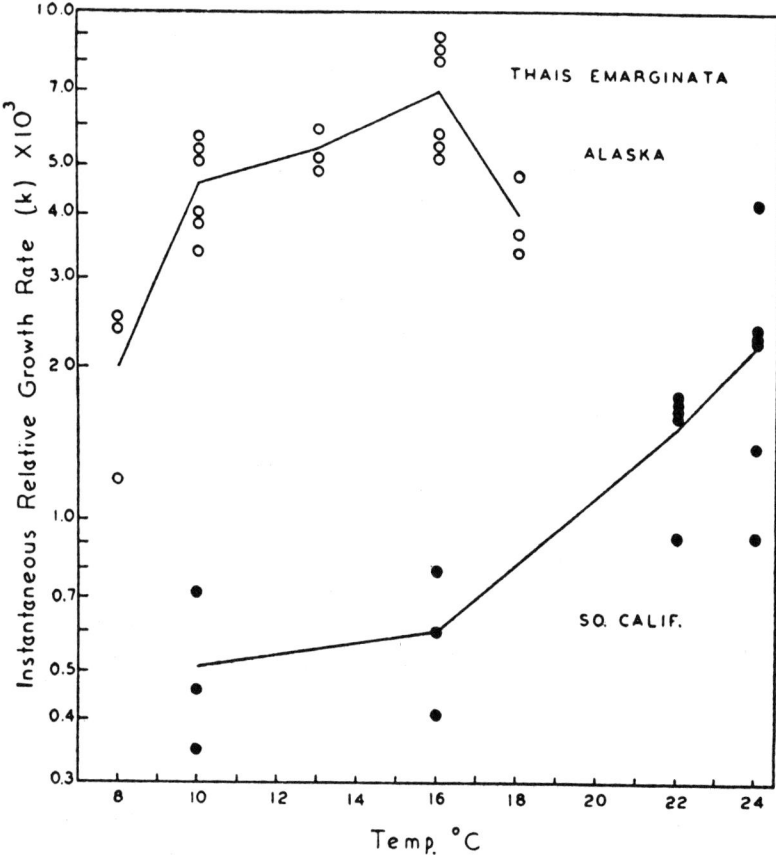

Fig. 4. Growth rates of shell length of veliger larvae of *Thais emarginata* as a function of temperature for populations from Mount Edgecombe, Alaska and Big Rock, California. From Dehnel (1955), by permission of University of Chicago Press.

Sound, where the temperature failed to reach 25°C.; yet only adults were transplanted, not spat (Loosanoff and Nomejko, 1951). When spat of another species (*Ostrea lurida*) were transplanted from Puget Sound to Long Island Sound, survival and breeding were possible only for those oysters brought into the laboratory during the winter (Davis, 1955). Similarly, the oyster drills, *Urosalpinx cinerea* of Virginia and Delaware, have different temperature thresholds for spawning (Stauber, 1950). The effect of latitude on rate of growth of embryos of various marine invertebrates has been well summarized by Dehnel (1955) who added observations on certain species of snails from southern California and Alaska. When growth rate was measured at different temperatures in the laboratory, the Alaskan snails grew significantly faster at corresponding stages (Fig. 4). In the middle of the physiological temperature range the northern $Q_{10}$ values were higher in those snails which showed no overlap in rates, lower in the others. Similar findings have been reported earlier for European marine invertebrates (Fox, 1938; Thorson, 1951). These differential growth rates can partly account for the relatively greater productivity of colder waters. It appears that the Alaskan and southern Californian populations of the same species of snails are genetically different, yet no transplantation experiments were performed and the animals at intermediate latitudes were not compared. It is likely that clines with respect to the effect of temperature on development of marine invertebrates may occur and that between the ends of such clines major genetic differences exist. However, there is need for transplantation of larval stages and acclimation over a life cycle.

Among insects, several clines have been described where temperature seems to be critical for development. In the moth *Lymantria dispar*, Goldschmidt (1940) showed in his classic studies that the rate of development in south European races is slower than in north European populations. In pine-needle wasps (*Diprion*) alpine races show faster development in the egg and lower $Q_{10}$'s, especially for conversion from pupa to adult, than do populations from warmer regions (Elens, 1953).

*Rana pipiens* from northern United States breed at lower temperatures, develop faster, have lower temperature coefficients of development than frogs from extreme southern states and lowlands of Mexico (Moore, 1949). Slight morphological differences and a zone of hybridization confirm the conclusion that the northern and southern *Rana pipiens* are genetic races or subspecies; in fact, hybrid abnormalities are so great that species status is approached (Volpe, 1954).

In several species of *Drosophila*, e.g., *pseudoobscura* and *persimilis*, natural populations contain individuals with several distinct arrangements of chromosome bandings. The proportion of each pattern may vary with season and altitude, and clines are observed with respect to certain inversions. Controlled breeding in cages at constant temperature and humidity shows that the percentage of certain types is in part temperature-dependent. For example, when several types of *D. persimilis* were reared together, the high-altitude type became predominant at 16° C. but not at 25° C. (Spiess, 1950). When two gene arrangements of *D. pseudoobscura* from the same locality were bred through many generations, the heterozygote became more common than the homozygotes. However, if the same two forms from different localities were bred there was sometimes stabilization with one homozygote predominating. Hence geographic differences in gene contents must exist separate from the differences in chromosome pattern (Dobzhansky, 1954). The selective effects of temperature on the chromosome polymorphism have been demonstrated to occur in both pupae and adults (Levine, 1952), and physiological studies are needed to learn the bases for the temperature adaptedness.

Both genetic and nongenetic differences are shown in the following experiments on fish in which the zones of temperature tolerance were tested after acclimation to different temperatures. Fourteen species of freshwater fish from different latitudes (Ontario, Tennessee, and Florida) were compared (Hart, 1952). The tolerance polygons showed genetic differences for three of these, the common shiner (*Notropis cornutus*), the largemouthed bass (*Micropterus salmoides*), and *Gambusia affinis*. Morphological differences between northern and southern forms of the first two are such that they are considered subspecies. Several of the others such as golden shiner and blacknose dace are morphological subspecies but they fail to show differences in their temperature tolerance polygons.

In *Paramecium aurelia* there are eight varieties which are really natural species (Sonneborn, 1950); two of these are distinguished by the temperatures optimal for their growth and reproduction (Sonneborn and Dippell, 1943).

*Oxygen.* A considerable amount of nongenetic variation with respect to oxygen requirement is possible. For example, intestinal parasites and such free-living animals as *Tubifex* and *Daphnia* are able to alter their metabolism or their oxygen transport system according to the environmental oxygen. It is entirely possible that some of the effects on freshwater poikilotherms attributed to temperature are in fact effects of oxygen, since in winter the oxygen in stagnant ponds where there is no photosynthesis may be very low. Altitudinal effects attributable to oxygen lack have been reported only for birds and mammals where the phenotypic response of increase in blood hemoglobin is well known. Of two species of a Russian mouse, *Apodemus,* one, *sylvaticus,* lives in both mountains and plains, the mountain populations having higher hemoglobin concentrations, while the other, *agrarius,* is restricted to the plains. Apparently at high altitudes members of the latter species fail to acclimate by increasing their blood hemoglobin (Kalabuchov, 1937).

Oxygen gradients do not in general provide good geographic series, but microecologically oxygen supply may be important in the isolation of populations. For example, in a series of chironomid larvae, the stream species survived anoxia only a short time whereas species from low-oxygen ditches survived anoxia for about 100 hours (Walshe, 1948). Gammarids and planaria from fast well-aerated water have higher rates of oxygen consumption than do related species from sluggish water poor in oxygen. Some snails and numerous parasitic helminths get along well on the energy of glycolysis and excrete the acids which are formed. In fact, several patterns of repayment of oxygen debt are recognized (von Brand, 1952). The known genotypic variations recognized with respect to oxygen are largely at the species level.

*Foods and Toxins.* There are marked interspecific differences in digestive enzymes according to natural diets and a few instances of slight differences in amino acid requirements, but in general, nutritional requirements are so similar and so general that subspecific differences are slight. Selection of specific food plants by herbivorous insects appears to be based on taste preferences for certain specific organic components of the plants which are repellent or neutral to other insects. Thorpe (1939) pointed out that selection of food plants can be developed by habituation, as by repeatedly exposing the larvae to a particular food, and he has experimentally habituated *Drosophila* to a medium containing peppermint and the ichneumid *Nemeritis* from *Ephestia* to *Meliphora*. *Dixippus morosus* was "forced" to feed on ivy, initially an unfamiliar food, instead of privet; initial mortality was high, but after several generations the progeny accepted the ivy readily (Sladden, 1934). Eventually natural selection can result in genetic fixation of food preferences which are initially phenotypic variants established by habituation. A population of sawfly, *Pontania salicis*, which had been feeding on *Salix andersonia*, was forced to eat only *S. rubra*, until after some four years the sawflies failed to make galls on *S. andersonia* when this species was planted near-by (Harrison, 1927). This interesting experiment deserves confirmation under carefully controlled conditions.

Most spectacular are tolerances of natural alkaloids by some insects, such as those feeding on tobacco, and the tolerance of modern insecticides by various strains. In some insects, resistance to DDT seems to be carried by a single gene but in most it is multifactorial. In many the genetic pattern necessary for resistance was present in the population before the specific stress of insecticides appeared (DDT resistance in *Musca*, Lichtwardt *et al.*, 1955). In a few the resistance appeared during many generations of selection (Crow, 1954). The complexities of resistance to insecticides have been recently reviewed (Metcalf, 1955). Resistant strains do provide a basis for natural selection of a particular race, which may ultimately lead to speciation, since areas where the toxin exists are spatially separable from those without the toxin and are well maintained.

Examples of species with very minor morphological differences but isolated by food preferences are: the homopterans *Psylla mali* on apple and *peregrina* on hawthorn (Lal, 1934); and two species of the butterfly *Colias, eurytheme* on alfalfa and *philodice* on red clover (Hovanitz, 1949).

Other environmental factors which may be important as isolating agents are such sensory stimuli as light and mechanical stimuli. For example, blind races of the Mexican characin, *Anoptichthys jordani,* inhabit dark caves where lack of vision is not a disadvantage (Breder, 1943). Wind and water currents, also substrate composition, limit the distribution of a variety of animals.

*Biotic Selection.* Related to food plant preference as an isolating mechanism is the selection of host organism or of symbiont. Termites which are very similar morphologically are clearly distinguished by the species of staphylinid beetle living in their nests (Emerson, 1935). *Ascaris lumbricoides* is morphologically similar in man, pig, and monkey, yet attempts at reciprocal infection have been unsuccessful (Faust, 1949); these ascarids may be true sibling species. Probably the sibling species most difficult to understand are those of the *Anopheles maculipennis* complex. Eight different strains are recognizable by their selection of man or other mammals as blood sources, also by their ability to develop in saline water. These strains are morphologically similar except for minor differences in egg structure. However, laboratory breeding experiments demonstrate that in reality the "strains" are sibling species (Hackett and Missiroli, 1935; Bates, 1949). An American mosquito, *Culex pipiens,* has two strains, *pipiens,* which for egg production requires a blood meal, and *molestus,* which can produce eggs without it (Kitzmiller, 1953).

Serological specificity distinguishes closely related groups and reveals incompatibility between more distant groups. "Blood types" have been analyzed in detail in man, in pigeons, and in cattle. In man, certain serological types, also certain hemoglobin types, are correlated with other genetic factors, and each type is carried at a level characteristic of the ethnic group (Boyd, 1950). Present suggestions of linkage of blood type with incidence of certain chronic diseases indicate the need for greatly expanded medical statistics.

*Behavior.* Fully as important as physical and chemical factors in maintaining (possibly not in initiating) isolation between populations are behavioral differences, particularly the behavior associated with reproduction. For example, six similar species of the *willistoni* group of *Drosophila* show mating habits so characteristic that no interspecific mating occurs (Spieth, 1947, 1949). Two sibling species of solitary wasps *Ammophila campestris* group, are distinguished mainly by their mode of opening the burrow and providing prey for their larvae (Adriaanse, 1947). Eastern and western American meadowlarks can be distinguished by their calls,* as can European chiffchaffs, several species of crickets, and also two southern frogs in regions of overlap (Blair, 1955). These behavioral traits could act as primary isolating characters after becoming established by genetic drift as in "island" populations, or they may secondarily separate species in which the initial isolating agents are no longer active.

## Mechanisms of Nongenetic Variation of the Phenotype

A stress physiologist is continually impressed by the phenotypic lability of individual organisms. In regulating animals, acclimation shifts the environmental limits for a stable internal state. In conforming animals, acclimation raises or lowers the tolerable internal variation or introduces some compensating mechanism, for example, metabolic adaptation in poikilotherms. Since phenotypic variation may be important in initial stages of speciation and since numerous population differences turn out to be nongenetic, it is important in a discussion of the species problem to point out possible mechanisms of phenotypic adaptation.

In describing any change induced in an individual by environmental stress, the time during which the change occurs must be precisely stated. For example, mammals exposed to cold show initial constriction of surface blood vessels and piloerection. Next they may shiver and then increase the activity of thyroid and adrenal cortex, and they may maintain elevated metabolism for weeks. Long-range adaptation, however, consists of increased in-

* A recent paper (Lanyon, W. E., 1956. Ecological aspects of the sympatric distribution of meadowlarks in the north-central states. *Ecology*, **37**, 98-108) suggests that the primary mechanism isolating the eastern and western species of meadowlark is indirectly related to moisture, particularly spring precipitation.

sulation in hair or subcutaneous fat. Thus, short-term regulation and long-term acclimation are accomplished by very different mechanisms. Long-term changes are often morphological—tadpoles have larger gills when reared in low oxygen, certain mosquito larvae have larger anal papillae in freshwater than in salt water. There may also be behavioral changes, such as aggregation by hive bees in winter.

In addition, many physiological and biochemical adaptations are known. Lungfish in estivation shift their products of nitrogen excretion to urea from the ammonia they excrete when actively swimming. *Daphnia* and *Artemia* reared in low oxygen develop high levels of hemoglobin whereas in high oxygen their hemolymph is colorless. Also, in low oxygen there is an increase in tissue cytochrome (Fox, 1955). The amount of "free" water diminishes in cold-hardened insects and molluscs.

We have referred above to the increased metabolic rate in aquatic poikilotherms acclimated to cold either seasonally or latitudinally. That this change is at least partly cellular is shown by the fact that tissue slices from certain organs of cold-acclimated fish and crustaceans consume more oxygen than do those from warm-water individuals (Piess and Field, 1950; Freeman, 1950; Roberts, 1953).

It appears likely, therefore, that environmental stresses can result in biochemical adaptations at the cellular level. The enzymes of thermophilic bacteria show higher temperatures of inactivation than corresponding enzymes of mesophilic bacteria, and the inactivation temperature differs for different enzymes from the same organism; these differences are clearly genetic. However, Precht and his associates (Christophersen and Precht, 1950, 1951) report that the highest temperature of activity in yeast measured by oxygen consumption and by methylene blue reduction is higher (65° C.) for yeast reared at 41° C. than it is (60° C.) for yeast reared at 20° C.; also the $Q_{0_2}$ of 20° yeast is higher than that of 41° yeast when measured at the same temperature. The experiments with yeast agree with data from poikilothermic animals that acclimation temperature may change the upper limiting temperature, the absolute level of oxygen consumption and, in some instances, the temperature characteristic of the total metabolic system. The explanation of such acclimation effects is by no means clear. It is unlikely that qualitative changes in enzyme proteins occur, since protein structure appears to be genetically fixed. There may, however, be quantitative changes in enzymes, a sort of enzyme induction.

Among microorganisms enzyme synthesis can be induced by forcing the use of unfamiliar substrates, for example sugars which the organism had not been able to metabolize. Enzyme induction has been demonstrated in animals—the synthesis of tryptophane peroxidase in rat liver induced by injection of L-tryptophane (Knox, 1951; Lee and Williams, 1953). There is no evidence, however, that physical factors such as salinity, temperature, and oxygen partial pressure can similarly induce enzyme synthesis. The following hypothesis might explain the effects of physical agents in terms of enzyme induction: It is recognized but not sufficiently emphasized that there are numerous alternate paths in digestion and in intermediary metabolism. Some of these paths are followed by certain tissues and animals and at certain stages in a life cycle more than are other paths. Parasitic helminths in hypoxia switch to glycolytic systems and excrete the acids formed. *Cecropia* pupae use a cyanide-insensitive oxidative pathway (presumably a flavoprotein) until cytochrome oxidase is activated by hormones of metamorphosis. If one path is affected more than another by temperature, by oxygen pressure, or by salinity, some of its substrates might pile up, and these might serve as intermediate substrates of an alternate path. These substances might then induce synthesis of some limiting enzyme of the alternate path. The limiting temperatures and the temperature characteristics of the overall metabolism might be changed according to relative importance of given pathways. In fact, a change in slope of the rate-temperature curve (change in temperature characteristic) could effectively increase or decrease the absolute rate of oxygen consumption at a given temperature.

What is needed is a quantitative assay of a variety of known alternate paths under different acclimation states, a very difficult task. An enzymatic analysis of adaptive phenotypic variation is as important as the study of the relations between genes and enzymes.

## Conclusions

Examination of physiological characters, particularly by stress tests, can provide much useful information about variation in natural populations. When populations of animals which differ markedly in their responses to a given stress are properly acclimated to each other's environment, the differences often are seen to be nongenetic. In fact, the functional lability of animals is very great. Part of this individual adaptability is behavioral, part occurs at the organ system level, part of it is cellular, and part may result from enzyme induction.

Some examples of genetic fixation of adaptive variations in mutant strains are known, for example, insecticide-resistant insects. Variants may become established as ecotypes or races, particularly in clinal situations; as an example, animals with different temperature dependence of rate functions. More commonly, when adaptive physiological differences between two populations are genetically established, the differentiation is sufficient for reproductive isolation; that is, speciation occurs. Many sibling species or morphologically similar species, which differ in physiological adaptability, are recognized, and stress tests applied to populations would probably reveal numerous natural species. In summary, most functional variation among animal populations appears to be either nongenetic or specific; relatively little is racial.

Neutral morphological variations are sometimes genetically carried with physiological characters by linkage or pleiotropism. Also, neutral characters may provide the basis for distinguishing typological species. Most adaptive functional characters of animals appear to be multifactorial; adaptation resides not in single genes but in the balanced genotype; the compensatory reactions to deviations from a "normal" physiological state make recognition of small genetic changes difficult.

There is urgent need for two kinds of research in the area of physiological variation. First, there is need for more sensitive functional tests of variation and for cellular and enzymatic analyses of the ways in which the environment induces phenotypic change. Second, to better understand natural variation and to distinguish nongenetic from genetic differences there is need for careful acclimation experiments on populations collected over the full range of a species, the examination of physiological clines, in experiments extending over complete life cycles. The use of physiological responses to various environmental stresses as criteria of natural variation permits a close approach to the key problem of evolution, the development of adapted types of plants and animals, if not to the origin of species.

## REFERENCES

Adriaanse, M. S. 1947. *Ammophila campestris* Latr. und *Ammophila adriaanus* Wilcke. Ein Beitrag zur vergleichenden Verhaltungsforschung. *Behaviour*, 1, 1-34.

Adolph, E. F. 1943. *Physiological Regulations*. Jaques Cattell Press, Lancaster, Pa.

Aird, I., and H. H. Bentall. 1953. A relationship between cancer of stomach and the ABO blood groups. *Brit. Med. J.*, 1, 799-801.

Aird, I., H. H. Bentall, and J. A. Mehigan. 1954. *Brit. Med. J.*, 2, 315-21.

Anderson, J. D., and C. L. Prosser. 1953. Osmoregulating capacity in populations occurring in different salinities. *Biol. Bull.*, 105, 369.

Bates, M. 1949. *The Natural History of Mosquitoes*. The Macmillan Company, New York, N.Y.

Blair, W. F. 1955. Mating call and stage of speciation in the *Microhyla olivacea-M. carolinensis* complex. *Evolution*, 9, 469-80.

Bläsing, I. 1953. Experimentelle Untersuchungen über den Umfang der ökologischen und physiologischen Toleranz von *Planaria alpina* Dana und *Planaria gonocephala* Dugès. *Zool. Jahrb. Abt. allg. Zool. u. Phys.*, 64, 111-52.

Boyd, W. C. 1950. *Genetics and the Races of Man*. Little, Brown and Co., Boston, Massachusetts.

Brand, T. v. 1952. *Chemical Physiology of Endoparasitic Animals*. Academic Press, New York, N.Y., Chapters 10, 13, 14.

Breder, C. M. 1943. Problems in the behavior and evolution of a species of blind cave fish. *Trans. N.Y. Acad. Sci.*, **5**, 168-76.

Brett, J. R. 1944. Some lethal temperature relations of Algonquin Park fishes. *Publ. Ontario Fish. Research Lab.*, **63**, 1-49.

Bullock, T. H. 1955. Compensation for temperature in the metabolism and activity of poikilotherms. *Biol. Revs. Cambridge Phil. Soc.*, **30**, 311-42.

Christophersen, J., and H. Precht. 1950. Fermentative Temperaturadaptation. *Biol. Zentr.*, **69**, 240-56.

Christophersen, J., and H. Precht. 1951. *Biol. Zentr.*, **70**, 261-74.

Crow, J. F. 1954. Analysis of a DDT resistant strain of *Drosophila*. *J. Econ. Entomol.*, **47**, 393-98.

Davis, H. C. 1955. Mortality of Olympia oysters at low temperatures. *Biol. Bull.*, **109**, 404-6.

Dehnel, P. A. 1955. Rates of growth of gastropods as a function of latitude. *Physiol. Zool.*, **28**, 115-44.

Delacour, J. 1954. *The Waterfowl of the World*. Country Life, London.

Dobzhansky, T. 1954. Evolution as a creative process. *Caryologia 6, suppl.*, **1**, 435-49.

Edwards, G. A., and L. Irving. 1943. The influence of temperature and season upon the oxygen consumption of the sand crab, *Emerita talpoida* Say. *J. Cellular Comp. Physiol.*, **21**, 169-81.

Elens, A. A. 1953. Étude écologique des lophyres en campine (Belgique) II. *Agricultura (Louvain)* n. s. **1**, 19-32; from *Biol. Abst.*, 1954.

Emerson, A. E. 1935. Termitophile distribution and quantitative characters as indicators of physiological speciation in British Guiana termites (*Isoptera*). *Ann. Entomol. Soc. Amer.*, **28**, 369-95.

Faust, E. C. 1949. *Human Helminthology*, Lea and Febiger, Philadelphia, Pennsylvania.

Fox, H. M. 1938. The activity and metabolism of poikilothermal animals in different latitudes. III. *Proc. Zool. Soc. London*, **108**, 501-5.

Fox, H. M. 1955. The effect of oxygen on the concentration of haem in invertebrates. *Proc. Roy. Soc. (London)*, **B143**, 203-14.

Fox, H. M., and C. A. Wingfield. 1937. The activity and metabolism of poikilothermal animals in different latitudes. II. *Proc. Zool. Soc. London*, **107**, 275-82.

Freeman, J. A. 1950. Oxygen consumption, brain metabolism and respiratory movement of goldfish during temperature acclimatization, with special reference to lowered temperatures. *Biol. Bull.*, **99**, 416-24.

Fromm, P., and R. E. Johnson. 1955. The respiratory metabolism of frogs as related to season. *J. Cellular Comp. Physiol.*, **45**, 343-59.

Fry, F. E., J. R. Brett, and G. H. Clawson. 1942. Lethal limits of temperature for young goldfish, *Rev. can. biol.*, **1**, 50-56.

Gause, G. F. 1947. Problems of evolution. *Trans. Connecticut Acad. Arts and Sci.*, **37**, 17-68.

Goldschmidt, R. 1940. *The Material Basis of Evolution.* Yale University Press, New Haven, Connecticut.

Hackett, L. W., and A. Missiroli. 1935. The varieties of *Anopheles maculipennis* and their relation to the distribution of malaria in Europe. *Riv. malariol.*, **14**, 45-109.

Harrison, J. W. H. 1927. Experiments on the egg-laying instincts of the sawfly *Pontania salicis* and their bearing on the inheritance of acquired characters. *Proc. Roy. Soc. (London)*, **B101**, 115-26.

Hart, J. S. 1952. Geographic variations of some physiological and morphological characters in certain freshwater fish. *Publ. Ontario Fish. Research Lab.*, **72**, 1-79.

Heuts, M. J. 1947. Experimental studies on adaptive evolution in *Gasterosteus aculeatus. Evolution*, **1**, 89-102.

Heuts, M. J. 1949. *J. Genetics*, **49**, 183-91.

Hovanitz, W. 1949. Increased variability in populations following natural hybridization. In Jepson, Mayr, and Simpson, *Genetics, Paleontology and Evolution.* Princeton University Press, Princeton, New Jersey, Chapter 18.

Johnson, D. S. 1952. A thermal race of *Daphnia atkinsoni* Baird and its distributional significance. *J. Animal Ecol.*, **21**, 118-19.

Kalabuchov, N. J. 1937. Some physiological adaptations of the mountain and plain forms of the woodmouse (*Apodemus sylvaticus*) and of other species of mouse-like rodents. *J. Animal Ecol.*, **6**, 254-72.

Kinne, O. 1954. Die Gammarus Arten der Kieler Bucht. *Zool. Jahrb. Abt. Syst. Okol.*, **82**, 405-24.

Kitzmiller, J. B. 1953. Mosquito genetics and cytogenetics. *Rev. brasil. malariol. e. doenças trop.*, **5**, 285-359.

Knox, W. E. 1951. Two mechanisms which increase in vivo the liver tryptophan peroxidase activity: specific adaptation and stimulation of the pituitary-adrenal system. *Brit. J. Exptl. Path.*, **32**, 462-69.

Krug, J. 1954. The influence of seasonal environmental changes upon the metabolism, lethal temperature and rate of heart beat of *Gammarus limnaeus* (Smith) taken from an Alaskan lake. *Biol. Bull.*, **107**, 397-410.

Lal, K. B. 1934. *Psyllia peregrina* Forst, the hawthorn race of the apple sucker *P. mali* Schmidb. *Ann. Appl. Biol.*, **21**, 641-48.

Lee, N. D., and R. H. Williams. 1953. Protein turnover and enzymatic adaptation. *J. Biol. Chem.*, **204**, 477-86.

Levine, R. P. 1952. Adaptive responses of some third chromosome types of *Drosophila pseudoobscura*. *Evolution*, **6**, 216-43.

Lichtwardt, E. T., W. M. Luce, G. C. Decker, and W. N. Bruce. 1956. A genetic test of DDT resistance in field house-flies. *Ann. Entomol. Soc. Amer.*, **48**, 205-10.

Loosanoff, V. L., and C. A. Nomejko. 1951. Existence of physiologically different races of oysters, *Crassostrea virginica*. *Biol. Bull.*, **101**, 151-56.

Metcalf, R. L. 1955. Physiological basis for insect resistance to insecticides. *Physiol. Revs.*, **35**, 197-232.

Moore, J. A. 1949. Geographic variation of adaptive characters in *Rana pipiens*. *Evolution*, **3**, 1-21.

Piess, C. N., and J. Field. 1950. The respiratory metabolism of excised slices of warm and cold-adapted fishes. *Biol. Bull.*, **99**, 213-24.

Pittendrigh, C. S. 1950. The ecoclimatic divergence of *Anopheles bellator* and *A. homunculus*. *Evolution*, **4**, 43-63, 64-78.

Prosser, C. L. 1955. Physiological variation in animals. *Biol. Revs., Cambridge Phil. Soc.*, **30**, 229-62.

Rao, K. P. 1953. Rate of water propulsion in *Mytilus californianus* as a function of latitude. *Biol. Bull.*, **104**, 171-81.

Rao, K. P., and T. H. Bullock. 1954. $Q_{10}$ as a function of size and habitat temperature in poikilotherms. *Am. Naturalist*, **88**, 33-44.

Ribbands, C. R. 1944. Differences between *Anopheles melas* and *Anopheles gambiae*. *Ann. Trop. Med. Parasitol.*, **38**, 87-99.

Roberts, J. L. 1953. Studies on thermal acclimatization in the lined shore crab, *Pachygrapsus crassipes* Randall. *XIX Intern. Physiol. Congr.*, 706-7.

Schlieper, C. 1929. Ueber die Einwirkung niederer Salzkonzentrationen auf marine Organismen. *Z. vergleich. Physiol.*, **9**, 478-514.

Schlieper, C. 1953. Zur Frage der Beziehungen zwischen osmotischer Resistenz und Grundumsatz bei euryhalinen Meeresvertebraten. *Naturwissenschaften*, **20**, 538-39.

Schlieper, C. 1955. Ueber die Physiol. Wirkungen des Brackwassers. *Kiel Meeresforsch.*, 11, 22-33.

Scholander, P. F., W. Flagg, V. Walter, and L. Irving. 1953. Climatic adaptation in arctic and tropical poikilotherms. *Phys. Zool.*, 26, 67-92.

Segal, E., K. P. Rao, and T. W. James. 1953. Rate of activity as a function of intertidal height within populations of some littoral molluscs. *Nature*, 172, 1108-9.

Sheppard, P. M. 1953. Polymorphism, linkage and the blood groups. *Am. Naturalist*, 87, 283-94.

Sladden, D. E. 1934. Transference of food habit from parent to offspring. *Proc. Roy. Soc. (London)*, B114, 441-49.

Sladden, D. E. 1935. *Proc. Roy. Soc. (London)*, 119, 31-46.

Smith, R. I. 1955a. On the distribution of *Nereis diversicolor* in relation to salinity in the vicinity of Tvärminne, Finland and the Isefjord, Denmark. *Biol. Bull.*, 108, 326-45.

Smith, R. I. 1955b. Comparison of the level of chloride regulation by *Nereis diversicolor* in different parts of its geographical range. *Biol. Bull.*, 109, 453-74.

Sonneborn, T. M. 1950. *Paramecium* in modern biology. *Bios.*, 21, 31-43.

Sonneborn, T. M., and R. V. Dippell. 1943. Sexual isolation, mating types and sexual response to diverse conditions in variety 4, *Paramecium aurelia*. *Biol. Bull.*, 85, 36-43.

Spiess, E. B. 1950. Experimental populations of *Drosophila persimilis* from an altitudinal transect of the Sierra Nevada. *Evolution*, 4, 14-33.

Spieth, H. T. 1947. Sexual behavior and isolation in *Drosophila*. *Evolution*, 1, 17-31.

Spieth, H. T. 1949. *Evolution*, 3, 67-81.

Spooner, G. M. 1947. The distribution of *Gammarus* species in estuaries. *J. Marine Biol. Assoc. United Kingdom*, 27, 1-52.

Spooner, G. M. 1951. *J. Marine Biol. Assoc. United Kingdom*, 30, 129-47.

Stauber, L. A. 1950. The problem of physiological species with special reference to oysters and oyster drills. *Ecology*, 31, 107-18.

Sumner, F. B., and U. N. Lanham. 1942. Studies of the respiratory metabolism of warm and cool spring fishes. *Biol. Bull.*, 82, 313-27.

Thorpe, T. H. 1939. Further studies in pre-imaginal olfactory conditioning in insects. *Proc. Roy. Soc. (London)*, B127, 424-33.

Thorpe, T. H. 1937. *Proc. Roy. Soc. (London)*, **124**, 56-81; and **126**, 370-97.
Thorson, G. 1951. Zur jetzigen Lage der marinen Bodentier Ökologie. *Verhandl. deut. Ges. Wilhelmshaven*, 276-327.
Volpe, S. P. 1954. Hybrid inviability between *Rana pipiens* from Wisconsin and Mexico. *Tulane Studies Zool.*, **1**, 110-23.
Walshe, B. M. 1948. The oxygen requirements and thermal resistance of chironomid larvae from flowing and from still waters. *J. Exptl. Biol.*, **25**, 35-44.
Wells, N. A. 1935. Variations in respiratory metabolism of the Pacific killifish, *Fundulus parvipinnis*, due to size, season and continued constant temperature. *Physiol. Zool.*, **8**, 318-336.
Wells, N. A. 1935. *Biol. Bull.*, **69**, 361-71.

# DIFFICULTIES AND IMPORTANCE OF THE BIOLOGICAL SPECIES CONCEPT

ERNST MAYR: MUSEUM OF COMPARATIVE ZOOLOGY,
HARVARD COLLEGE, CAMBRIDGE, MASSACHUSETTS

To bring into focus the picture that emerges from this symposium one must consider very diverse material on which the seven contributors have based their discussions—living organisms and fossils, animals and plants, freshwater and terrestrial organisms, vertebrates and invertebrates.* It would not have been surprising if the seven speakers had represented seven entirely different viewpoints. This did not happen. Indeed, the general agreement among the speakers was quite far-reaching; for instance, all speakers with one exception have emphatically endorsed the biological species concept. Yet it is evident that we have not yet reached a true synthesis. The approach of every worker in this field is still largely colored by his intimate knowledge of the material with which he himself works. Let me illustrate this with a few examples. When an ornithologist speaks of "species" he has a phenomenon in mind in which hybridization or lack of sexuality are of no consequence. He has great difficulty in determining whether or not species in birds and species in plants are the same kind of phenomenon. Or let us take another case. The population geneticist who has emerged in the last thirty years from the typological thinking of mutationism is likely to consider everything as new that he himself is learning about species. He is unaware of the fact that thinking in terms of populations, and indeed the whole biological species concept, came from systematics into genetics rather than the reverse (Mayr, 1955). Population genetics has, often quite independ-

* The revised version of T. M. Sonneborn's contribution was not available when this discussion was prepared.

ently, rediscovered much that has been considered axiomatic in population systematics for seventy to eighty years.

Our thinking on these questions is one-sided not only because it is so strongly affected by our working material, but also because it is based on a limited number of selected examples. An author who works with a hybrid complex will be impressed by the importance of hybridization, another one who works with insular forms by the importance of geographic isolation. We now have reached the point where we are badly in need of comparative systematics and of a strictly quantitative approach. Analyses such as were done by Verne Grant (this symposium) should be done for as many groups of animals and plants as possible. Obviously, this method can be applied only where the group has reached a considerable degree of taxonomic maturity, but there are now many such groups in the vertebrates, insects, and plants. The need for such comparative systematics emphasizes the importance of sound orthodox taxonomic monographs.

**Difficulties in the Application of Species Concepts**

In the introduction I attempted to describe the three basic philosophical species concepts which play a role in systematics. Much of the discussion of this symposium dealt with the problem of the utilization of such concepts in the taxonomic practice. It may be useful at this point to say a word or two on the basic method of the application of concepts. In particular it must be emphasized that a concept is not necessarily invalidated if it cannot be applied in an individual case. The concept tree is unquestionably valid, yet one may have doubts whether or not to include in this concept such plants as a spreading juniper, a dwarf willow, a giant cactus, and a strangler fig. Some of our most universally accepted concepts encounter the same difficulties as the species concept, namely, borderline cases or insufficient information. Child and adult are two concepts which are not invalidated by the fact that the adolescent is a borderline stage. Father is a completely valid concept, but its application sometimes encounters difficulties as is evident in paternity suits.

The student who attempts to apply the species concept to concrete situations in nature faces, as all the speakers have emphasized, numerous difficulties. At first sight there appears to be a bewildering diversity of perplexities, but these can be classified into some major groups as was shown by Grant (this symposium), and a study of the various classes of difficulties helps considerably in an understanding of the species problem. Basically all obstacles in the application of a biological species concept are due either to a lack of pertinent information on some essential property of the investigated material or to its evolutionary intermediacy.

## Lack of Information

Different kinds of information are needed to permit the correct assignment of individuals to species. Most commonly the question arises whether certain morphologically rather distinct individuals belong to the same species or not. The long list of synonyms, characteristic for some groups of animals and plants, are a concrete expression of this difficulty (Mayr, Linsley, and Usinger, 1953). This difficulty is particularly acute in three branches of animal taxonomy. In many families of insects, particularly hymenoptera, males and females are so different that separate classifications for males and females have to be adopted until the proper associations have been made. Even more different are the larval stages in many groups of insects and aquatic organisms. The same is true for parasites for which it is likewise sometimes necessary to have two sets of names, one for larval and one for adult stages, until association can be demonstrated through elucidation of the life cycle. The difficulties caused by sexual dimorphism, age differences, or nongenetic habitat differences which the neontologist faces in his work must be emphasized because some taxonomists seem to believe that paleontologists are the only ones who have to cope with the difficulty of having to draw inferences from morphological types. That this difficulty is particularly acute in paleontology no one will deny. The worker who finds two or more essentially similar, yet somewhat different, morphological types in a single sample of fossils is forced to make a somewhat arbitrary decision whether to consider them as variants within a single interbreeding population or rather as several similar species. There is no automatic solution. The splitting of every bimodal curve into two morphological species would lead to the separation of males and females or of age classes into separate species and to other equally unbiological conclusions. Sylvester-Bradley (1956) and Imbrie (this symposium) have pointed to the fallacy of this approach. "The paleontologist who classes members of a single interbreeding community in more than one species is liable to confuse even himself when he refers to hybridism. The distinction, in fact, between morphological species and bio-species can only be overlooked at the peril of utter confusion." Paleontologists cannot afford to forget that fossils are nothing but the remains of formerly living organisms and that these organisms when still alive occurred in the form of genetically defined populations exactly as the species still living today. It is these populations which the paleontologist attempts to classify with the help of fossil remains, and morphological criteria are used merely as a means to an end. Paleontologists classify their material on the basis of inferences. The better they understand the nature of biological populations, the relations of genotype and phenotype, and the results of developmental physiology, the more skillfully they can interpret the available morphological evidence. There is no justification for abandoning the biological approach merely because it is sometimes difficult to decide whether or not several morphological types in a population are conspecific.

A second type of difficulty is introduced when there is insufficient information on the reproductive isolation of populations that are not in contact with each other. Some paleontologists have insisted that the classification of samples of discontinuous vertical sequences introduces a new element in the evaluation of populations, not appreciated by the neontologist. This is not altogether true. The taxonomic, genetic, and ecological problems of the multidimensional species are quite the same whether one deals with a series of populations in a chronological series, as does the paleontologist, or with geographically isolated populations, as does the neontologist. This has been recognized correctly by the paleontologist Gwynn Thomas (1956): "The essential similarity of temporal and spatial variation in fossils also makes the paleontological species concept for practical purposes almost identical with the neontological in fragmentary lineage segments." In either case we are dealing with isolated samples, and in either case we have to base a somewhat arbitrary decision as to species status on a good deal of indirect evidence. Again the difficulty is not with the yardstick but with its application.

There is a third group of cases where we lack information on the reproductive isolation of individuals, namely, all forms of asexuality. However, in this case more is involved than mere lack of information because here we are not dealing with populations. This then is a much more formidable and more fundamental obstacle to the application of the species concept, and the discussion of this difficulty shall therefore be postponed to a later section. (See below.)

## Evolutionary Intermediacy

A species definition postulates at least a discontinuity and, under the most favorable conditions, a triad of characteristics: reproductive isolation, ecological difference, and morphological distinguishability. However, evolution is a gradual process and so is the multiplication of species. As a consequence, there are many instances where a population is on the borderline and has acquired some but not yet all the attributes of a distinct species. The gradual nature of the speciation process raises the following difficulties for the student of multidimensional species.

*Evolutionary Continuity.* Species that are widespread in space or time may have terminal populations which satisfy the criteria of distinct species, yet are connected by an unbroken chain of populations. Dr. Moore has described such a situation for contemporary populations of *Rana pipiens;* they are indeed not rare in polytypic species. Theoretically this should be the standard condition in paleontology, yet one has difficulty finding such cases in the literature because breaks in the fossil record are sufficiently frequent to prevent the piecing together of unbroken lineages. One of the best substantiated ones is that of *Micraster* discussed by Imbrie (this symposium). On the other hand, there are conspicuous gaps in most of the celebrated cases of so-called unbroken lineages such as for instance Brinkmann's Kosmoceras. In some cases of real continuity between morphological species the differences are so slight that most contemporary neontologists would not hesitate to consider these forms merely subspecies of a single polytypic species. Even though the number of cases that cause real difficulties is very small, the fact remains that an objective delimitation of species in a multidimensional system is an impossibility.

*Acquisition of Reproductive Isolation without Equivalent Morphological Change.* The reconstruction of the genotype which is responsible for the reproductive isolation between two species takes place sometimes without visible effect on the phenotype. The resulting "sibling species" qualify as biological species, but not as morphological species. Sonneborn has shown that the varieties of *Paramecium* belong in this class. In plants most sibling species appear to be autopolyploids. Where such sibling species clearly differ in their ecology, nothing is gained by ignoring them merely because the morphological difference is slight. On the other hand, where morphological and ecological differences are not discernible, as seems to be the case in some instances of autopolyploidy in plants or where the differences can be established only by breeding in the laboratory such as the varieties of *Paramecium,* it would seem impractical to separate these forms as species in routine taxonomic work.

*Strong Morphological Differences without Reproductive Isolation.* A number of genera of animals and plants are known where even strikingly different populations interbreed freely wherever they come in contact. Grant (this symposium) has discussed this situation in detail and has given numerous illustrations. The attitude of calling every morphologically distinct population a species, which was widespread among classical taxonomists, is definitely losing ground. Yet to combine all morphological species that freely hybridize in zones of contact also leads to absurdity. Full agreement as to where to compromise between these two extremes has not yet been reached. Such situations occur also in the animal kingdom as, for instance, in the snail genus *Cerion* (Mayr and Rosen, 1956). When the morphological differences between such populations far exceed those normally found between good species in related genera, the taxonomist is reluctant to unite them into a single species. These cases are in a way the exact opposite to sibling species. The acquisition of reproductive isolating mechanisms is lagging far behind the general genetic divergence of these populations as indicated by their morphological divergence.

*Deficiencies in the Isolating Mechanisms.* One, if not the most important, attribute of a species is its possession of isolating mechanisms. These isolating mechanisms are usually composite and mutually reinforcing, but often they maintain only partial isolation between populations. This becomes evident when a temporary isolation between geographical isolates breaks down, resulting in allopatric, or if the isolation was primarily ecological, in sympatric hybridization. Numerous cases are known where natural populations acted toward each other like good species in areas of contact as long as their habitats were undisturbed. Yet when the characteristics of these habitats were suddenly changed in a drastic manner, usually by the interference of man, the reproductive isolation broke down. Before this breakdown everyone would have agreed that these populations were species, but after the breakdown and after the loss of reproductive isolation, they agreed better with the specifications of conspecific populations.

The frequency of hybridization, that is, the susceptibility to a secondary breakdown of partial reproductive isolation, is very different in different groups of animals and plants. In most higher animals hybridization is sufficiently rare (or else the hybrids sterile) not to cause any serious difficulties in species delimitation. In cases like the snail genus *Cerion*, discussed above, it is a major source of difficulty. Grant has discussed in detail (in this symposium) the effect of this phenomenon on the species problem in plants and has supplied quantitative data indicating its relative importance in different genera. It is obvious that the delimitation of species becomes a serious problem in genera where species hybridize freely with each other and where introgression is a major factor.

The role of isolating mechanisms for the species problem is brought into sharp focus by a consideration of hybridization. Moore has paid special attention to this side of the species problem in his discussion of the origin of isolating mechanisms (this symposium). He concludes that they originate as a consequence of the genetic differences which accumulate among isolated populations during adaptation to local conditions. This thesis seems well substantiated and corresponds indeed to my own analysis of the situation. Yet part of his evidence may have to be interpreted in a different manner, namely, the significance of the climatic races of *Rana pipiens*. Moore has shown that the various geographic races of this species are adapted in their embryonic development to prevailing local water temperatures. When individuals of a cold-adapted race are crossed with ones of a warm-adapted race, a more or less inviable hybrid will result. Moore concludes from this "that the northern and southern forms have developed isolating mechanisms, in reference to each other, to such a degree that they could coexist and remain distinct." I am not convinced that this conclusion is warranted. It is quite possible, if not probable, that much of this local adaptation is purely ecotypic and essentially reversible. If a warm climate race would reinvade a cool climate, its developmental rates and temperature tolerance would have to be modified by selection to permit survival in the cooler waters. If it should subsequently come into contact with a local cool water race, it might hybridize with it and produce harmonious viable zygotes. That this is not pure speculation is indicated by the *Rana pipiens* population in the mountains of Costa Rica, which in the cool waters has acquired the developmental properties of Vermont frogs and when crossed with them produces normal viable zygotes. Differences that are purely ecotypic local adaptations are not necessarily good isolating mechanisms because they tend to disappear as soon as the environmental differences disappear. This does not preclude the possibility of other incipient isolating mechanisms in previously isolated populations.

*Different Levels of Speciation in Different Local Populations of the Same Polytypic Species.* The amount of genetic divergence may differ in various isolated populations of a polytypic species. This inequality in the level of speciation takes different forms. Particularly spectacular are the cases of circular overlap, an increasing number of which are cited in the literature. Other cases are sympatric species, which are completely distinct at certain localities but hybridize freely at others. Lorkovicz (1953) has suggested broadening the term semispecies (Mayr, 1940) to include all isolated populations which on the basis of some criteria have reached species level, but not on the basis of others. Grant (this symposium) has made a similar suggestion to broaden the application of the term semispecies. All these difficulties demonstrate the fact that the clear-cut alternatives of the nondimensional situation are absent in a multidimensional system.

The species indicates a discontinuity above the level of the individual, but a new difficulty is introduced by the fact that there may be several such discontinuities, not all of them being species. If we designate as isolate any more or less isolated population or population segment, we can distinguish in sexually reproducing organisms between geographical, ecological, and reproductive isolates of which only the latter are species. Among asexually reproducing organisms every clone and, in fact, every individual is an isolate. Here, obviously, the species and the isolate can even less be synonymized with each other than in sexually reproducing organisms.

## Asexuality and the Species Problem

The essence of the biological species concept is discontinuity due to reproductive isolation. Without sexuality this concept cannot be applied. Asexuality then is the most formidable and most fundamental obstacle of a biological species concept. In truly asexual organisms there are no "populations" in the sense in which this term exists in sexual species nor can "reproductive isolation" be tested. Students of the species problem have neglected asexual situations for a number of reasons. The geneticist is interested only in sexual species because it is the recombining of genetic factors through the sexual process which permits formal genetics. The only event of genetic interest that happens in asexually reproducing organisms is an occasional mutation the effects of which will be invisible unless it is dominant or the organism is haploid. Absence of sexuality in existing organisms is almost certainly a secondary, derived phenomenon (Dougherty, 1955) and consequently does not require the setting up of a primarily different species category. Finally, widespread though it is in certain groups of organisms, asexuality is an exception rather than the rule, and species can be defined and delimited in most groups of organisms without any reference to loss of sexuality.

What can one do if one wants to apply the species concept to asexual organisms in view of the breakdown of the usual criteria? A number of different solutions have been proposed. The first is to find a species definition which would be equally suitable for sexual and asexual organisms. Du Rietz (1930) thought that the following definition was satisfactory: "The smallest natural populations permanently separated from each other by a distinct discontinuity in the series of biotypes are called species." This forgets that asexual organisms do not form "natural populations" and that in asexual organisms every individual and every clone is such a distinct biotype. The clear realization of this difficulty has led some authors to go one step farther and abandon the biological concept altogether, because of its inapplicability to asexual situations. Frankly, it appears to me that there is nothing to recommend this solution. In exchange for the biological species with all its advantages, it reintroduces the morphological species with all its weaknesses pointed out by most of the speakers at this symposium.

A second solution is to restrict the term species to sexually reproducing organisms and use it only in the sense of biological species. This proposal comes closer to a satisfactory solution, but it still leaves open the question how to classify morphologically differing individuals in asexual organisms. It has been suggested to use for them a neutral term, such as the term binom, mentioned by Grant (this symposium). The suggestion overlooks the fact that the word species has not only the biological meaning of a reproductively isolated population but also the purely formal meaning "kind of," simply a classifying unit. The term "agamospecies" has been used to designate "totally asexual populations" but, as stated above, an "asexual population" is a biological impossibility.

The most satisfactory solution in taxonomic practice has been a frankly dualistic one. It consists in defining the term species biologically in sexual organisms and morphologically in asexual ones. There is more justification in this procedure than a mere pragmatic one. The growing elucidation of the relations between genotype and phenotype also justify this approach. Reproductive isolation is effected by physiological properties which have a genetic basis. Morphological characters are the product of the same gene complex. Once this is clearly understood, a new role can be assigned to morphological differences associated with reproductive isolation, namely that of indicators of specific distinctness. This permits the assumption that the amount of genetic difference which, in a given taxonomic group, results in reproductive isolation will be correlated with a certain amount of morphological difference. If this is true, it is permissible to conclude from the degree of morphological difference on the probable degree of reproductive isolation. To base this inference on genetic reasoning is new; the method itself, however, of determining empirically with the help of morphological criteria whether or not a population has reached species status goes back to classical taxonomy. This inference method is by no means a return to a morphological species concept since reproductive isolation always remains the primary criterion and degree of morphological difference only a secondary indicator, which will be set aside whenever it comes in conflict with the biological evidence.

It is possible to use the same kind of inference to classify asexual organisms into species. Those asexual individuals are included in a single species that display no more morphological difference from each other than do conspecific individuals in related sexual species. Criteria must be adjusted to individual

382  DIFFICULTIES AND IMPORTANCE OF THE CONCEPT

situations since there is great diversity in the forms of asexuality. Where sexuality is abandoned only temporarily, as in the cases of seasonal parthenogenesis in *Daphnia* and aphids, there is no problem. It is customary and biologically sound to consider such temporary clones as portions of the total gene pool of a species. Nor is there any major difficulty where sexuality is lost in a single species of a genus or in a number of lines which can be clearly traced back to a common ancestor and where the morphological differences are still slight enough to justify combining these "microspecies" into a single collective species (Mayr, 1951). By far more difficult is the situation in many microorganisms where sexual reproduction or its genetic equivalents are totally absent in large groups or at best highly sporadic. To include all descendants of a common ancestor in a single collective species is also impossible in groups like the bdelloid rotifers which, apparently without sexual reproduction, have grown to an order with four families, some twenty genera, and several hundred "species" all reproducing parthenogenetically and possibly all descendants from a single ancestor. If such a group were a complete morphological continuum, any attempt to break it up into species would be doomed to failure. Curiously enough there seem to be a number of discontinuities which make taxonomic subdivision possible. The most reasonable explanation of this phenomenon is that the existing types are the survivors among a great number of produced forms, that the surviving types are clustered around a limited number of adaptive peaks, and that ecological factors have given the former continuum a taxonomic structure. Each adaptive peak is occupied by a different "kind" of organism, and it is legitimate to call each of these clusters of biotypes a species.

The large list of difficulties which the application of the species concept faces may seem to confirm the opinion of those who consider the species as something purely subjective and arbitrary. To counterbalance this impression it must be emphasized (*a*) that none of these difficulties of application invalidates the three basic concepts of the species; (*b*) that these difficulties are infrequent in most groups of animals and higher plants and that the frequency and significance of their occurrence can be determined rather accurately; (c) that such difficulties are usually of minor importance in nondimensional situations which are those most frequently encountered by the taxonomist; and (d) that in spite of these difficulties it is usually possible to classify doubtful entities into taxonomic species which satisfy at least one or the other species concept. To use the words of G. G. Simpson (1943): "A taxonomic species is an inference as to the most probable characters and limits of the morphological species from which a given series of specimens is drawn."

## The Practical Importance of Species

Those who maintain that species are something purely subjective, vague, and arbitrary sometimes ask: "What is gained by recognizing species?" The answer is that the attempt to determine species status has led in many cases not only to a more precise formulation of a biological problem but very often also to its solution. As an illustration I will call attention to only three areas of biological research where this is true.

*The Clarification and Simplification of Classification.* The need for classifying the morphological diversity of nature into biological species forces an unequivocal decision how to handle morphological variants. Accepting the biological species concept no longer permits either describing all variants as morphological species or listing the more pronounced ones as species and the less distinct ones as varieties. Now only those variants are given species rank which satisfy the criterion of the biological species concept, namely, reproductive isolation. The result is a simplification of classification which is not only of practical help to the working taxonomist but also actually aids the understanding of distribution, ecology, and phylogeny. In ornithology it has permitted a reduction in the number of species from nearly 20,000 to 8,600, and a similar simplification is apparent throughout zoology. There are still some authors who resent having to make such decisions, some of which by necessity will be incorrect, and who prefer to classify all organisms into meaningless but comparable morphological species. It is quite evident that these workers fight a losing battle because biologically trained taxonomists are unhappy to be degraded into pebble sorters. They would much rather make an occasional mistake than be burdened forever with a multitude of purely morphologically defined pigeon holes.

*Fossil Species.* The need to evaluate fossil specimens in terms of biological species has led and is leading to a new outlook in paleontological classification (Sylvester-Bradley, 1956). It forces the paleontologist to make clear decisions: Different specimens found in the same exposure (the same sample) must be either different species or intrapopulation variants, "for two geographical subspecies cannot come from the same locality, and two chronological subspecies cannot come from the same horizon" (Sylvester-Bradley, *op. cit.*). The recognition of subspecies and polytypic species in paleontology leads to the same simplification and greater precision as it has in neontology. In all this work the evidence is largely morphological, but the interpretation is based on biological concepts. An occasional error of interpretation in the synthesis of polytypic species in paleontology is vastly to be preferred to the chaotic accumulation of morphologically defined entities without biological meaning.

*Biological Races.* The shift of emphasis from morphological difference to reproductive isolation has necessitated a reanalysis of the whole complex of phenomena loosely referred to as "biological races." These are of great practical importance since most of these so-called biological races were found during the study of disease vectors or of injurious animals. Here again the need for a clear decision, species or not, has led to clarification and simplification. The study of sibling species which has been the major outcome of this analysis has been, in many cases, of great practical importance in applied biology.

The species is, however, of more than purely practical importance. It has a very distinct biological significance which has been described above (Introduction). Finally, a new field of research is growing around the species.

## The Science of the Species

It has become apparent within recent years that a new branch of biology is developing, the science of species. It is devoted to the many-sided aspects of the species and to the understanding of the level of integration which is denoted by the term species. This branch of biology is as legitimate as is cytology, devoted to the level of the cell, or histology, dealing with the level of the tissue. The species is an important unit in evolution, in ecology, in the behavioral sciences, and in applied biology.

The study of the biological properties of species has within recent decades led to the development of several new fields in biology, each being a borderline field between the science of species on the one hand and some other branch of biology, e.g., genetics, ecology, animal psychology on the other. It would lead too far to paint a detailed picture of this recent development, and I will content myself to outline it with a few bold strokes. These new fields are:

*The Study of Speciation.* Darwin, as we saw, had fallen down in his attempt to explain the multiplication of species because he was not fully aware of the multiple aspects of species. Nor did he realize that multiplication of species, an origin of discontinuities, is not the same as simple evolutionary change. The study of the mechanisms and modes of speciation has become an interesting borderline field between systematics, genetics, and evolution.

*The Study of Isolating Mechanisms.* The appreciation of the fact that in addition to sterility there are many other factors which safeguard the genetic integrity of species has led to the development of a new field of study. As far as animals are concerned, contact was established with the field of animal psychology because the ethological isolating mechanisms are among the most conspicuous manifestations of animal behavior. The study of courtship behavior has importantly contributed to the concepts of Lorenz and Tinbergen and has led to a reinterpretation of many facts which Darwin had grouped together under the heading "Sexual Selection."

*The Study of Intraspecific and Interspecific Competition.* The realization that there is a subtle difference between the competition that goes on among conspecific individuals and that among individuals belonging to different species has had a very stimulating effect in the field of ecology. That interspecific competition is an important centrifugal factor in evolution has been demonstrated by various recent authors. Attempts to determine and to measure such competition more precisely have led to a much more detailed study of the ecology and the population structure of species than was attempted by the ecologists of preceding decades.

*Study of the Genetic Structure of Species.* The realization that local populations belong to a broader system, the species, and that the total genetic content of these intercommunicating populations are the gene pool of the species has had a profound effect on population genetics. It has led to new questions and has facilitated the understanding of certain intrapopulation phenomena. Studies of the balance of the gene complex, of the balance between local adaptation and compatibility with gene dispersal and of hybridization between species have led to a broadening of genetics which transcends considerably the genetics of the early Mendelian period.

*The Study of Physiological Species Differences.* In physiology a development is taking place which parallels that in genetics. It involves the realization that the ecotypic physiological differences of local populations are of a different order of magnitude from that existing between good species. Each species is a separate physiological system, and these physiological differences are of different kinds as pointed out by Prosser (in this symposium). With physiological characteristics of populations, special care must be exercised to distinguish between genetically controlled characters and others. Prosser has shown that nongenetic acclimatizations are particularly frequent in marine animals with pelagic larval stages. Such organisms must be capable of coping with the water conditions in the area where they settle. It would be a mistake, however, to conclude from this special case that all infraspecific population differences of physiological characters are nongenetic. There is not only the excellent work of the Stanford group (Clausen, Hiesey, and Keck) and their associates, which clearly establishes the genetic nature of physiological differences between local races of plants, and similar results have been obtained elsewhere, but also a large body of fact demonstrating geographic variation of physiological characters in terrestrial animals. The ultimate differences between species are in part built up from such population differences. However, a distinction must be made between purely ecotypic local adaptation to purely local conditions and a more general physiological divergence which is almost always found in isolated populations (Mayr, 1956). Physiological differentiation sometimes proceeds more rapidly than morphological change, and this is the explanation for the occurrence of sibling species in many groups.

The study of physiological adaptations and limitations is still very much at the beginning. This seems to be one of the most promising branches of comparative physiology. A study of the physiological differences between species, particularly aquatic species, with respect to rates of development, temperature adaptation, and temperature tolerance has already yielded most interesting findings.

These are only a few indications of newly developing branches of biology specifically dealing with the species level. They refer to the development I had in mind when I said that we were witnessing at the present time the growth of a "science of species."

## Conclusion

The discussions of this symposium show clearly that species are still a stimulating subject and that the species problem still offers a challenge. There are many difficulties in applying the species concept to the vast variety of discontinuities found in organic nature. Yet many biological phenomena would make no sense if individuals were not organized into populations, and these populations into species. Species are an important biological phenomenon, important to every biologist because every biologist works with species.

*Note:* For references, see those in the Introduction.

# INDEX

Acclimation, 340 ff., 353, 361
Adaptation, 352
   ecotypic, 378, 387
   physiological, 340 ff., 349, 361
Adolph, E. F., 343, 364
Agameon, 61
Agamic complex, 70
Agamospecies, 61, 381
*Agoseris*, 47
*Allophyllum*, 47
*Anadara*, 135
Anderson, E., 71, 77
*Anopheles*, 351
Apomicts, 36
Archetype, 12
Arkell, W. J., 6, 19
*Artemisia tridentata*, 54
*Asclepias*, 53
Asexuality, 379 ff.
Asexual organisms, 283, 286, 288
Associations of species, 104, 115
*Asterias*, 350
Austin, M. L., 179, 315
Autogamy, 188

von Baer, K. E., 7, 19
Banta, A. M., 86, 121
Barrel, J., 143, 152
Bateson, W., 5
Beale, G. H., 182, 214, 235, 315
Behavior
   relation to isolation, 359
   reproductive, 343
Berg, L. S., 111, 121, 122
Bessey, C. E., 4, 19
*Beyrichia*, 140-141
Binom, 46, 61, 380
Biological races, 384
Biological species, 160
   characteristics of, 50
   criteria of, 200
   difficulties of concept, 57
   in paleontology, 129

Biological units, 39
Birge, E. A., 83, 122
Brauer, F., 8, 19
Breeding system, 296
   of *Paramecium aurelia*, 202, 206, 234 ff.
   of *Paramecium bursaria*, 245
   of *Paramecium caudatum*, 239
   in Protozoa, 155-315
   of *Tetrahymena pyriformis*, 267
Brett, J. R., 340, 365
*Bufo*, 337
Bullock, B. H., 125, 152
Burma, B. H., 6, 19, 44, 78, 125, 152
Butzel, H. M., Jr., 168, 315, 316

Cain, A. J., 10, 19
Calkins, G. N., 292, 316
*Callinectes*, 350
Camp, W. H., 10, 19, 44, 61, 78
*Carbonicola*, 128
Carson, H. L., 35, 37
Caryonide, 223
   definition, 186
Categories
   collective, 16
*Ceanothus*, 47, 53
*Cerion*, 377
Characters
   neutral, 347
   physiological, 348, 363
Chen, T. T., 241, 242, 316
Ciliates, 157 ff.
   life cycle, 187
Cladocera, 104
Classification, 383
Cleveland, L. R., 280, 316, 317
Clines, 355
Clones, 159, 223
Coenospecies, 65
Colbert, E. H., 133, 134, 152
*Colpidium truncatum*, 277
Commiscuum, 65

Comparium, 65
Competition
  interspecific, 386
Concepts, 10 ff.
  application of, 132, 372
  biological, 42, 45, 57, 74, 126, 132, 155, 296, 371
  conflicting, 6
  critique of, 42
  difficulties of, 371-383
  evolution of, 41
  genetic, 16
  and hybridization, 62
  importance of, 383-388
  morphological, 126
  multidimensional, 16
  of naturalist, 41
  nominalistic, 44
  nondimensional, 14-16
  philosophical, 11
  typological, 11, 42, 43, 45, 126
Conjugation
  definition, 188
Conodonts, 125
Constancy, 2, 7
Continuity
  evolutionary, 375
Copepods, 104
*Coregonus*, 110-120
  hybridization in, 118
  spawning behavior, 118
*Coregonus albula*, 111
*Coregonus lavaretus*, 111
*Crassostrea*, 354
*Crepis*, 70
*Crinia signifera*, 331
Criteria
  biological, 9
  serologic, 285
*Crurithyris*, 141
*Cubitostrea*, 145
Cuvier, G., 7
Cytoplasm, 185

Danser, B. H., 65, 78
*Daphnia*, 82-110
  characters of species, 105
  distribution of, 82
  helmet, 97
  hybridization, 106 ff.
  life cycle, 85
  phenotypic variation, 90 ff.
  reproduction modes, 86
*Daphnia dubia*, 106
*Daphnia galeata mendotae*, 106
*Daphnia longiremis*, 99 ff.
*Daphnia longispina*, 83
*Daphnia middendorffiana*, 107, 108
*Daphnia parvula*, 98
*Daphnia pulex*, 83, 84, 107, 108
*Daphnia retrocurva*, 93-99
*Daphnia schødleri*, 90, 91, 107
Darlington, C. D., 30, 37
Darwin, C., 3, 4, 7, 19, 325
Davidson, J. F., 44, 65, 78
Definitions, 16
  biological, 17, 46
  secondary speciation, 60
De Garis, C. F., 158, 317
Delimitation, 51
Delineation
  borderlines, 54
  difficulties of, 53
Deme, 28
De Vries, H., 5
Differences
  interspecific, 357
  morphological, 12, 376
  nongenetic, 363
  physiological, 386
*Difflugia*, 286
Dimorphism
  sexual, 373
Dippell, R., 167, 175, 177, 179, 189, 216, 219, 317, 323
Discontinuities
  kinds, 379
  morphological, 47
Dispersal
  consequences of, 218
Dobzhansky, T., 17, 18, 20, 26, 59, 72, 78, 283, 294, 296, 317, 326, 337
Dougherty, E. C., 19, 380
*Drosophila*, 35, 356
  behavior, 360

*Drosophila persimilis*, 57, 336
*Drosophila pseudoobscura*, 57, 336
*Drosophila robusta*, 35
Du Rietz, G. E., 5, 17, 20, 51, 78, 380

Eagar, R. M. C., 125, 128, 129, 152
*Eidos*, 11
Eimer, G. H. T., 15, 20
Elliott, A. M., 256, 317, 318
Embryologist's viewpoint, 325
Engler, A., 42
Environment
    influence on freshwater animals, 97
Enzymes, 361
*Euplotes*, 271-276, 351
    breeding system, 274
    mating hormones, 273
    mating types, 273
    species and varieties, 271
*Euplotes patella*, 272

Fish, 356
    hybridization in, 110
Flagellates, 280-283
    sexual processes, 280 ff.
Fossil animals, 125-151, 384
    application of biological concept, 132
    incompleteness of record, 142
    size-frequency distribution of population, 141
    species, 384
Frequency
    of good species, 52
Freshwater animals, 81-123
Frog, see *Rana*

*Gammarus*, 350
*Gasterosteus*, 351
Gene pool, 14, 26
Gene recombination, 23-37; see also *Recombination*
Genes
    drift, 360
    flow, 32

Genotype, 24
Giese, A. C., 190-192, 209, 318
*Gilia*, 48-50, 62-64
*Gilia capitata*, 62
*Gilia leptantha*, 47
*Gilia millefoliata*, 62
*Gilia tenuiflora-latiflora*, 63
Gillis, C. L., 10, 19
Gill rakers, 112, 117
Gilman, L. C., 236, 239, 240, 318
*Ginkgo biloba*, 65
Ginsburg, I., 12, 20
Gloger, C. L., 9, 20
Godron, D. A., 7, 20
Grant, V., 372, 376, 381
Greene, E. L., 3, 20
Gregg, J. R., 6, 20
Groupings, 42, 67
Gruchy, D. F., 256, 318

Habituation, 358
Haploidy, 202
Hayes, B. E., 257, 317
Hermaphroditism, 70
Heuts, M. J., 351, 366
History of problem, 1
Hiwatashi, K., 236, 240, 318, 319
Hoare, C. A., 283, 319
Homogamic complex, 67
Hooijer, D. A., 133, 134, 152
Hubbs, C. L., 110, 122
Human species, 34
Hybridization, 32, 55, 59, 62, 70-72, 106, 118, 327, 377
    in fishes, 110
    introgressive, 110, 112
    reasons for, 74
Hybrids, 330
    methods of identification, 58
    sterility of, 44

Identification
    methods of, for hybrids, 58
Illiger, J. C. W., 8, 20
Inbreeders, 206 ff.
Inbreeding, 297 ff.
    indications of, 220 ff.
    obligatory, 280

Intermediacy
   evolutionary, 375
Introgression, 55, 109, 119
Isolation
   agents, 359
   mechanisms, 72, 167, 232 ff., 326, 330, 332 ff., 359, 377, 378, 385
   microgeographic, 235
   relation to behavior, 359
   reproductive, 15, 47, 374, 376, 377

Jennings, H. S., 158, 159, 233, 241, 245, 248, 250, 253, 286, 319, 320
Jordan, K., 5, 15, 17, 20

Karyotype, 216
Kermack, K. A., 141, 148-152
Kimball, R. F., 252, 271 ff., 320
Kinne, O., 351, 366
Kiser, R. W., 84, 122
Koopman, K. F., 336, 337
Kosmoceras, 376
Kotaka, T., 135, 152
Kurtén, B., 141, 152

Lederberg, J., 285, 320
Leitch, D., 128, 152
Levine, M., 167, 320
Linnaean species, 42
Linnaeus, 2, 3, 15
Lorkovicz, Z., 20, 379
Lotsy, J. P., 42, 67, 79

Margolin, P., 180, 320
Mason, H. L., 44, 60, 79
Mating
   *Paramecium aurelia,* 164
   reactions, intervarietal, 165
   types, 158, 164
Maupas, E., 157, 187, 321
Mayr, E., 10, 17, 21, 50, 68, 79, 80, 289, 321, 326, 337, 382, 387
McKerrow, W. S., 138, 153
Meadowlark, 360
Meglitsch, P. A., 10, 14, 21
Melvin, J. B., 167, 321

Mendelian population, 26
Merging
   of species, 60
Metz, C. B., 242, 321
*Micraster,* 141, 147-151
Micronuclei, 178
Microspecies, 70, 382
Moore, J. A., 378
Morphological concept, 126
Morphological differences, 12, 376
Morphological types, 373
Munz, P. A., 53, 77
*Mustela*
   comparison of osteological characteristics, 133
*Mytilus,* 350, 352, 353

Nanney, D. L., 186, 227, 229, 251, 267, 321
*Nereis,* 350
*Neurospora,* 348
Newell, N. B., 125, 129, 143, 144, 153
Niggli, P., 11, 21

Objectivity, 2
Oeder, G. L., 7
*Oenothera,* 56
Oken, L., 9
*Opuntia,* 56, 70
*Ornithella,* 139
Outbreeders, 206 ff.
Outbreeding, 297 ff., 347
   indications of, 220 ff.
Overlaps, 48
   circular, of races, 50
*Oxytrichia bifaria,* 277 ff.
   breeding system, 278
Oysters (*Crassostrea*), 354

Paleontology, 127
   biological species in, 129
   typological, 127
*Paramecium,* 158 ff.
   autogamy, 188
   characters of species, 161
   conjugation, 188

*Paramecium (continued)*
  isolating mechanism, 167
  killer strains, 179
  life cycle, 187
  macronuclear differentiation, 227
  sexual processes, 203
*Paramecium aurelia*, 162-236
  breeding system, 202, 206, 234 ff.
  characteristics, 197
  conjugation, 188
  cytoplasmic exchange, 224
  distribution of varieties, 169
  evolution of groups A and B, 226
  fission rate, 176
  geographical distribution, 231
  geographical variation, 216
  identification of varieties, 173
  immaturity, 190
  inheritance of mating types, 185
  karyotypes, 216
  killers, 234
  life cycle differences, 187
  mating types, 163 ff., 227
  maturity, 191
  morphological differences, 177
  mortality of intervarietal crosses, 172
  races, 235
  senility, 193
  serotypes, 180, 230
  taxonomic status of varieties, 195
  temperature range, 171
  temperature tolerance, 175
  varieties, 15, 16, 163 ff., 173, 315
  varieties as species, 199
*Paramecium bursaria*
  breeding system, 245
  characteristics of varieties, 243
  distribution of varieties, 243
  karyotypes, 244
  mating types, 241, 244, 250
  reproductive isolation, 242
  self-differentiation, 248
  speciation, 256
*Paramecium calkinsi*, 162, 276
*Paramecium caudatum*, 236-241
  breeding system, 239
  varieties, 236, 237

*Paramecium multimicronucleatum*, 161, 163, 314
*Paramecium trichium*, 162, 276
Parasites, 373
Parkinson, D., 142, 153
Parthenogenesis, 86
Pennak, R. W., 84, 122
Phenotype, 24, 360
  effect of temperature, 97
  plasticity of, 114
Phylogenetic tree, 130
Physiologist's viewpoint, 339-364
Physiology
  comparative, 287
*Pinus muricata*, 60
*Pinus remorata*, 60
Pittendrigh, C. H., 352, 367
Plant species, 39-77
Plate, L., 8, 21
Plato, 11
Polyploidy, 54
Populations
  competition between, 334
  local, 27, 31
  Mendelian, 26
  recombination in, 29
  size-frequency distribution of fossils, 141
  in space, 32
  in time, 32
Poulton, E. B., 2, 21
Preer, J. R., Jr., 179, 182, 322
Pringle, C. R., 182, 208, 210, 211, 238, 322
Prosser, C. L., 386
Properties, 14
Protozoa, 155-315
  asexual, 280, 299
  reproductive methods, 155-315
  variety in, 160 ff.

Races
  biological, 384
  osmotic, 350
  physiological, 347, 353
Ramsbottom, J., 3, 21
*Rana*, 328 ff.
*Rana pipiens*, 329, 355, 375, 378

394 INDEX

Rate functions, 343
Rawson, D. S., 103, 122
Ray, C., Jr., 268, 322
Ray, J., 7, 15, 21, 41
Recognition of species, 105
Recombination, 23-37, 202
    amount of, 33
    in man, 34
    systems, 33, 36
Redi, F., 3
Regan, C. T., 127, 153
Regulation, 342
Rensch, B., 17, 21
Reproduction
    asexual, 36, 56, 61, 70, 202
    sexual, 203
    uniparental, 88
*Rhizomys*, 134
Rhizopods, 286
Robinet, C., 3
Rotifers, 104
    bdelloid, 382
Rowe, A. W., 147, 153

Sagebrush, 55
Salmonidae, 110
Schopenhauer, A., 3, 21
Self-pollination, 56
Semispecies, 68, 379
Senility
    of *Paramecium aurelia*, 193
Serotypes, 180, 359
Sexual dimorphism, 373
Sexuality
    loss of, 281
Sibley, C. G., 68, 80
Sibling species, 13, 54, 57, 69, 291, 293, 348, 359, 376, 384
Siegel, R. W., 179, 191, 234, 277 ff., 322
Simpson, G. G., 14, 15, 17, 21, 125, 127, 134, 153, 383
Sirks, M. J., 3, 22
Sonneborn, T. M., 251, 322, 323
Spallanzani, L., 3
*Spathidium spathula*, 292
Speciation, 69, 256, 346, 375, 385
    allopatric, 49, 326, 344

    in asexual organisms, 286
    ecotypic, 387
Species
    boundaries of, 150
    genetic structure of, 386
    multidimensional, 375
    nondimensional, 131
    secondary, 60
    successional, 131, 147 ff.
    typological, 127
Spencer, H., 292, 324
Spjeldnaes, N., 140-141, 153
Spontaneous generation, 2
Spooner, G. M., 351, 368
Stebbins, G. L., Jr., 36, 38, 72, 80
Stenzel, H. B., 145, 153
Sterility, 72
Strains
    amicronucleate, 263
    killer, 179, 234
Stresemann, E., 17, 22
Stress, 340
*Strophodonta*, 141
*Stylonychia putrina*, 277
Superspecies, 50, 67
Survival, 340
Svärdson, G., 111-120, 122, 123
Sylvester-Bradley, P. C., 125, 129, 153, 374, 384
Sympatry, 49
Synclone, 223
Syngameon, 64, 66, 67, 71, 76
Syngen, 201, 289, 295
    asexual, 290
Systematics
    comparative, 372

Temperature, 97, 352
    adaptations to, 328
    tolerance, 175, 356
*Tetrahymena pyriformis*, 256-271
    breeding systems, 267
    characteristics, 258
    life cycle, 261
    mating types, 259
    varieties, 257
*Tetrahymena rostrata*, 270
*Thais emarginata*, 354

Thoday, J. M., 33, 38, 291, 323
Thomas, G., 375
Transitions between species, 59
Trueman, A. E., 129
Trypanosomes, 284
Turbulence, 97
Turesson, G., 65, 80

*Uroleptus mobilis,* 292
*Urosalpinx,* 355

Valentine, D. H., 10, 22
Variation
  geographical, 98
  nongenetic, 360
  overlapping, 48
  phenotypic, 90 ff., 112, 138
  physiological, 340-343, 345, 349, 350, 352, 357, 359
  polytypic, 43, 384
  seasonal, 94
Variety, 4
Voigt, F. S., 9, 22
Volpe, G. P., 337, 338

Wagler, E., 83, 123
Ward, G. H., 55, 80
Weir, J., 127, 153
Westoll, T. S., 138, 153
White, M. J. D., 30, 38
Whitefish, see *Coregonus*
Wichterman, R., 162, 276, 323, 324
Wild type, 25
Woltereck, R., 99, 102, 123
Woodruff, L. L., 292, 324

Zirkle, C., 3

# NATURAL SCIENCES IN AMERICA

*An Arno Press Collection*

Allen, J[oel] A[saph]. **The American Bisons,** Living and Extinct. 1876

Allen, Joel Asaph. **History of the North American Pinnipeds:** A Monograph of the Walruses, Sea-Lions, Sea-Bears and Seals of North America. 1880

**American Natural History Studies:** The Bairdian Period. 1974

**American Ornithological Bibliography.** 1974

Anker, Jean. **Bird Books and Bird Art.** 1938

Audubon, John James and John Bachman. **The Quadrupeds of North America.** Three vols. 1854

Baird, Spencer F[ullerton]. **Mammals of North America.** 1859

Baird, S[pencer] F[ullerton], T[homas] M. Brewer and R[obert] Ridgway. **A History of North American Birds:** Land Birds. Three vols., 1874

Baird, Spencer F[ullerton], John Cassin and George N. Lawrence. **The Birds of North America.** 1860. Two vols. in one.

Baird, S[pencer] F[ullerton], T[homas] M. Brewer, and R[obert] Ridgway. **The Water Birds of North America.** 1884. Two vols. in one.

Barton, Benjamin Smith. **Notes on the Animals of North America.** Edited, with an Introduction by Keir B. Sterling. 1792

Bendire, Charles [Emil]. **Life Histories of North American Birds** With Special Reference to Their Breeding Habits and Eggs. 1892/1895. Two vols. in one.

Bonaparte, Charles Lucian [Jules Laurent]. **American Ornithology:** Or The Natural History of Birds Inhabiting the United States, Not Given by Wilson. 1825/1828/1833. Four vols. in one.

Cameron, Jenks. **The Bureau of Biological Survey:** Its History, Activities, and Organization. 1929

Caton, John Dean. **The Antelope and Deer of America:** A Comprehensive Scientific Treatise Upon the Natural History, Including the Characteristics, Habits, Affinities, and Capacity for Domestication of the Antilocapra and Cervidae of North America. 1877

Contributions to American Systematics. 1974

Contributions to the Bibliographical Literature of American Mammals. 1974

Contributions to the History of American Natural History. 1974

Contributions to the History of American Ornithology. 1974

Cooper, J[ames] G[raham]. Ornithology. Volume I, Land Birds. 1870

Cope, E[dward] D[rinker]. The Origin of the Fittest: Essays on Evolution and The Primary Factors of Organic Evolution. 1887/1896. Two vols. in one.

Coues, Elliott. Birds of the Colorado Valley. 1878

Coues, Elliott. Birds of the Northwest. 1874

Coues, Elliott. Key To North American Birds. Two vols. 1903

Early Nineteenth-Century Studies and Surveys. 1974

Emmons, Ebenezer. American Geology: Containing a Statement of the Principles of the Science. 1855. Two vols. in one.

Fauna Americana. 1825-1826

Fisher, A[lbert] K[enrick]. The Hawks and Owls of the United States in Their Relation to Agriculture. 1893

Godman, John D. American Natural History: Part I — Mastology and Rambles of a Naturalist. 1826-28/1833. Three vols. in one.

Gregory, William King. Evolution Emerging: A Survey of Changing Patterns from Primeval Life to Man. Two vols. 1951

Hay, Oliver Perry. Bibliography and Catalogue of the Fossil Vertebrata of North America. 1902

Heilprin, Angelo. The Geographical and Geological Distribution of Animals. 1887

Hitchcock, Edward. A Report on the Sandstone of the Connecticut Valley, Especially Its Fossil Footmarks. 1858

Hubbs, Carl L., editor. Zoogeography. 1958

[Kessel, Edward L., editor]. A Century of Progress in the Natural Sciences: 1853-1953. 1955

Leidy, Joseph. The Extinct Mammalian Fauna of Dakota and Nebraska, Including an Account of Some Allied Forms from Other Localities, Together with a Synopsis of the Mammalian Remains of North America. 1869

Lyon, Marcus Ward, Jr. Mammals of Indiana. 1936

Matthew, W[illiam] D[iller]. Climate and Evolution. 1915

Mayr, Ernst, editor. The Species Problem. 1957

Mearns, Edgar Alexander. Mammals of the Mexican Boundary of the United States. Part I: Families Didelphiidae to Muridae. 1907

Merriam, Clinton Hart. **The Mammals of the Adirondack Region,** Northeastern New York. 1884

Nuttall, Thomas. **A Manual of the Ornithology of the United States and of Canada.** Two vols. 1832-1834

Nuttall Ornithological Club. **Bulletin of the Nuttall Ornithological Club:** A Quarterly Journal of Ornithology. 1876-1883. Eight vols. in three.

[Pennant, Thomas]. **Arctic Zoology.** 1784-1787. Two vols. in one.

Richardson, John. **Fauna Boreali-Americana;** Or the Zoology of the Northern Parts of British America, Containing Descriptions of the Objects of Natural History Collected on the Late Northern Land Expeditions Under Command of Captain Sir John Franklin, R. N. Part I: Quadrupeds. 1829

Richardson, John and William Swainson. **Fauna Boreali-Americana:** Or the Zoology of the Northern Parts of British America, Containing Descriptions of the Objects of Natural History Collected by the Late Northern Land Expeditions Under Command of Captain Sir John Franklin, R. N. Part II: The Birds. 1831

Ridgway, Robert. **Ornithology.** 1877

**Selected Works By Eighteenth-Century Naturalists and Travellers.** 1974

**Selected Works in Nineteenth-Century North American Paleontology.** 1974

**Selected Works of Clinton Hart Merriam.** 1974

**Selected Works of Joel Asaph Allen.** 1974

**Selections From the Literature of American Biogeography.** 1974

Seton, Ernest Thompson. **Life-Histories of Northern Animals:** An Account of the Mammals of Manitoba. Two vols. 1909

Sterling, Keir Brooks. **Last of the Naturalists:** The Career of C. Hart Merriam. 1974

Vieillot, L. P. **Histoire Naturelle Des Oiseaux de L'Amerique Septentrionale,** Contenant Un Grand Nombre D'Especes Decrites ou Figurees Pour La Premiere Fois. 1807. Two vols. in one.

Wilson, Scott B., assisted by A. H. Evans. **Aves Hawaiienses:** The Birds of the Sandwich Islands. 1890-99

Wood, Casey A., editor. **An Introduction to the Literature of Vertebrate Zoology.** 1931

Zimmer, John Todd. **Catalogue of the Edward E. Ayer Ornithological Library.** 1926